Advanced Materials in Drug Release and Drug Delivery Systems

Advanced Materials in Drug Release and Drug Delivery Systems

Editor

Katarzyna Winnicka

MDPI • Basel • Beijing • Wuhan • Barcelona • Belgrade • Manchester • Tokyo • Cluj • Tianjin

Editor
Katarzyna Winnicka
Department of Pharmaceutical Technology,
Faculty of Pharmacy,
Medical University of Bialystok
Poland

Editorial Office
MDPI
St. Alban-Anlage 66
4052 Basel, Switzerland

This is a reprint of articles from the Special Issue published online in the open access journal *Materials* (ISSN 1996-1944) (available at: https://www.mdpi.com/journal/materials/special_issues/drug_release_delivery).

For citation purposes, cite each article independently as indicated on the article page online and as indicated below:

LastName, A.A.; LastName, B.B.; LastName, C.C. Article Title. *Journal Name* **Year**, *Volume Number*, Page Range.

ISBN 978-3-0365-1058-3 (Hbk)
ISBN 978-3-0365-1059-0 (PDF)

Contents

About the Editor

Katarzyna Winnicka chief of the Department of Pharmaceutical Technology, Faculty of Pharmacy, Medical University of Bialystok. Current research interests concern the development of novel drug dosage forms (orodispersible forms, multicompartment formulations) and the utilization of multifunctional polymers to design pharmaceutical dosage forms with model drugs.

materials

MDPI

Editorial

Special Issue: Advanced Materials in Drug Release and Drug Delivery Systems

Katarzyna Winnicka

Department of Pharmaceutical Technology, Faculty of Pharmacy, Medical University of Bialystok, Kilinskiego 1, 15-089 Bialystok, Poland; kwin@umb.edu.pl; Tel.: +48-85-7485615

Development of new drug molecules is costly and requires longitudinal, wide-ranging studies; therefore, designing advanced pharmaceutical formulations for existing and well-known drugs seems to be an attractive device for the pharmaceutical industry. Properly formulated drug delivery systems can improve pharmacological activity, efficacy and safety of the active substances. Advanced materials applied as pharmaceutical excipients in designing drug delivery systems can help solve problems concerning the required drug release—with the defined dissolution rate and at the determined site. Novel drug carriers enable more effective drug delivery, with improved safety and with fewer side effects. Investigations concerning advanced materials represent a rapidly growing research field in material/polymer science, chemical engineering and pharmaceutical technology. Exploring novel materials or modifying and combining existing ones is now a crucial trend in pharmaceutical technology.

The Special Issue "Advanced Materials in Drug Release and Drug Delivery Systems" was established to present the most recent insights into utilization of different materials with promising potential in drug delivery and into different formulation approaches that can be used in the design of pharmaceutical formulations. The brief summary of eleven articles included in the issue is presented below.

The papers [1–3] present development in the biomedical field using nano- or microparticulate carriers. Wang et al. provide information about developing an effective and relatively simple one-pot technique to load graphene quantum dots (GQDs) into nanoparticles intended for fluorescent imaging offering a new way to visualize the distribution and metabolism of nanoparticles in vivo without radioactive damage. It was shown that designed GQD-loaded optical magneto ferroferric oxide@polypyrrole (Fe_3O_4@PPy) core shell nanoparticles are characterized by high potential applicability in monitoring the distribution and metabolism of nanoparticles and in long-term in vivo real-time tracking [1]. Zheng et al. [2] describe information about the preparation and characterization of methoxy poly(ethylene glycol)-poly(lactide) copolymer (mPEG-PLA) microspheres as carriers of alkaloids from *Alstonia scholaris* leaves—a traditional Chinese drug used in the treatment of chronic respiratory diseases. The limiting factor connecting administration of these alkaloids is short half-life, so the authors focused on the development of a multicompartment dosage form providing sustained drug release. mPEG-PLA microspheres with alkaloids from *Alstonia scholaris* leaves were obtained by w/o/w double emulsion method. Designed particles significantly prolonged drug release, were not cytotoxic in HL-7702 cells, and inhibited auricle and pedal swelling evaluated in mice and rats. Wasilewska et al. [3] report data on the exploitation of ethylcellulose microparticles containing rupatadine fumarate in the development of orodispersible minitablets—a novel solid dosage form connecting benefits of both liquid and solid formulations. As rupatadine fumarate—a second generation antihistamine drug—is characterized by bitter taste, an attempt was made to formulate "patient friendly" orodispersible minitablets with acceptable taste. To mask rupatadine fumarate's bitterness, the drug was enclosed in microparticles obtained with widely used in the pharmaceutical technology, non soluble in water ethylcellulose. In

check for updates

Citation: Winnicka, K. Special Issue: Advanced Materials in Drug Release and Drug Delivery Systems. *Materials* **2021**, *14*, 1042. https://doi.org/10.3390/ma14041042

Received: 15 February 2021
Accepted: 22 February 2021
Published: 23 February 2021

this work, orodispersible minitablets containing rupatadine fumarate were formulated by direct compression of commercially ready to use blends and microparticles prepared with an aqueous ethylcellulose dispersion. Developed formulations of minitablets possessed beneficial pharmaceutical parameters and provided a taste-masking effect evaluated by three alternative techniques (e-tongue, human taste panel, and the in vitro drug release). It was indicated that ethylcellulose microparticles can be successfully used in the direct compression. A more detailed description of ethylcellulose applicability in drug dosage forms development is presented in the review paper by Wasilewska et al. [4]. Ethylcellulose is a multifunctional, hydrophobic cellulose derivative possessing advantageous and unique properties to be used in pharmaceutical formulations. This water insoluble polymer is regarded as safe, biocompatible, is degraded into nontoxic products, and shows gastroresistance. Ethylcellulose is widely utilized in the pharmaceutical industry, it is a valuable excipient to modify drug release profiles and to create controlled release dosage forms. In this article, particular attention was placed on its role in the oral and topical formulations.

The papers [5,6] present data about multilayer films and electrospun matrices as advanced materials used in drug delivery systems. Nie et al. [5] reported on the versatile layer by layer highly stable multilayer films prepared from polymer (hyaluronic acid) and inorganic–mesoporous silica nanoparticles for the delivery of FITC-labeled short peptides. The described method improves the stability of the multilayer structure by forming covalently cross-linked super strata on the outlying areas of the films. Moreover, the super stratum functions as a "nano net" reducing nanoparticles diffusion and is simultaneously permeable for small molecules, which can be used to develop multifunctional drug delivery systems based on mesoporous silica as the active substance depot. Mesoporous silica nanoparticles enhance the stability of peptides, retention properties of peptides can be modified by varying capping-layer numbers, and the release profile of small molecules can also be adjusted by adequate carrier arrangements. Nazarkina et al. describe sirolimus matrices for coating vascular stents [6]. Development of stent coatings is a valid issue, as stent implantation is a frequently performed medical procedure which can initialize various reactions stimulating an inflammatory response. In this work, an attempt was made to design matrices with sirolimus-drug possessing immunosuppressive and antiproliferative activity and approved both in Europe and United States as stent-coating component. Matrices were formulated by electrospinning polycaprolactone, human albumin, hexafluoroisopropanol, and dimethyl sulfoxide, and the optimal composition, the most applicable for bare-metal stent coating, ensuring prolonged sirolimus release to provide its concentration preferable for antiproliferative activity, was indicated.

The next three papers are devoted to the ever-expanding technological approaches in solid oral dosage forms like tablets [7,8] and capsules [9]. Jamróz et al. focus on the deep analysis of the relationship between structure and pharmaceutical characteristics in 3D printed tablets. Application of 3D printing in pharmaceutical technology is relatively novel, requires proper selection of excipients, and to date, only one medicinal product obtained by this technique is commercially available (Spritam, by Aprecia Pharmaceuticals, registered in the US). In this work, the composition of the 3D printed tablets and the relationship between the tablet geometry, tablet shape, and drug release profile was discussed. To evaluate the influence of a degree of infill on the printing reproducibility and to analyze the tablet structure, micro-computed tomography was used. Authors developed 3D printed immediate release tablets with liquid crystal-forming itraconazole and indicated that optimal tablet characteristics (reproducibility during printing, optimal drug release) was assured when filament formulation of 20% itraconazole, 76% polyvinyl alcohol, and 4% crospovidone was utilized [7]. Supercritical fluid method is a relatively new technique with potential application in the pharmaceutical industry to reduce particle size and to create inclusion complexes or solid dispersions. Antosik-Rogóż et al. report on the utility of the supercritical fluid method (using supercritical CO_2) in the development of dosage forms with poorly soluble bicalutamide-drug from the antiandrogens group used in prostate cancer treatment [8]. The influence of supercritical CO_2 on the characteristics of binary systems with bicalutamide and polymeric excipients (Macrogol 6000, Poloxamer 407), and tablets obtained from solid dispersions was deeply described. It was shown that this process affected particle size and shape, improved the in vitro bicalutamide release profile, and it evoked changes in the drug crystallinity. The paper by Maciejewski et al. [9] provides data on the gastro resistant soft gelatin films and capsule formulation from binary-gelatin/cellulose acetate phthalate and ternary-gelatin/cellulose acetate phthalate/carrageenan films. They suggest that the discreet kinetically-limited phase separation was the crucial factor affecting the disintegration of the films. In medium, at acidic pH, gelatin undergoes swelling and dissolving processes, and exposes the cellulose acetate phthalate skeleton, which is insoluble in acidic environments. Moreover, addition of carrageenan improved this acid-resistant effect. The dissolution study performed with diclofenac sodium as a model drug from the lab-scale soft capsule formulations indicated that the proposed fillings were compatible with the films, and the drug was not released at acidic pH, but it was released at pH 6.8.

In the paper by Chyzy et al. [10], the latest scientific reports on hydrogel applications as drug dosage forms were discussed. The authors provide an in-depth description of the main formulation approaches, classification of hydrogels, their physicochemical characteristics, and their exploitation in the biomedicine and drug delivery fields (in oral, dermal, ocular, vaginal, and parenteral administration). This review also provides information on stimuli responsive hydrogels and modern approaches to improve treatment efficiency by using hydrogel formulations. Readers can achieve an interdisciplinary vision into the advances in the design, optimization, and application of hydrogel delivery systems [10]. Faizan et al., in turn, reviewed recent literature reports concerning carbon monoxide-releasing materials providing carbon monoxide (CO) for medicinal purposes [11]. At higher concentration, CO has a toxic effect, but in lower, strictly controlled doses, it can be treated as a valuable agent participating in cell signaling, with potential treatment application. The authors discuss the therapeutic properties of CO and problems connected with designing and development of effective and safe CO-releasing molecules and CO-releasing materials. They thoroughly describe the possibilities of delivering CO by using a great variety of structures and strategies available for biological research and treatment applications (e.g., copolymer assembles, micelles, nano fiber gels, nanoparticles, nanosheets, metal organic frameworks, conjugation with proteins, peptides, metallodendrimers, vitamins, nanodiamonds), and consider toxicological aspects of CO delivery involved in the development of CO-releasing materials-based pharmaceuticals.

Conflicts of Interest: No conflict of interest is declared.

References

1. Wang, Y.; Xu, N.; He, Y.; Wang, J.; Wang, D.; Gao, Q.; Xie, S.; Li, Y.; Zhang, R.; Cai, Q. Loading Graphene Quantum Dots into Optical-Magneto Nanoparticles for Real-Time Tracking In Vivo. *Materials* **2019**, *12*, 2191. [CrossRef] [PubMed]
2. Zheng, X.; Li, H.; He, Y.; Yuan, M.; Shen, M.; Yang, R.; Jiang, N.; Yuan, M.; Yang, C. Preparation and In Vitro Release of Total Alkaloids from *Alstonia scholaris* Leaves Loaded mPEG-PLA Microspheres. *Materials* **2019**, *12*, 1457. [CrossRef] [PubMed]
3. Wasilewska, K.; Ciosek-Skibińska, P.; Lenik, J.; Srčič, S.; Basa, A.; Winnicka, K. Utilization of Ethylcellulose Microparticles with Rupatadine Fumarate in Designing Orodispersible Minitablets with Taste Masking Effect. *Materials* **2020**, *13*, 2715. [CrossRef] [PubMed]
4. Wasilewska, K.; Winnicka, K. Ethylcellulose–A Pharmaceutical Excipient with Multidirectional Application in Drug Dosage Forms Development. *Materials* **2019**, *12*, 3386. [CrossRef] [PubMed]
5. Nie, K.; Yu, X.; Kumar, N.; Zhang, Y. Versatile Layer-By-Layer Highly Stable Multilayer Films: Study of the Loading and Release of FITC-Labeled Short Peptide in the Drug Delivery Field. *Materials* **2019**, *12*, 1206. [CrossRef] [PubMed]
6. Nazarkina, Z.K.; Chelobanov, B.P.; Chernonosova, V.S.; Romanova, I.V.; Karpenko, A.A.; Laktionov, P.P. Sirolimus-Eluting Electrospun-Produced Matrices as Coatings for Vascular Stents: Dependence of Drug Release on Matrix Structure and Composition of the External Environment. *Materials* **2020**, *13*, 2692. [CrossRef] [PubMed]
7. Jamróz, W.; Pyteraf, J.; Kurek, M.; Knapik-Kowalczuk, J.; Szafraniec-Szczęsny, J.; Jurkiewicz, K.; Leszczyński, B.; Wróbel, A.; Paluch, M.; Jachowicz, R. Multivariate Design of 3D Printed Immediate-Release Tablets with Liquid Crystal-Forming Drug—Itraconazole. *Materials* **2020**, *13*, 4961. [CrossRef] [PubMed]

8. Antosik-Rogóż, A.; Szafraniec-Szczęsny, J.; Chmiel, K.; Knapik-Kowalczuk, J.; Kurek, M.; Gawlak, K.; Danesi, V.P.; Paluch, M.; Jachowicz, R. How Does the CO_2 in Supercritical State Affect the Properties of Drug-Polymer Systems, Dissolution Performance and Characteristics of Tablets Containing Bicalutamide? *Materials* **2020**, *13*, 2848. [CrossRef] [PubMed]
9. Maciejewski, B.; Arumughan, V.; Larsson, A.; Sznitowska, M. Prototype Gastro-Resistant Soft Gelatin Films and Capsules—Imaging and Performance In Vitro. *Materials* **2020**, *13*, 1771. [CrossRef] [PubMed]
10. Chyzy, A.; Tomczykowa, M.; Plonska-Brzezinska, M.E. Hydrogels as Potential Nano-, Micro- and Macro-Scale Systems for Controlled Drug Delivery. *Materials* **2020**, *13*, 188. [CrossRef] [PubMed]
11. Faizan, M.; Muhammad, N.; Niazi, K.U.K.; Hu, Y.; Wang, Y.; Wu, Y.; Sun, H.; Liu, R.; Dong, W.; Zhang, W.; et al. CO-Releasing Materials: An Emphasis on Therapeutic Implications, as Release and Subsequent Cytotoxicity Are the Part of Therapy. *Materials* **2019**, *12*, 1643. [CrossRef]

materials

MDPI

Communication

Loading Graphene Quantum Dots into Optical-Magneto Nanoparticles for Real-Time Tracking *In Vivo*

Yu Wang [1,2], Nan Xu [1,2], Yongkai He [1,2,3], Jingyun Wang [4], Dan Wang [1,2], Qin Gao [1,2], Siyu Xie [1,2], Yage Li [1,2], Ranran Zhang [3,*] and Qiang Cai [1,2,3,*]

1 State key Laboratory of New Ceramics and Fine Processing, School of Materials Science and Engineering, Tsinghua University, Beijing 100084, China
2 Key Laboratory of Advanced Materials of Ministry of Education of China, Tsinghua University, Beijing 100084, China
3 Tsinghua Shenzhen International Graduate School, Tsinghua University, Shenzhen 518055, China
4 Shenzhen Geim Graphene Center, Tsinghua-Berkeley Shenzhen Institute, Tsinghua University, Shenzhen 518055, China
* Correspondence: biomnano@163.com (R.Z.); caiqiang@mail.tsinghua.edu.cn (Q.C.); Tel.: +86-010-6277116 (Q.C.)

Received: 21 May 2019; Accepted: 3 July 2019; Published: 8 July 2019

Abstract: Fluorescence imaging offers a new approach to visualize real-time details on a cellular level *in vitro* and *in vivo* without radioactive damage. Poor light stability of organic fluorescent dyes makes long-term imaging difficult. Due to their outstanding optical properties and unique structural features, graphene quantum dots (GQDs) are promising in the field of imaging for real-time tracking *in vivo*. At present, GQDs are mainly loaded on the surface of nanoparticles. In this study, we developed an efficient and convenient one-pot method to load GQDs into nanoparticles, leading to longer metabolic processes in blood and increased delivery of GQDs to tumors. Optical-magneto ferroferric oxide@polypyrrole (Fe_3O_4@PPy) core-shell nanoparticles were chosen for their potential use in cancer therapy. The *in vivo* results demonstrated that by loading GQDs, it was possible to monitor the distribution and metabolism of nanoparticles. This study provided new insights into the application of GQDs in long-term *in vivo* real-time tracking.

Keywords: GQDs; real-time tracking; optical-magneto nanoparticles; *in vivo*

1. Introduction

Cancer is known to be one of the leading causes of death in almost every country of the world [1]. The ideal therapy for cancer is to deliver suitable treatment (chemotherapeutic drugs, genetic drugs, nanoparticles, etc.) to the right place at the right time. Various delivery systems are designed to track the release and infiltration process of nanoparticles or drugs at the tumor site, guiding local treatment and monitoring the effects after treatment [2–4]. Due to low temporal or spatial resolution, the contemporary methods available are unable to track the kinetics of drugs or nanoparticles *in vivo* for a sustained period of time [5]. Optical imaging presents non-ionizing, non-invasive, and non-destructive features with high precision and efficiency [6]. Small molecule organic dyes, including near-infrared fluorescent dyes sulfo-cyanine7 (Cy7), indocyanine green (ICG), Dye800 and green fluorescent dyes fluorescein isothiocyanate (FITC), are employed as imaging probes for their excellent luminescence, while properties such as poor light stability severely limit their biomedical applications [7,8]. Semiconductor quantum dots have excellent luminescence and light stability, and are considered as alternatives to organic dyes. However, high toxicity and poor water solubility would hinder their application in biomedical fields. Moreover, the relatively large size of semiconductor quantum dots and the scintillating fluorescence are both obstacles to successful imaging at the molecular level [9]. Therefore, the development of new fluorescent materials is of great significance to biomedical imaging.

Graphene quantum dots (GQDs) have drawn more attention in recent years due to their outstanding properties, including excellent light stability and water solubility. Other noteworthy properties of GQDs include their ultra-small size, highly adjustable photoluminescence, as well as excellent multiphoton excitation, high luminosity, and chemical inertness [10,11]. Moreover, various nanomaterials including GQDs and GQD-based nanomaterials are reported to yield reactive oxygen species (ROS) in cells, leading to the promise of photodynamic therapy for cancer [12,13]. GQDs have shown even more promising applications in fluorescence imaging, imaging-guided surgery and biosensing [6,14]. Studies pay close attention to the biocompatibility of GQDs when they are applied in biomedical fields. *In vitro* studies illustrate that GQDs show excellent biocompatibility when co-incubated with various cells; *in vivo* studies also demonstrate the biocompatibility of GQDs [15]. N-doped graphene quantum dots exhibit low cytotoxicity when concentrations are under 200 μg/mL [16]. These studies indicate the promising biomedical applications of GQDs and GQD-based materials.

To achieve multiple functions, it is common to combine GQDs with other nanomaterials. There are many studies on the combination of GQDs and nanomaterials by dipping or sticking GQDs on the surface of nanoparticles for drug delivery and imaging [17–19]. Our study proposed an efficient and convenient one-pot approach to load GQDs into Fe_3O_4@PPy nanoparticles, which could avoid complicated binders and be easily extended to other nanoparticles. Fe_3O_4@PPy nanoparticles were chosen as model nanoparticles since they were proven to be excellent tumor diagnosis and treatment platforms, performing multiple diagnostic functions (such as magnetic resonance imaging and photoacoustic imaging), and have been utilized in various therapies (magnetic hyperthermia and photothermal therapy) [20]. Combination of the GQDs and magnetic nanoparticles enabled the function of real-time tracking in the optical-magneto nanoparticles. Moreover, when loading GQDs into PPy-coated Fe_3O_4 nanoparticles, longer metabolic processing in the blood and increased delivery of GQDs were achieved, laying the foundation for the application of GQDs *in vivo*.

2. Materials and Methods

2.1. Preparation and Characterization

Polyvinyl alcohol (Sinopharm Chemical Reagent Co., Ltd, Shanghai, China) (PVA, 7.5 g) was dissolved in 100 mL of deionized water. Next, 0.1 g of ferroferric oxide (Sinopharm Chemical Reagent Co., Ltd, Shanghai, China) was dissolved in 10 mL of deionized water and then mixed with 40 mL of PVA solution, followed by 30 minutes of stirring. GQDs (0.15 g) were dissolved in 5 mL of deionized water, then mixed with the above solution and stirred for 30 minutes for uniform dispersion. The synthesis of used OH–GQDs was carried out according to a previous work of Wang et al. [21]. Subsequently, 0.2 mL of pyrrole liquid was slowly added to the above mixed solution and stirred for 10 min. Then, 0.1 g of ferric chloride hexahydrate (Sinopharm Chemical Reagent Co., Ltd, Shanghai, China) was added and stirred at room temperature for three hours. Eventually, the solution was centrifuged five times, washed, and finally dried at 70 °C. The obtained nanoparticles were referred to as GQD-NPs. The as-prepared GQD-NPs were characterized and analyzed by a high resolution transmission electron microscope (HRTEM; JEM-2010F, JEOL, Tokyo, Japan), X-ray diffraction (XRD, Bruker D8 Advance X-ray diffractometer, Bruker, Karlsruhe, Germany), Raman spectroscopy analysis (HORIBA, Montpellier, France), and X-ray photoelectron spectroscopy (XPS; Escalab 250Xi, Thermo Scientific, Waltham, MA USA). Fluorescent spectra were measured at room temperature with a fluorescence spectrometer (FLSP920, Edinburgh Instruments, Edinburgh, UK). Magnetic properties were detected by a vibrating sample magnetometer (Lake Shore VSM 7307, Lakeshore, Columbus, OH, USA).

2.2. Cell Toxicity

Murine fibrosarcoma L929 cells were purchased from Peking Union Medical College and cultured in RPMI 1640 basic medium (Gibco, Carlsbad, CA, USA) supplied with 10% fetal bovine serum (Gibco, Carlsbad, CA, USA) and 1% penicillin–streptomycin (Gibco, Carlsbad, CA, USA). Cells were seeded in 96-well plates at 5×10^3 cells per well for 24 h. Then, GQD-NPs with gradient concentrations (50 μg/mL, 100 μg/mL, 160 μg/mL, 250 μg/mL, 500 μg/mL, and 1000 μg/mL) were added into each well. The doses of GQD-NPs were selected according to previous studies [3,22–24]. After incubation for 24 h or 48 h, cell viability was measured using a cell counting kit-8 (CCK-8; Dojindo, Kyushu, Japan) following the manufacturer's instructions. The cell viability values were all normalized to control groups (untreated cells after incubation for 24 h or 48 h). The experiments were conducted in triplicate and the data were expressed as means ± standard deviation. Student's t-test was used to determine the level of significance. Differences with $p < 0.05$ and $p < 0.01$ were considered statistically significant and highly significant, respectively.

2.3. In Vivo Tracking

Animal experiments were conducted under the guidance of the Animal Testing Center of Tsinghua University in accordance with strict animal ethical standards. The BALB/c female mice (5–6 weeks old and 16–18 g) used in our experiments were purchased from Beijing Weitong Lihua Experimental Animal Technology Co., Ltd. (Beijing, China). To establish the tumor model, 0.2 mL human breast cancer cell line MCF-7 cell suspension (5×10^6 cells) was injected subcutaneously into the hind limb of the nude mice. After the tumor grew to 75–100 mm^3, the nude mice were used in the animal experiments. Each mouse was injected with 0.15 mL of 1 mg mL^{-1} solution intravenously. Subsequently, the mice were anesthetized at 6 h, 24 h, 48 h and then placed in supine position. For image acquisition, the Cellvizio®dual band imaging system (Mauna Kea Technologies, Paris, France) was used to obtain real-time images of the blood vessels (imaged by tail vein injection of Evans blue under 660 nm excitation), and GQD-NPs (imaged under 488 nm excitation) at the tumor site.

3. Results and Discussion

The structure of the obtained samples could be clearly seen in the HRTEM image (Figure 1a) since the outside layer and inside nanoparticles had different contrasts. The HRTEM image (shown in Figure 1b,c) further illustrated the amorphous coating of the polymer and the inside nanoparticles with a clear lattice structure. The energy dispersive spectroscopy (EDS) map (Figure 1d–i) demonstrated that the inside nanoparticles mainly consisted of Fe and O, and the outer layer mainly contained C and N. GQDs were hardly identified in the HRTEM and EDS images (Figure S1).

X-ray diffraction was employed to further validate the content of the nanocomposites. Peaks in Figure 2a were ascribed to Fe_3O_4 [3]. The Raman spectrum of the GQD-NPs was shown in Figure 2b. The D peak at 1372 cm^{-1} was generally considered to be the disordered vibration peak of GQDs, caused by lattice vibrations leaving the center of the Brillouin zone, which characterized the defects or edges of the GQDs. The peak present at 1582 cm^{-1} was the G peak, which was a characteristic peak of GQDs. The G peak was higher than the D peak, indicating the more edged structure of the GQDs [21]. The XPS spectra of GQD-NPs were shown in Figure 2c–g. From Figure 2c, it could be seen that the nanocomposites consisted of C, N, O and Fe. The C=C peak of C1s in Figure 2d indicated that there were a large number of conjugated structures in the nanocomposites. The appearance of the C–O peak revealed that there were both amino groups and hydroxyl groups on the surface of GQDs. In the spectrum of O1s in Figure 2e, the C–O and O–H peaks further confirmed the presence of hydroxyl groups on the surface of GQDs. The peak of N–H in Figure 2f was presumed to be the N–H in the nitrogen-containing ring in PPy. The Fe2p in Figure 2g illustrated the presence of Fe_3O_4. The XPS spectra further demonstrated that there were Fe_3O_4, PPy and GQDs in the nanocomposites [3,21,25]. Studies have shown that excellent fluorescent luminescence performance is one of the most outstanding properties of graphene quantum dots [11,26,27]. Therefore, after loading the GQDs, the optical performance of the nanocomposite would be of concern. As shown in Figure 2h, the GQD-NPs presented an absorption peak under the excitation wavelength of 490 nm, which was the characteristic peak of the GQDs. In Figure 2i, the GQD-NPs retained a slightly decreased saturation magnetization compared to bare Fe_3O_4, allowing them to perform as a contrast agent for magnetic resonance imaging.

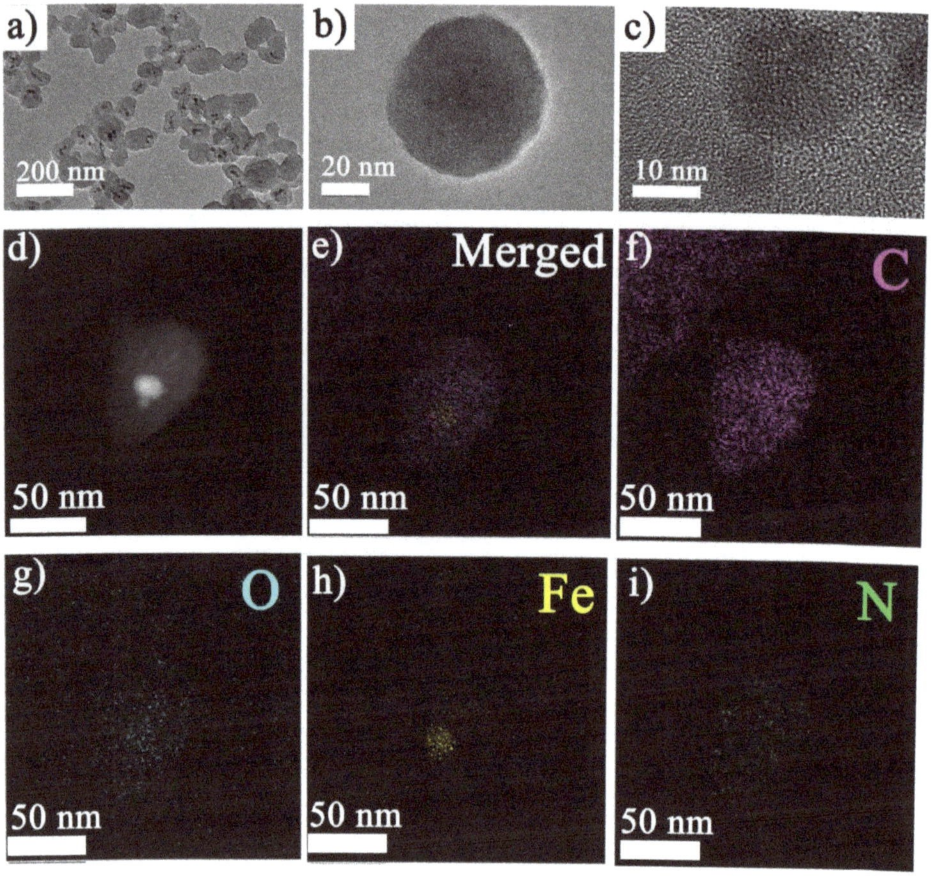

Figure 1. Morphology and elemental analysis of graphene quantum dots (GQDs) loaded nanoparticles: (**a**~**c**) high resolution tranmission electron microscopy (HRTEM) images at different magnifications (bright field image); (**d**~**i**) energy dispersive spectroscopy (EDS) analysis of the nanocomposite; (**d**) graph representation, (**e**) EDS analysis of the nanocomposite (merged), (**f**) EDS C elemental map, (**g**) EDS O elemental map, (**h**) EDS Fe elemental map, (**i**) EDS N elemental map.

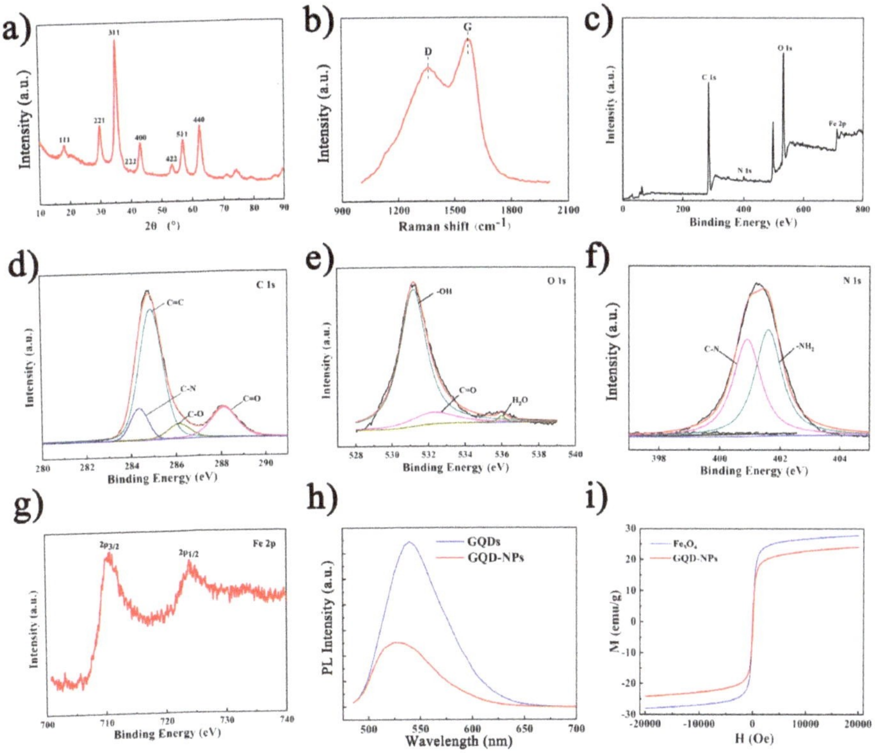

Figure 2. Properties of GQD-nanoparticles (NPs): (**a**) X-ray diffraction (XRD) patterns; (**b**) Raman spectrum, the ordered G band and disordered D band were indicated; (**c**) X-ray photoelectron spectrum (XPS); (**d–g**) XPS analysis of C, O, N and Fe, respectively; (**h**) Photoluminescence (PL) spectra with the pulsed laser excitations at 490 nm; (**i**) Magnetizing curve, magnetization (M), magnetic field strength (H).

The viability of L929 cells incubated with gradient concentrations of GQD-NPs (50 μg/mL, 100 μg/mL, 160 μg/mL, 250 μg/mL, 500 μg/mL, and 1000 μg/mL) for 24 h or 48 h was detected by CCK8. As shown in Figure 3, after co-incubation for 24 h, there were significant differences between the groups whose concentrations of GQD-NPs were higher than 50 μg/mL and the control group; after co-incubation for 48 h, high significant differences were found between all the groups treated with GQD-NPs and the control group. The outcome illustrated that GQD-NPs exhibited excellent biocompatibility when the concentration was lower than 250 μg/mL. This particular outcome was in line with previous studies which investigated the cell toxicity of GQDs [28] or Fe_3O_4@PPy [22].

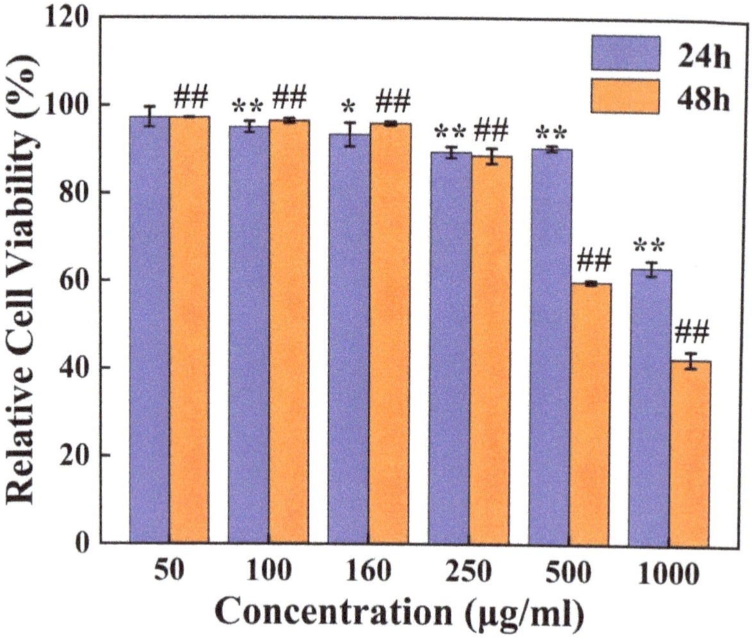

Figure 3. Cell viability of L929 cells incubated with GQD-NPs. The experiments were carried out in triplicate. Data were presented as mean ± standard deviation (SD) ($n = 3$). The cell viability values were all normalized to control groups (untreated cells after incubation for 24 h or 48 h). Asterisk (*) and double asterisks (**) refer to statistical significance of $p < 0.05$ and $p < 0.01$, respectively, compared with control groups between the cell viability values after co-incubation for 24 h; double pounds (##) refer to a statistical significance of $p < 0.01$ compared with control groups between the cell viability values after co-incubation for 48 h.

Evans blue dye strongly binds to hemoglobin in blood, and is therefore widely used to track the presence of blood vessels [29]. As seen in Figure 4, prior to injection, a clear vascular system was displayed at the tumor site under excitation with the 660 nm laser. A few green fluorescence lights were seen under 488 nm, since some substances in the visible light band would also emit green fluorescence under visible-light excitation. GQDs were employed in the tracking of nanoparticles *in vivo*. When the GQD-NPs were injected for six hours, green fluorescence indicated the presence of GQD-NPs, while the blood vessels at the tumor site were still clearly visible (red). In the merged image, it could be seen that a large number of GQD-NPs apeared in the tumor blood vessels, and a small amount of GQD-NPs penetrated into the surrounding tissues through the tumor blood vessel wall. When the GQD-NPs were injected for 24 h, the blood vessels at the tumor site were still clearly visible (red), and there were still many GQD-NPs within the tumor blood vessels (green). As time passed (from 6 h to 24 h), a large amount of GQD-NPs passed through the tumor vessel wall and penetrated into the surrounding tissues as for the enhanced permeability and retention (EPR) effect, achieving enrichment in the tumor tissues. When the GQD-NPs were injected for 48 h, there were still fluorescence signals of the GQD-NPs (green) in the tumor tissues and the blood vessels, but the fluorescence signals in the tumor tissues were significantly weakened (relative to 24 h). This indicated that more GQD-NPs had been metabolized. Thus, reduced enrichment of the GQD-NPs at the tumor site was found. The results illustrated that by loading GQDs into nanoparticles, the dynamic changes of nanoparticles could be tracked *in vivo*.

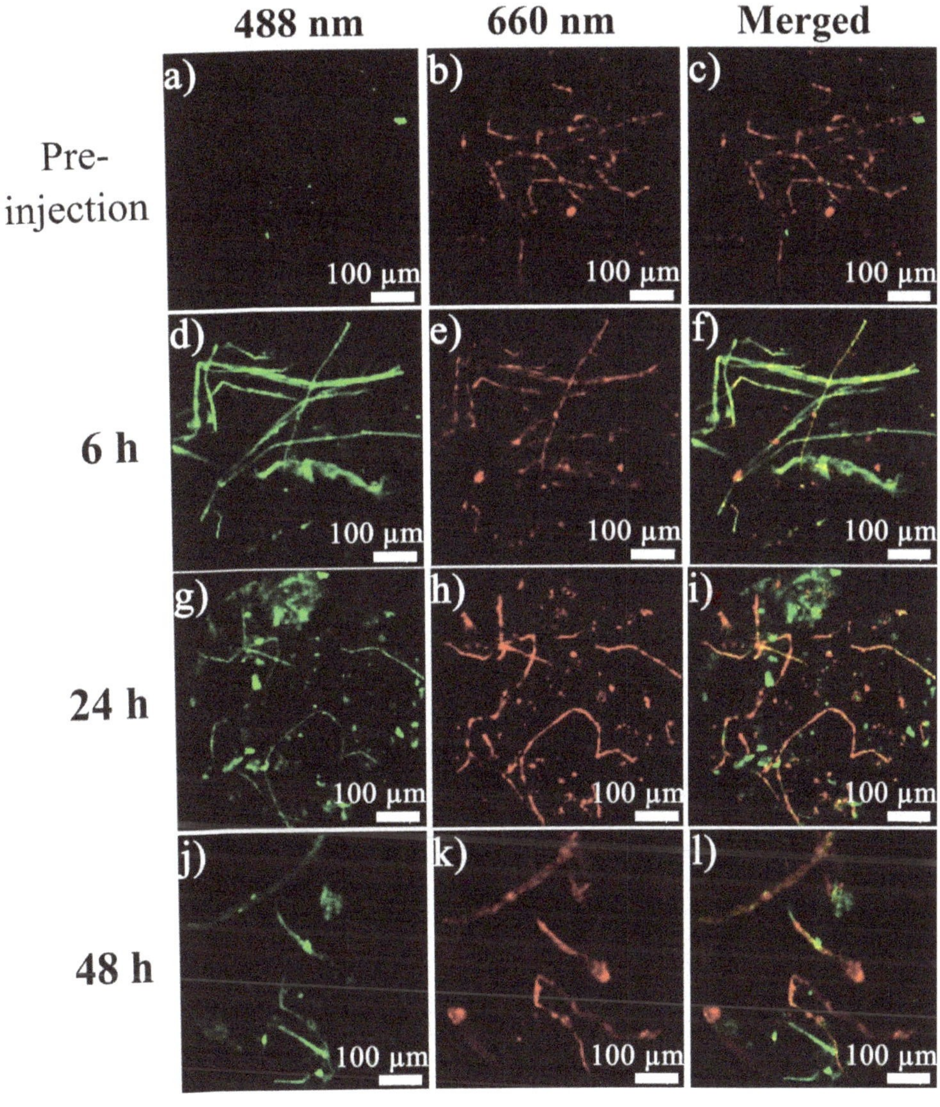

Figure 4. Fibered confocal fluorescence microscopic (FCFM) images of GQD-NPs (imaged under 488 nm excitation, green) and blood vessels (imaged by tail vein injection of Evans blue under 660 nm excitation, red) in tumor tissues over time.

4. Conclusions

The one-pot method of loading graphene quantum dots (GQDs) was proposed as a means to modify nanoparticles for fluorescent imaging. GQDs were loaded into the optical-magneto Fe_3O_4@PPy nanoparticles instead of on the surface. The morphology, components, optical and magnetic properties, biocompatibility and biomedical application of the GQD-loaded optical-magneto nanoparticles were investigated. The GQD-loaded optical-magneto nanoparticles performed as an excellent tacking agent, highlighting the potential application of GQDs in tracking the real-time distribution and metabolism of nanoparticles *in vivo*.

Supplementary Materials: The following are available online at http://www.mdpi.com/1996-1944/12/13/2191/s1, Figure S1: Hydrodynamic size of GQDs-NPs nanocomposites.

Author Contributions: Conceptualization, writing—review and editing, R.Z. and Q.C.; Writing—original draft preparation, Y.W.; Material preparation and characterization Y.W., Y.H.; Cell experiment, N.X., In vivo experiment D.W., Q.G.; Figures, J.W.; Data analysis S.X., Y.L.

Funding: This research was funded by National Key Research & Development Program of China (2016YFC0304502).

Acknowledgments: We gratefully acknowledge the measurement support from State key Laboratory of New Ceramics and Fine Processing, School of Materials Science and Engineering, Tsinghua University, Beijing 100084, China.

Conflicts of Interest: The authors declare no conflict of interest.

References

1. Bray, F.; Ferlay, J.; Soerjomataram, I.; Siegel, R.L.; Torre, L.A.; Jemal, A. Global Cancer Statistics 2018: Globocan Estimates of Incidence and Mortality Worldwide for 36 Cancers in 185 Countries. *Ca A Cancer J. Clin.* **2018**, *68*, 394–424. [CrossRef] [PubMed]
2. Wang, Q.; Zhao, X.; Yan, H.; Kang, F.; Li, Z.; Qiao, Y.; Li, D. A Cross-Talk Egfr/Vegfr-Targeted Bispecific Nanoprobe for Magnetic Resonance/Near-Infrared Fluorescence Imaging of Colorectal Cancer. *Mrs Commun.* **2018**, *8*, 1008–1017. [CrossRef]
3. Yan, H.; Zhao, L.; Shang, W.; Liu, Z.; Xie, W.; Qiang, C.; Xiong, Z.; Zhang, R.; Li, B.; Sun, X.; et al. General Synthesis of High-Performing Magneto-Conjugated Polymer Core-Shell Nanoparticles for Multifunctional Theranostics. *Nano Res.* **2017**, *10*, 704–717. [CrossRef]
4. Wang, C.; Xu, H.; Liang, C.; Liu, Y.; Li, Z.; Yang, G.; Cheng, L.; Li, Y.; Liu, Z. Iron Oxide @ Polypyrrole Nanoparticles as a Multifunctional Drug Carrier for Remotely Controlled Cancer Therapy with Snergistic Antitumor Effect. *Acs Nano* **2013**, *7*, 6782–6795. [CrossRef] [PubMed]
5. Ogata, G.; Ishii, Y.; Asai, K.; Sano, Y.; Nin, F.; Yoshida, T.; Higuchi, T.; Sawamura, S.; Ota, T.; Hori, K.; et al. AMicrosensing System for the *in vivo* Real-Time Detection of Local Drug Kinetics. *Nat. Biomed. Eng.* **2017**, *1*, 654–666. [CrossRef] [PubMed]
6. Cai, Y.; Wei, Z.; Song, C.; Tang, C.; Han, W.; Dong, X. Optical Nano-Agents in the Second Near-Infrared Window for Biomedical Applications. *Chem. Soc. Rev.* **2019**, *48*, 22–37. [CrossRef] [PubMed]
7. Resch-Genger, U.; Grabolle, M.; Cavaliere-Jaricot, S.; Nitschke, R.; Nann, T. Quantum Dots Versus Organic Dyes as Fluorescent Labels. *Nat. Methods* **2008**, *5*, 763–775. [CrossRef]
8. Wang, J.; Qiu, J. A Review of Organic Nanomaterials in Photothermal Cancer Therapy. *Cancer Res. Front.* **2016**, *2*, 67–84. [CrossRef]
9. Ponomarenko, L.A.; Schedin, F.; Katsnelson, M.I.; Yang, R.; Hill, E.W.; Novoselov, K.S.; Geim, A.K. Chaotic Dirac Billiard in Graphene Quantum Dots. *Science* **2008**, *320*, 356–358. [CrossRef]
10. Liu, R.; Wu, D.; Feng, X.; Mullen, K. Bottom-Up Fabrication of Photoluminescent Graphene Quantum Dots with Uniform Morphology. *J. Am. Chem. Soc.* **2011**, *133*, 15221–15223. [CrossRef]
11. Yan, Y.; Gong, J.; Chen, J.; Zeng, Z.; Huang, W.; Pu, K.; Liu, J.; Chen, P. Recent Advances on Graphene Quantum Dots: From Chemistry and Physics to Applications. *Adv. Mater.* **2019**, *31*, 1808283. [CrossRef]
12. Tabish, T.A.; Scotton, C.J.; J Ferguson, D.C.; Lin, L.; der Veen, A.V.; Lowry, S.; Ali, M.; Jabeen, F.; Ali, M.; Winyard, P.G.; et al. Biocompatibility and Toxicity of Graphene Quantum Dots for Potential Application in Photodynamic Therapy. *Nanomedicine* **2018**, *13*, 1923–1937. [CrossRef] [PubMed]
13. Saquib, Q.; Faisal, M.; Al-Khedhairy, A.A.; Alatar, A.A. *Cellular and Molecular Toxicology of Nanoparticles*; Springer: Cham, Switzerland, 2018; Volume 1048.
14. Iannazzo, D.; Pistone, A.; Celesti, C.; Triolo, C.; Patané, S.; Giofré, S.; Romeo, R.; Ziccarelli, I.; Mancuso, R.; Gabriele, B.; et al. A Smart Nanovector for Cancer Targeted Drug Delivery Based on Graphene Quantum Dots. *Nanomaterials* **2019**, *9*, 282. [CrossRef] [PubMed]
15. Zhang, X.; Wei, C.; Li, Y.; Yu, D. Shining Luminescent Graphene Quantum Dots: Synthesis, Physicochemical Properties, and Biomedical Applications. *Trac Trends Anal. Chem.* **2019**, *116*, 109–121. [CrossRef]
16. Şenel, B.; Demir, N.; Büyükköroğlu, G.; Yıldız, M. Graphene Quantum Dots: Synthesis, Characterization, Cell Viability, Genotoxicity for Biomedical Applications. *Saudi Pharm. J.* **2019**. [CrossRef]

17. Chen, T.; Yu, H.; Yang, N.; Wang, M.; Ding, C.; Fu, J. Graphene Quantum Dot-Capped Mesoporous Silica Nanoparticles through an Acid-Cleavable Acetal Bond for Intracellular Drug Delivery and Imaging. *J. Mater. Chem. B* **2014**, *2*, 4979. [CrossRef]
18. Su, X.; Chan, C.; Shi, J.; Tsang, M.; Pan, Y.; Cheng, C.; Gerile, O.; Yang, M. A Graphene Quantum Dot@Fe_3O_4@SiO_2 Based Nanoprobe for Drug Delivery Sensing and Dual-Modal Fluorescence and Mri Imaging in Cancer Cells. *Biosens. Bioelectron.* **2017**, *92*, 489–495. [CrossRef]
19. Wang, X.; Sun, X.; Lao, J.; He, H.; Cheng, T.; Wang, M.; Wang, S.; Huang, F. Multifunctional Graphene Quantum Dots for Simultaneous Targeted Cellular Imaging and Drug Delivery. *Colloids Surf. B Biointerfaces* **2014**, *122*, 638–644. [CrossRef]
20. Tian, Q.; Wang, Q.; Yao, K.X.; Teng, B.; Zhang, J.; Yang, S.; Han, Y. Multifunctional Polypyrrole@Fe_3O_4 Nanoparticles for Dual-Modal Imaging and In Vivo Photothermal Cancer Therapy. *Small* **2014**, *10*, 1063–1068. [CrossRef]
21. Wang, L.; Wang, Y.; Xu, T.; Liao, H.; Yao, C.; Liu, Y.; Li, Z.; Chen, Z.; Pan, D.; Sun, L.; et al. Gram-Scale Synthesis of Single-Crystalline Graphene Quantum Dots with Superior Optical Properties. *Nat. Commun.* **2014**, *5*, 5357. [CrossRef]
22. Feng, W.; Zhou, X.; Nie, W.; Chen, L.; Qiu, K.; Zhang, Y.; He, C. Au/Polypyrrole@Fe_3O_4 Nanocomposites for Mr/Ct Dual-Modal Imaging Guided-Photothermal Therapy: An *in vitro* Study. *ACS Appl. Mater. Inter.* **2015**, *7*, 4354–4367. [CrossRef] [PubMed]
23. Yan, H.; Shang, W.; Sun, X.; Zhao, L.; Wang, J.; Xiong, Z.; Yuan, J.; Zhang, R.; Huang, Q.; Wang, K.; et al. "All-in-One" Nanoparticles for Trimodality Imaging-Guided Intracellular Photo-Magnetic Hyperthermia Therapy Under Intravenous Administration. *Adv. Funct. Mater.* **2018**, *28*, 1705710. [CrossRef]
24. Yao, W.; Ni, T.; Chen, S.; Li, H.; Lu, Y. Graphene/Fe_3O_4@Polypyrrole Nanocomposites as a Synergistic Adsorbent for Cr(Vi) Ion Removal. *Compos. Sci. Technol.* **2014**, *99*, 15–22. [CrossRef]
25. Wang, J.; Yan, H.; Liu, Z.; Wang, Z.; Gao, H.; Zhang, Z.; Wang, B.; Xu, N.; Zhang, S.; Liu, X. Langmuir–Blodgett Self-Assembly of Ultrathin Graphene Quantum Dot Films with Modulated Optical Properties. *Nanoscale* **2018**, *10*, 19612–19620. [CrossRef] [PubMed]
26. Iannazzo, D.; Ziccarelli, I.; Pistone, A. Graphene Quantum Dots: Multifunctional Nanoplatforms for Anticancer Therapy. *J. Mater. Chem. B* **2017**, *5*, 6471–6489. [CrossRef]
27. Bacon, M.; Bradley, S.J.; Nann, T. Graphene Quantum Dots. *Part. Part. Syst. Charact.* **2014**, *31*, 415–428. [CrossRef]
28. Chong, Y.; Ma, Y.; Shen, H.; Tu, X.; Zhou, X.; Xu, J.; Dai, J.; Fan, S.; Zhang, Z. The In Vitro and In Vitro Toxicity of Graphene Quantum Dots. *Biomaterials* **2014**, *35*, 5041–5048. [CrossRef]
29. Han, Z.; Shang, W.; Liang, X.; Yan, H.; Hu, M.; Peng, L.; Jiang, H.; Fang, C.; Wang, K.; Tian, J. An Innovation for Treating Orthotopic Pancreatic Cancer by Preoperative Screening and Imaging-Guided Surgery. *Mol. Imaging Biol.* **2019**, *21*, 67–77. [CrossRef]

materials

MDPI

Article

Preparation and In Vitro Release of Total Alkaloids from *Alstonia Scholaris* Leaves Loaded mPEG-PLA Microspheres

Xiangyu Zheng [1], Hongli Li [1], Yi He [1], Mingwei Yuan [1], Meili Shen [2], Renyu Yang [1], Nianfeng Jiang [1], Minglong Yuan [1,*] and Cui Yang [1,*]

1 National and Local Joint Engineering Research Center for Green Preparation Technology of Biobased Materials, Yunnan Minzu University, Kunming 650500, China; zhengxiangyu1993@163.com (X.Z.); honglili@vip.163.com (H.L.); heyi_sichuan@163.com (Y.H.); yuanmingwei@163.com (M.Y.); yangrenyu1995@163.com (R.Y.); jnf1514@163.com (N.J.)

2 National and Local Joint Engineering Laboratory for Synthesis Technology of High Performance Polymer, Jilin University, Changchun 130012, China; sml199211@163.com

* Correspondence: yml@188.com (M.Y.); yangcui@ynni.edu.cn (C.Y.); Tel.: +86-871-6591-4825 (M.Y. & C.Y.)

Received: 3 April 2019; Accepted: 28 April 2019; Published: 6 May 2019

Abstract: Total alkaloids of *Alstonia scholaris* leaves (ASAs) are extracted from the lamp leaves, which have positive anti-inflammatory activity and remarkable effects in treating bronchitis. Due to its short half-life, we used a degradable mPEG-PLA to physically encapsulate the total alkali of the lamp stage, and prepared a sustained-release microsphere by double-emulsion method. The ASAs-loaded $mPEG_{10000}$-PLA microspheres were screened for better performance by testing the morphology, average particle size, embedding rate and drug loading of different molecular weight mPEG-PLA microspheres, which can stably and continuously release for 15 days at 37 °C. The results of cytotoxicity and blood compatibility indicated that the drug-loaded microspheres have beneficial biocompatibility. Animal experiments showed that the drug-loaded microspheres had a beneficial anti-inflammatory effect. These results all indicated that mPEG-PLA is a controlled release carrier material suitable for ASAs.

Keywords: total alkaloids from *Alstonia scholaris* leaves; mPEG-PLA; microspheres; drug release; biocompatibility

1. Introduction

Respiratory diseases are common and frequently occurring diseases whose incidence has increased in recent years. Inhalation of dust and irritating gases are the main causes of respiratory diseases, which cause tremendous harm to human health [1]. *Alstonia scholaris* leaves, a traditional Chinese medicine from the Dai nationality, are mainly used to treat chronic respiratory diseases [2]. *A. scholaris* leaves are also used as a traditional medicine to treat respiratory diseases in India, Malaysia, the Philippines, and Thailand [2]. *A. scholaris* leaf extracts, especially alkaloids, are widely used to treat bronchitis and have significant efficacy [3–7]. Luo et al. studied *A. scholaris* leaves and found an alkaloid component that beneficial effects respiratory diseases. This component has a positive effect on bronchitis and post-cold infections [8,9]; however, its half-life is short, which limits its clinical application.

Research and development of sustained-release formulations have always been a focus in drug research and have gradually begun to solve the problem of short half-lives. In recent years, polymer nanoparticles have shown to be efficient drug carriers for encapsulating drugs [10]. At present, designing and selecting drug nanocomposites has mainly focused on amphiphilic block copolymers. Polyethylene glycol (PEG) is often used in the hydrophilic segment of the copolymers, while many homopolymers, copolymers and derivatives of degradable polymers, such as polylactic acid (PLA), polyglycolic acid (PGA) and polycaprolactone (PCL), are used in the hydrophobic segment [11–14]. Methoxy poly(ethylene glycol)-poly(lactide) copolymer (mPEG-PLA) is an amphiphilic polymer formed by grafting mPEG onto PLA. mPEG-PLA has excellent biocompatibility, a low molecular weight, and many hydroxyl groups and is nontoxic and widely used as a coating material in drug-delivery systems [15,16]. PEG blocks can improve the polymer's hydrophilicity and flexibility, prevent protein adsorption and avoid recognition and phagocytosis by the reticuloendothelial system (RES). PEG encapsulates the drug to form the "core", while the "shell" is formed by the outside hydrophobic segment, which constitutes the typical "core-shell" structure, with great advantages for drug release [17–19]. Therefore, in this work, total alkaloids from *A. scholaris* leaves (ASAs) were studied, and mPEG-PLA with different molecular weights was used as the carrier material. A series of drug-loaded microspheres were prepared by the water-oil-water (W/O/W) double-emulsion technique. Based on the morphology, particle size, encapsulation efficiency (EE), drug-loading efficiency (LE), and in vitro drug release of the prepared microspheres, the best drug-loaded materials were screened out, and then the biological properties of its drug-loaded microspheres such as cytotoxicity, blood compatibility and anti-inflammatory activity were studied.

2. Materials and Methods

2.1. Materials

L-lactide was purchased from Purac Co., Ltd. (Shanghai, China). Dichloromethane (DCM), ethanol, tert-butyl methyl ether, and isopropanol were purchased from Tianjin Damao Chemical Reagent Co., Ltd. (Tianjin, China) and were all analytical grade. ASAs were purchased from Kunming Institute of Botany, Chinese Academy of Sciences, batch number 20180501 (Kunming, China). ASAs is a mixture of brown solid powder. Two of main active ingredients are Vallesamine and Picrinine, which were detected by HPLC with peak times of 43.133 min and 72.190 min. NaH_2PO_4, KH_2PO_4, and NaCl were purchased from Chengdu Chron Chemicals Co., Ltd. (Chengdu, China). Stannous octoate (95%, analytical grade) and polyvinyl alcohol (PVA; Mw = 75,000 Da and 88% alcoholysis degree) were purchased from Shanghai Jingchun Chemicals Co., Ltd. (Shanghai, China). mPEG (Mw 5000) was purchased from Shanghai Seebio Biotechnology Co., Ltd. (Shanghai, China). HL-7702 cells (normal human hepatocytes) were purchased from Procell Life Sciences and Technology Co., Ltd. (Wuhan, China). High glucose and 1640 medium were provided by HyClone Company. The 3-(4,5-dimethyl-2-thiazolyl)-2,5-diphenyl-2-H-tetrazolium bromide (MTT) was purchased from APExBIO (Houston, TX, USA). High-quality fetal bovine serum (FBS) was purchased from Shanghai ExCell Bio Co., Ltd. (Shanghai, China). Ethylenediaminetetraacetic acid (EDTA)-trypsin solution, cyan-streptomycin, and cisplatin were provided by Solarbio (Shanghai, China). Kunming mice, Sprague-Dawley (SD) rats, and rabbits were purchased from Kunming Medical University (Kunming, China). Dengtaiye Pian (DP) was purchased from Yunnan Datang Hanfang Pharmaceutical Co., Ltd. (Kunming, China). Aspirin was purchased from Bayer HealthCare Manufacturing Co., Ltd. (Beijing, China).

2.2. Preparation of mPEG-PLA

mPEG-PLA was prepared by ring-opening polymerization. L-LA (18 g) and mPEG (2 g) were putted in reaction flask as raw materials. Protected by nitrogen, catalyzed by stannous octoate (0.02 g), the mPEG-PLA was polymerized at 130–160 °C for 24 h. Using recrystallization, the product was precipitated by methylene chloride and ethanol. Product was filtered and washed with hot water to remove the mPEG bulk polymer. The obtained product was dried with P_2O_5 for 48 h to obtain a white copolymer. Figure 1 shows the synthesis process.

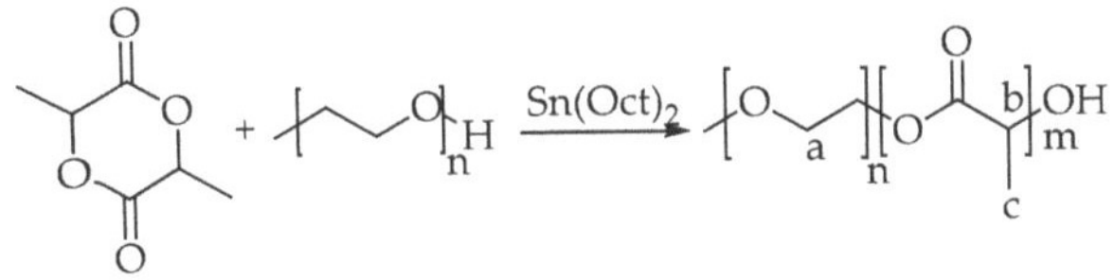

Figure 1. The synthesis process of mPEG-PLA.

2.3. *Determination of Molecular Mass*

3 mg copolymer were dissolved in 1 mL chromatography grade THF and was filtered by using a 0.45 μm nylon 66 filter membrane. Then using gel permeation chromatography (GPC) which was purchased from Waters Inc. (Milford, MA, USA) to define the molecular weight and distribution of the compounds, and THF was used as the eluent. The system is equipped with a column (7.8 × 300 mm, Waters Styragel, Waters Inc., Milford, MA, USA), a Waters 515 pump and a Waters 2414 refractive index detector. When the column temperature of GPC is 40 °C, the flow rate is 1 mL/min, and the baseline is smooth. The filtrate was pulled into an injection needle, and the sample was slowly and uniformly injected into the sampler when the air was removed. All data were obtained under the same standard curve.

2.4. *Preparing the Blank and ASAs-Loaded mPEG-PLA Microspheres*

The ASAs-loaded mPEG-PLA microspheres were prepared using the W/O/W double-emulsion technique. First, 100 mg of ASAs were dissolved in 1 mL methanol solution, which was the internal water phase (W_1). The 500-mg mPEG-PLA copolymer material was weighed and completely dissolved in 10 mL of methylene chloride as the oil phase (O). The external water phase (W_2) was a 2% PVA solution. Next, the ASAs solution was injected into the oil phase in an ice bath at a constant rate. The mixture was emulsified at high speed (21,000 rpm) for 2 min to form the first emulsion. Similarly, the first emulsion was continuously dropped into 20 mL of 2% PVA (W_2) in an ice bath and emulsified at high speed (21,000 rpm) for 2 min to form a double emulsion. The double emulsion was poured into 400 mL of 5% isopropanol solution and stirred at low speed at room temperature for 6 h. After the organic solvent had completely evaporated, the ASAs-loaded microsphere solution was obtained. Finally, the ASAs-loaded microsphere solution was centrifuged (6500 rpm, 10 min), and the supernatant was discarded. The microspheres were collected (white solid), washed with pure water 3 times, and lyophilized at −50 °C for 24 h. The obtained microspheres were dried at −20 °C. Blank microspheres were prepared by the same method.

2.5. *Morphology, Particle size and Particle Size Distribution of the Microspheres*

The morphological characteristics of the ASAs-loaded mPEG-PLA microspheres were observed using SEM (NOVA NANOSEM-450, FEI, Hillsboro, OR, USA). Conductive adhesive tape was applied to the surface of the loading platform. A few lyophilized microspheres (white powder) were evenly spread on a conductive adhesive, and the surface was sprayed with gold to observe. The particle size distribution of the microspheres was measured using a laser particle-size analyzer (Mastersizer 3500, Microtrac, Malvern, UK). A small portion of the microsphere suspension was diluted in a 15 mL centrifuge tube until the solution became nearly transparent. The particle size distribution of the microspheres was determined using a particle-size analyzer.

2.6. *Measuring the EE of the Microspheres*

Twenty milligrams of lyophilized microspheres were weighed, and 500 μL of methylene chloride was added to completely dissolve them. Next, 2 mL of methanol solution was added. After centrifugation, the supernatant was filtered through a 0.45-μm microfiltration membrane. Next,

20 μL of filtered solution was injected via HPLC, and the area normalization method was used to calculate the EE% and LE% using the following equations.

$$LE\% = \frac{\text{weight of ASAs in microspheres}}{\text{total weight of microspheres}} \times 100\%$$

$$EE\% = \frac{\text{weight of encapsulated drug}}{\text{weight of initial drug loading}} \times 100\%$$

2.7. *In Vitro Microsphere Release*

Fifty milligrams of lyophilized microspheres were weighed and placed in a 15 mL centrifuge tube, then 5 mL of phosphate buffer saline (PBS) was added and shaken in a thermostatic oscillator at 37 °C. The tube was removed at set time intervals and centrifuged at 5000 rpm for 6 min. Next, 1 mL of the release solution was removed from the supernatant, an equal amount of fresh PBS was added to the tube, and the tube was placed back in the thermostatic oscillator and shaken at 37 °C. The ASAs content in the release solution was determined by HPLC, and the cumulative drug release amount was calculated using the following formula [20]. The amount (%) of the released ASAs-loaded microspheres was plotted against the release time (t, in days) to obtain the in vitro release curve of the microspheres.

$$Q = C_n \times V_1 + V_2 \sum C_{n-1}$$

Q: cumulative drug release (μg);
C_n: concentration of the release solution (μg/mL) removed at time t;
V_1: volume of the released medium (mL);
V_2: volume of the medium to be withdrawn each time (mL).

2.8. *In Vitro Microsphere Degradation*

Fifty milligrams of lyophilized microspheres were weighed and placed in a 15 mL centrifuge tube, and then, 5 mL of PBS was added and shaken at 37 °C. The tube was removed at set time intervals and centrifuged at 5000 rpm. The supernatant was collected to determine the pH changes in the microspheres. Buffer salts on the microsphere surface were washed with pure water and frozen at −50 °C for 24 h. The microspheres were weighed using an analytical balance, and the dry weight loss of the degraded microspheres was calculated. The molecular weight of 3 mg of the lyophilized microspheres was measured via GPC. The pH value, the microsphere weight and the relative molecular weight (Mn) were separately plotted against degradation time to obtain the in vitro degradation curve of the microspheres.

2.9. *Blood Compatibility Testing of The Microspheres*

All the trials were approved by the laboratory animals Ethics Committee of Yunnan Minzu University and were registered on the Kunming Science and Technology Bureau (SYXK (Yunnan) K2017-0001, 16 January 2017).

2.9.1. Hemolysis Experiment

The samples were dissolved in deionized water (DI water), and the solutions were prepared at concentrations of 10.6 μg/mL, 1.06 mg/mL, and 0.106 mg/mL. Eight milliliters of fresh anticoagulated rabbit blood (blood mixed with sodium citrate at a volume ratio of 9:1) was diluted with 10 mL of 0.9% sodium chloride solution. Two hundred microliters of the sample was placed in a test tube, and 5 mL of 0.9% sodium chloride solution was added. The tube was then kept in a 37 °C water bath for 30 min. Next, 100 μL of diluted blood was added, and the mixture was gently mixed and incubated for 60 min. The positive control was treated with 5 mL DI water and 100 μL blood (D = 0.8 ± 0.3). The negative control was treated with 5 mL 0.9% NaCl solution and 100 μL blood. After centrifugation at 800 r/min for 5 min, the supernatant was transferred into a cuvette, and an ultraviolet (UV) spectrophotometer was used to measure the absorbance at 540 nm. Hemolysis of the sample was calculated per the formula below. The final sample concentrations were 40 μg/mL, 4 μg/mL, and 0.4 μg/mL. The material was considered hemolyzed when the hemolysis rate was above 5% [21–27].

$$\text{Hemolysis rate}(\%) = \frac{\text{Abs}_{\text{sample}} - \text{Abs}_{\text{negative control}}}{\text{Abs}_{\text{positive control}} - \text{Abs}_{\text{negative control}}} \times 100\%$$

2.9.2. Coagulation Experiment

The samples were dissolved in distilled water to prepare solutions with concentrations of 245.2 μg/mL, 24.52 μg/mL, and 2.452 mg/mL. The final concentrations were 40 μg/mL, 4 μg/mL, and 0.4 μg/mL. Next, 200 μL of the solution was placed in a 15-mL centrifuge tube and kept at 37 °C for 5 min. Fifty microliters of fresh anticoagulated rabbit blood was added to the samples, which were and kept at a constant temperature for 5 min. Ten microliters of aqueous calcium chloride solution (0.2 mL/L) was added to the blood sample, and the centrifuge tube was shaken to evenly mix the calcium chloride and blood and kept at a constant temperature for 5 min. The centrifuge tube was then removed, 12 mL of DI water was added to the solution, and the supernatant was collected. The blood was measured at 540 nm using a UV spectrophotometer. The optical absorbance (i.e., the optical density [OD] of the free hemoglobin remaining in the beaker containing 50 μL of whole blood treated with 12 mL of DI water) was used as a reference. The average of 5 measurements was taken. The sample's anticoagulant activity was expressed as the relative absorbance [28,29]:

$$\text{BCI} = \frac{I_o}{I_w} \times 100\%$$

I_o: relative absorbance of the mixture of blood and calcium chloride after contact with the sample for a set period of time;
I_w: relative absorbance of blood mixed with a certain amount of DI water.

2.10. Cytotoxicity of The Microspheres

2.10.1. Cell Culture

Normal human hepatocytes (HL-7702) were inoculated into culture flasks, and RPMI-1640 medium (containing 10% FBS, 100 U/mL penicillin, and 100 U/mL streptomycin) was added. Cells were cultured in a 5% CO_2 incubator at 37 °C. Cells grew as monolayers adherent to glass and were passaged once every 3–5 days using 0.25% trypsin.

2.10.2. MTT Assay of Sample Inhibition on Tumor Cell Proliferation

HL-7702 cells (180 μL; 5 × 10^4 cells/mL) in the logarithmic growth phase were inoculated in a 96-well plate, and 20 μL of the sample was added to each well after incubating overnight. Three concentration gradients were set with 3 wells per concentration. After a 48-h incubation, 20 μL of 5 mg/mL MTT was added to each well, and the cells were continuously cultured for 4 h. The culture medium was aspirated and discarded, and 150 μL dimethylsulfoxide (DMSO) was added to terminate the reaction. The plate was shaken for 15 min in a shaker. The OD value at 490 nm was measured in a microplate reader [30,31], and the inhibition rate was calculated [32]:

$$\text{Cell viability} = \frac{\text{ODsample}}{\text{ODnegative control}} \times 100\%$$

2.11. Anti-Inflammatory Activity Testing of The Microspheres

To determine the influence of xylene-induced auricle swelling on the mice [33], 35 Kunming mice (19–22 g) were randomly divided into 7 groups by body weight and sex, with 5 mice per group. Except for the aspirin group, only one intragastric administration on the day of the experiment, the other groups were pre-administered once daily for 3 consecutive days. The control group was administered the same volume of 1% CMC-Na (carboxymethyl cellulose-sodium). The gavage volume was 20 mL/kg per group. Thirty minutes after the last gastric gavage, 0.05 mL of xylene was evenly applied on both sides of each mouse's right ear, while the left ear was used as the control. The mice were sacrificed via cervical dislocation 1 h after inflammation. The same area of both ears was cut off using a 10-mm-diameter puncher, and the weight difference between the two ears was used as the swelling degree.

To determine the influence of the egg-white-induced pedal swelling in the rats, 35 male SD rats (170–220 g) were randomly divided into 7 groups by body weight, with 5 rats per group. All groups were intragastrically administered the microspheres once daily for 3 consecutive days, except the aspirin group, which received the microspheres via intragastric administration on the day of the experiment. The control group received the same volume of 1% CMC-Na (carboxymethyl cellulose-sodium). The gavage volume was 10 mL/kg per group. Thirty minutes after the last gastric gavage, 0.1 mL of fresh egg white was injected subcutaneously into the pedal of the right rear foot of each rat to induce inflammation. The foot areas before inflammation and at 0.5, 1, 2, 3, 4, and 5 h after inflammation were measured. The difference in the foot area before and after inflammation was determined as the degree of swelling, and the swelling rate was calculated.

$$\text{Swell rate} = \frac{A_{t1} - A_{t0}}{A_{t0}} \times 100\%$$

A_{t0}: foot area before administration
A_{t1}: foot area after administration

3. Results and Discussion

3.1. Characterization of Copolymer mPEG-PLA

Figure 2 shows the nuclear magnetic resonance (NMR) spectra of the mPEG-PLA copolymer. The ^{1}H NMR image shows the characteristic peaks of hydrogen in the mPEG-PLA. The peaks at 5.19 ppm and 1.5 ppm correspond to the PLA's methine and methyl peaks, respectively. The methylene peak at 3.6 ppm is the repeating unit in the PEG. The peak at 3.39 ppm corresponds to the hydrogen of the CH_3O- group of the mPEG at the end of the copolymer. NMR spectroscopy was performed at 25 °C (Bruker 400 MHz, Karlsruhe, Germany).

Figure 3 shows the mPEG-PLA gel permeation chromatography (GPC) curve with the different molecular weights of mPEG. Data were processed using different molecular weights of polystyrene as the reference material. Table 1 shows the weight-average molecular weight (Mw), number-average molecular weight (Mn) and polydispersity index of the molecular weight distribution (PDI) of the copolymer. The obtained copolymers have relatively small PDIs and relatively uniform molecular weights. Four copolymers with different molecular weights were prepared by copolymerization of four mPEG with the same amount of L-LA. The GPC curves show that the different molecular weights of the four mPEG-PLA, they all have a single peak, indicating that no other copolymers were formed. The copolymer synthesized from mPEG of different molecular weights, wherein the larger the molecular weight of mPEG, the higher the molecular weight of the obtained copolymer and the shorter the peak time of the copolymer, which is consistent with the results obtained by the GPC test.

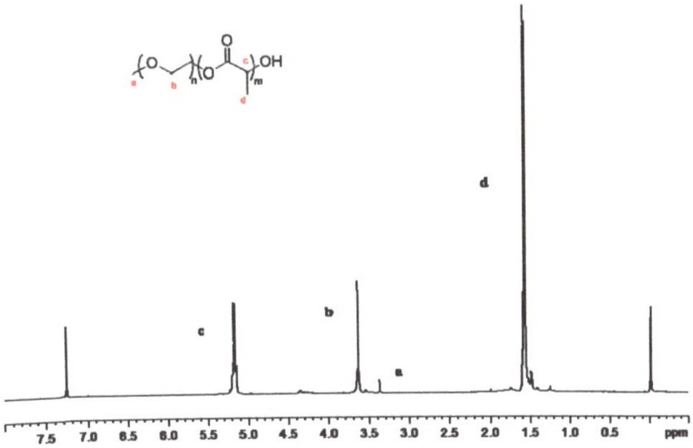

Figure 2. ^{1}H nuclear magnetic resonance (NMR) spectrum of mPEG-PLA.

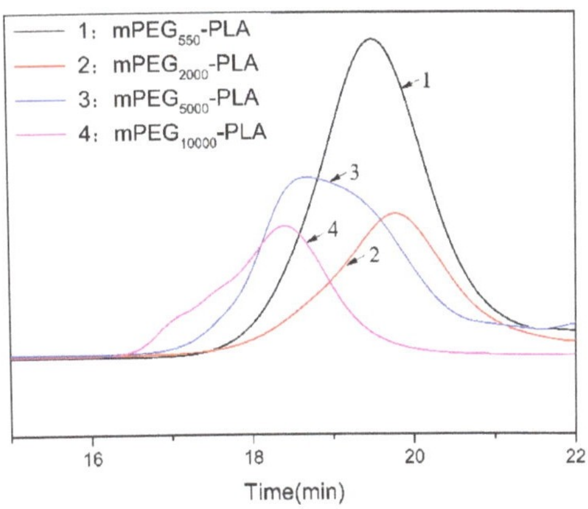

Figure 3. Gel permeation chromatography (GPC) of mPEG-PLA.

Table 1. The Mn, Mw and PDI of different molecular weight of mPEG-PLA.

Sample	Mn [1] ($g{\cdot}mol^{-1}$)	Mn [2] ($g{\cdot}mol^{-1}$)	Mw [1] ($g{\cdot}mol^{-1}$)	PDI [1]
$mPEG_{550}$-PLA	4013	4097	4540	1.13
$mPEG_{2000}$-PLA	5489	5504	6320	1.15
$mPEG_{5000}$-PLA	7378	7419	8637	1.17
$mPEG_{10000}$-PLA	11,307	11,416	12,889	1.14

[1] Determined by GPC via a universal calibration curve and appropriate Mark–Houwink parameters for THF. [2] Calculated by ^{1}H NMR. Mn: The number-average molecular weight; Mw: The weight-average molecular weight; PDI: polydispersity index of the molecular weight distribution.

3.2. Morphology and Size Distribution of the Microspheres

Figure 4 shows the scanning electron microscopy (SEM) images of the mPEG-PLA microspheres loaded with ASAs. Figure 4A shows $mPEG_{550}$-PLA microspheres, which have a small particle diameter, a round smooth spherical shape, and no obvious adhesion. Figure 4B shows $mPEG_{2000}$-PLA microspheres with small pores on the surface and varied particle sizes. Figure 4C shows $mPEG_{5000}$-PLA microspheres with a few small pores on the surface and with different pore sizes and numbers. Figure 4D shows $mPEG_{10000}$-PLA microspheres. Compared with the $mPEG_{550}$-PLA, $mPEG_{2000}$-PLA, and $mPEG_{5000}$-PLA microspheres, the $mPEG_{10000}$-PLA's particle size is larger, its surface has few small pores and its shape is smooth. The adhesion is due to the relatively large molecular weight of the mPEG-PLA copolymer. The pores on the surface are caused by water molecule diffusion during lyophilization. Figure 5 shows the particle size distribution of the mPEG-PLA microspheres loaded with ASAs using a laser particle-size analyzer (Mastersizer 3500, Microtrac, Malvern, UK). The average diameter of the $mPEG_{550}$-PLA microspheres was 1.803 ± 0.21 μm, and the PDI was 0.15. The average diameter of the $mPEG_{2000}$-PLA microspheres was 2.083 ± 0.17 μm, and the PDI was 0.14. The average diameter of the $mPEG_{5000}$-PLA microspheres was 2.631 ± 0.2 μm, and the PDI was 0.20. The average diameter of the $mPEG_{10000}$-PLA microspheres was 3.84 ± 0.30 μm, and the PDI was 0.18(Table 2). These results are nearly the same as those shown in the SEM images. The experimental results showed that the diameters of the 4 microspheres were smaller than 5 μm, indicating that these microspheres are suitable for oral of drug delivery, which is consistent with the study by Wei et al. [34].

Figure 4. Scanning electron microscopy (SEM) images of mPEG-PLA microsphere. (**A**) mPEG 550-PLA, (**B**) mPEG 2000-PLA, (**C**) mPEG5000-PLA, (**D**) mPEG10000-PLA. The magnification of SEM image was 2000.

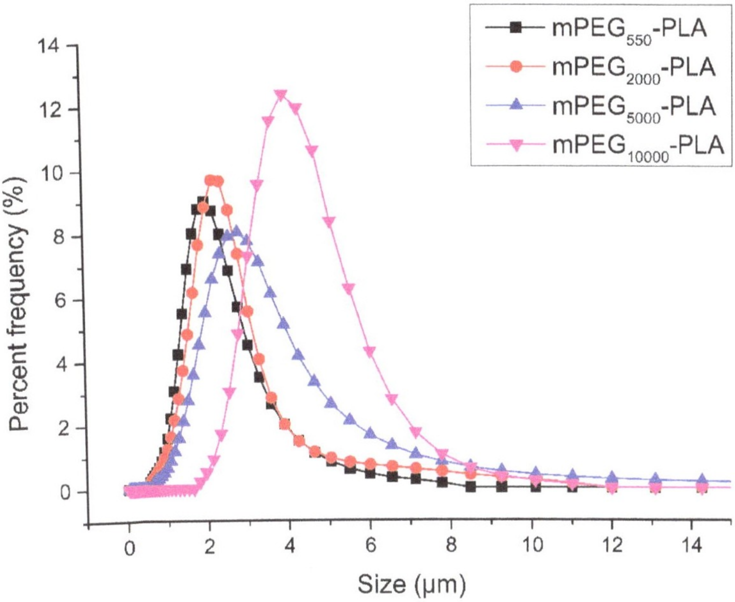

Figure 5. Size distribution of mPEG-PLA microsphere.

3.3. *Drug Loads*

mPEG-PLA microspheres loaded with ASAs were prepared using the W/O/W double-emulsion technique, and the EE and LE were measured by high-performance liquid chromatography (HPLC; Table 2). With the increased molecular weight, the microsphere's EE and LE increased, with trends similar to those of the mPEG-PLA microspheres loaded with recombinant human growth hormone (rhGH) prepared by Yi et al. [35]. While preparing the microspheres, a portion of the drug diffused into the external water phase due to solvent evaporation, resulting in drug loss. A copolymer with a high molecular weight has a large molecular gap, enabling the drug to diffuse outward, thus increasing drug loss.

Table 2. The average diameter, PDI, EE, LE of mPEG-PLA microspheres loaded ASAs.

Sample	Average Diameter (μm)	[1]PDI	[1]EE (%)	[1]LE (%)
$mPEG_{550}$-PLA	1.803 ± 0.21	0.15	53.97 ± 2.57	2.71 ± 0.21
$mPEG_{2000}$-PLA	2.083 ± 0.17	0.14	62.78 ± 3.44	3.05 ± 0.11
$mPEG_{5000}$-PLA	2.631 ± 0.20	0.20	70.28 ± 3.61	4.14 ± 0.12
$mPEG_{10000}$-PLA	3.840 ± 0.30	0.18	75.12 ± 4.25	5.17 ± 0.19

[1] PDI: polydispersity index of the particle size distribution. EE: encapsulation efficiency. LE: loading efficiency.

3.4. *Analysis of In Vitro Microsphere Release*

Figure 6 shows the release curve of mPEG-PLA microspheres loaded with ASAs in phosphate-buffered saline (PBS) (pH = 7.40). The cumulative amount of $mPEG_{550}$-PLA microspheres released on the first day was 24.32 ± 0.37%, and then, the release rate slowed, reaching 64.38 ± 1.21% after a two-week stable release. The cumulative amount of $mPEG_{2000}$-PLA microspheres released on the first day was 20.74 ± 0.29%, and then, the release rate slowed, reaching 57.09 ± 1.28% after a two-week stable release. The cumulative amount of $mPEG_{5000}$-PLA microspheres released on the first day was 19.54 ± 0.22%, and then, the release rate slowed, reaching 52.45 ± 1.28% after a two-week stable release. The cumulative amount of the $mPEG_{10000}$-PLA microspheres released on the first day was 15.03 ± 0.51%, and then, the release rate slowed, reaching 45.12 ± 1.04% after a two-week stable release. The relatively large amount of the drug released on the first day occurred because a small amount of the drug adhered to the microsphere surface when the microspheres encapsulated the drug. With the microsphere's increased molecular weight, in vitro release of the drug from the microspheres was relatively slow. The release time of the mPEG-PLA microspheres in this study was longer than that of the microspheres prepared by Zheng et al. [36] and Xiong et al. [37], indicating that the mPEG-PLA was well encapsulated on the ASAs.

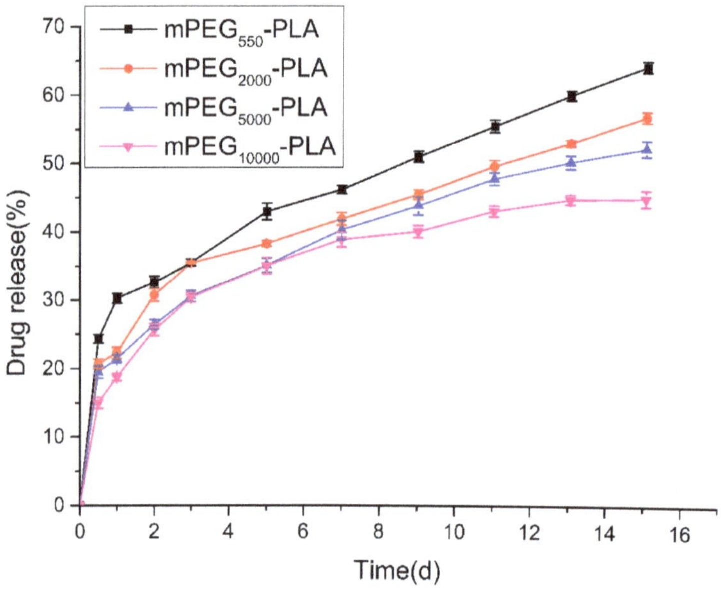

Figure 6. Release curve of mPEG-PLA microspheres loaded ASAs in PBS

3.5. *Analysis of In Vitro Microsphere Degradation*

Based on the morphology, particle size, EE, LE, and in vitro microsphere release, we selected $mPEG_{10000}$-PLA as the optimal microspheres for the study. Figure 7 shows the degradation process of the ASAs-loaded $mPEG_{10000}$-PLA microsphere. The figure shows the changes in the microsphere's dry weight, system pH and molecular weight over time. In PBS (pH = 7.40, 10 mM), the three curves all decreased with time. Within 60 days, the microspheric pH decreased from 7.40 to 5.89, the microspheres lost 47.16% of their dry weight, and their molecular weight dropped from 10307 Da to 8258 Da. As its degradation progressed, the mPEG-PLA was hydrolyzed to produce CO_2, which was dissolved in water to decrease the pH. mPEG-PLA is an amphiphilic material that absorbs water in PBS and breaks the hydrophilic segment, thus decreasing the molecular weight. As the hydrophilic segment breaks, the molecular weight of the copolymer decreases, resulting in a loss of copolymer quality. Li et al. [38] studied the degradation of mPEG-PLA nanoparticles and found that the mPEG-PLA degraded very slowly, and the Mn decreased by 27.6% within 30 days. Simone [39] et al. studied the polymer degradation conditions at pH and found that the pH decreased correspondingly over time, which is similar to the mPEG-PLA degradation trend in this study.

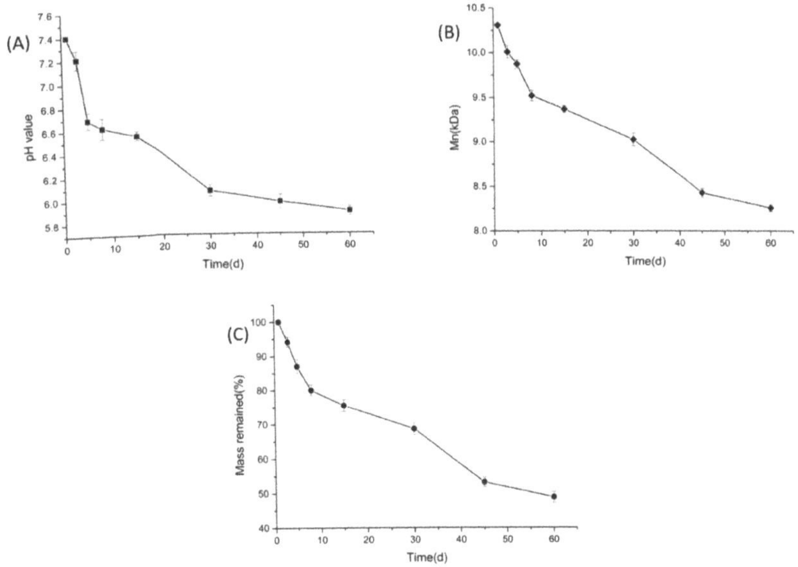

Figure 7. The degradation process of the ASAs-loaded mPEG$_{10000}$-PLA microspheres in the phosphate buffer saline (PBS) (pH = 7.40): the pH (**A**) Mn (**B**) and mass (**C**) changed with time.

3.6. *Analysis of In Vitro Hemolytic Properties of Microspheres*

Figure 8 shows the hemolysis and anticoagulation rates of the ASAs-loaded mPEG$_{10000}$-PLA microsphere, the mPEG$_{10000}$-PLA, and the ASAs. The hemolysis rates of the ASAs-loaded mPEG$_{10000}$-PLA microsphere, the mPEG$_{10000}$-PLA, and the ASAs increased as the concentration increased but did not exceed 5% at 40 μg/mL. Studies have shown that hemolysis rates below 5% indicate blood compatibility, which is consistent with the International Organization for Standardization (ISO) hemolysis standard, indicating applicability for intravenous injection. The anticoagulant chart shows that the blood clotting index (BCI) increases as the concentration increases, indicating that its anticoagulant activity increases as the sample concentration increases.

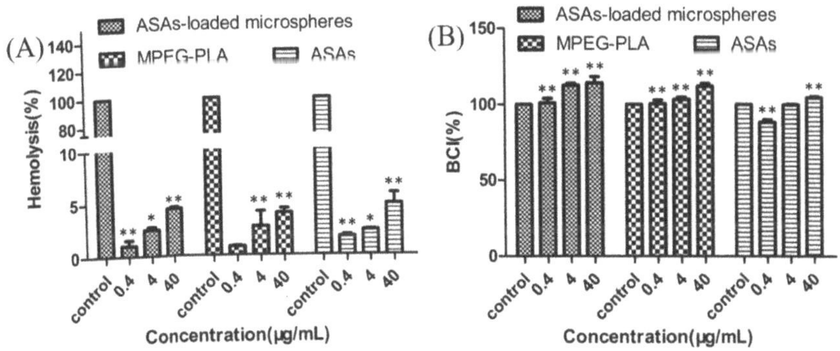

Figure 8. Hemolytic activity (**A**) and blood clotting index (**B**) of the ASAs-loaded mPEG$_{10000}$-PLA microsphere. * $P < 0.05$ and ** $P < 0.01$, compared with control, and the whole blood sample with the ASAs was used as a control.

3.7. *In Vitro Cytotoxicity Analysis of The Microspheres*

Figure 9 shows the cytotoxicity of the ASAs-loaded mPEG$_{10000}$-PLA microspheres, the mPEG$_{10000}$-PLA, and the ASAs to the HL-7702 (normal human hepatocyte) cell line. To study the cytotoxicity of the ASAs-loaded mPEG$_{10000}$-PLA microspheres, the 3-(4,5-dimethylthiazol-2-yl)-2,5-diphenyltetrazolium bromide (MTT) assay was used to detect HL-7702 cell viability under different concentrations of ASAs-loaded mPEG$_{10000}$-PLA microspheres, mPEG$_{10000}$-PLA, and ASAs. Figure 8 shows the toxicity at concentrations from 0.4 μg/mL to 40 μg/mL. The results showed that mPEG$_{10000}$-PLA microspheres could be used for drug loading. The microspheres prepared by Song et al. [40] showed that the HL-7702 cells were dose- and time-dependent, and as the concentration increased, the HL-7702 cell survival rate decreased, which is similar to the HL-7702 cell viability under the ASAs in this study.

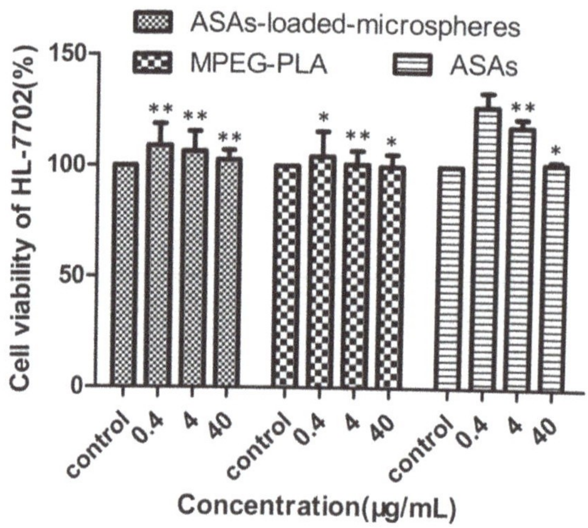

Figure 9. The cytotoxicity of the ASAs-loaded mPEG$_{10000}$-PLA microspheres, the mPEG$_{10000}$-PLA, and the ASAs to the HL-7702 cells. * $P < 0.05$ and ** $P < 0.01$, compared with control, and the whole blood sample with the ASAs was used as a control.

3.8. *Analysis of the Microspheric Anti-Inflammatory Activity*

Table 3 shows the effect of the ASAs on xylene-induced auricle swelling in mice. The ASAs-loaded microspheres significantly inhibited the xylene-induced auricle swelling in the mice. The effect of the ASAs-loaded microspheres at 10 mg/kg was comparable to that of the 0.3 g/kg DP, and the efficacies at 20 and 40 mg/kg were stronger than that of 10 mg/kg aspirin.

Table 3. Effect of the ASAs on xylene-induced auricle swelling in mice.

Group	Doses (/kg)	Auricular Swelling Degree ($\overline{x} \pm s$, mg)	Inhibition Rate (%)
Control	—	50.85 ± 3.61	—
Aspirin	10 mg	34.65 ± 2.90	31.86%
DP	0.3 g	37.65 ± 3.04	25.96%
ASAs	0.1 g	39.65 ± 0.78	22.03%
	10 mg	39.00 ± 2.69	23.30%
Loaded microspheres	20 mg	31.35 ± 6.86	38.35%
	40 mg	27.65 ± 7.57	45.62%

Table 4 shows the effect of the ASAs on egg-white-induced pedal swelling in rats. The ASAs-loaded microspheres significantly inhibited the egg-white-induced pedal swelling in the rats, and the action time of the ASAs-loaded microspheres at 20 mg/kg was longer than that of the DP, and the action time of ASAs-loaded microspheres at 80 mg/kg was similar to that of aspirin.

Table 4. Effect of the ASAs on egg-white-induced pedal swelling in rats.

Group	Doses (/kg)	Inhibitory Swelling Rate of Sole (%)				
		1h	2h	3h	4h	5h
Control	—	—	—	—	—	—
Aspirin	20 mg	66.77	62.07	57.02	52.17	49.55
DP	0.24 g	68.81	65.86	60.39	57.31	54.48
ASAs	0.16 g	78.97	69.99	68.29	60.14	57.87
Loaded microspheres	20 mg	51.73	47.12	43.72	38.97	35.14
	40 mg	59.62	55.47	46.64	44.61	41.78
	80 mg	75.37	73.22	65.14	58.07	55.27

4. Conclusions

ASAs-loaded mPEG-PLA microspheres were successfully prepared using the W/O/W double-emulsion technique. The comprehensive results for the microspheres, including particle size, EE, LE, in vitro release and other tests, showed that $mPEG_{10000}$-PLA microspheres performed excellently. The particle size of ASAs-loaded $mPEG_{10000}$-PLA microspheres was 3.84 ± 0.30 μm, the EE was 75.12 ± 4.25%, and the LE was 5.17 ± 0.19%. The microspheres sustained release for 15 days in a simulated human environment, with beneficial biocompatibility at 40 μg/mL, and showed no toxicity to HL-7702 cells. It significantly inhibited auricle swelling in mice and pedal swelling in rats. These results indicate that mPEG-PLA is a carrier material suitable for controlled ASAs release.

Author Contributions: M.Y. and C.Y. conceived and designed the experiments; X.Z., Y.H., R.Y. and N.J. performed the experiments; X.Z., M.S. and H.L. analyzed the data; M.Y., H.L. and M.Y. contributed reagents/materials/analysis tools; X.Z. wrote the paper. X.Z. and H.L. contributed equally to this work.

Funding: This work was supported by the National Natural Science Foundation of China (Project Nos. 31460247, 81460542, 81760644), the Biodegradable Materials Innovative Research Team (in Science and Technology) at the University of Yunnan Province and the Innovation Team Based on Research and Application of Biological Functional Materials of Yunnan Minzu University (2017HC034), Yunnan Science and Technology Project - Major Science and Technology Specialities of Biological Medicine (2018ZF008). Graduate Innovation Foundation of Yunnan Minzu University (2018YJCXS268).

Conflicts of Interest: The authors declare no conflicts of interest.

References

1. Mentese, S.; Mirici, N.A.; Otkun, M.T.; Bakar, C.; Palaz, E.; Tasdibi, D.; Cevizci, S.; Cotuker, O. Association between Respiratory Health and Indoor Air Pollution Exposure in Canakkale, Turkey. *Build. Environ.* **2015**, *93*, 72–83. [CrossRef]
2. Khyade, M.S.; Kasote, D.M.; Vaikos, N.P. *Alstonia scholaris* (L.) R. Br. and Alstonia macrophylla Wall. ex G. Don: A comparative review on traditional uses, phytochemistry and pharmacology. *J. Ethnopharmacol.* **2014**, *153*, 1–18. [CrossRef]
3. Atta-Ur-Rahman; Alvi, K.A. Indole alkaloids from *Alstonia scholaris*. *Phytochemistry* **1987**, *26*, 2139–2142. [CrossRef]
4. Atta-Ur-Rahman, A.; Asif, M.; Ghazala, M.; Fatima, J.; Alvi, K.A. Scholaricine, an alkaloid from *Alstonia scholaris*. *Phytochemistry* **1985**, *24*, 2771–2773. [CrossRef]
5. Xiang-Hai, C.; Zhi-Zhi, D.; Xiao-Dong, L. Unique monoterpenoid indole alkaloids from *Alstonia scholaris*. *Org. Lett.* **2007**, *9*, 1817–1820.
6. Cai, X.H.; Shang, J.H.; Tao, F.; Luo, X.D. Novel Alkaloids from *Alstonia scholaris*. *Zeitschrift Für Naturforschung B* **2010**, *65*, 1164–1168. [CrossRef]

7. Cai, X.H.; Tan, Q.G.; Liu, Y.P.; Feng, T.; Du, Z.Z.; Li, W.Q.; Luo, X.D. A cage-monoterpene indole alkaloid from *Alstonia scholaris*. *Org. Lett.* **2008**, *10*, 577–580. [CrossRef]
8. Zhao, Y.-L.; Shang, J.-H.; Pu, S.-B.; Wang, H.-S.; Wang, B.; Liu, L.; Liu, Y.-P.; Hong-Mei, S.; Luo, X.-D. Effect of total alkaloids from *Alstonia scholaris* on airway inflammation in rats. *J. Ethnopharmacol.* **2016**, *178*, 258–265. [CrossRef] [PubMed]
9. Shang, J.-H.; Cai, X.-H.; Zhao, Y.-L.; Feng, T.; Luo, X.-D. Pharmacological evaluation of *Alstonia scholaris*: Anti-tussive, anti-asthmatic and expectorant activities. *J. Ethnopharmacol.* **2010**, *129*, 293–298. [CrossRef] [PubMed]
10. Ho, D.; Leong, J.W.; Crew, R.C.; Norret, M.; House, M.J.; Mark, P.J.; Waddell, B.J.; Iyer, K.S.; Keelan, J.A. Maternal-placental-fetal biodistribution of multimodal polymeric nanoparticles in a pregnant rat model in mid and late gestation. *Sci. Rep.* **2017**, *7*, 2866. [CrossRef]
11. Anderson, J.M.; Shive, M.S. Biodegradation and biocompatibility of PLA and PLGA microspheres. *Adv. Drug Deliv. Rev.* **2012**, *64*, 72–82. [CrossRef]
12. Dong, C.M.; Guo, Y.Z.; Qiu, K.Y.; Gu, Z.W.; Feng, X.D. In vitro degradation and controlled release behavior of d, l-PLGA50 and PCL-b-d, l-PLGA50 copolymer microspheres. *J. Control. Release* **2005**, *107*, 53–64. [CrossRef]
13. Carreras, N.; Acuña, V.; Martí, M.; Lis, M.J. Drug release system of ibuprofen in PCL-microspheres. *Colloid Polym. Sci.* **2013**, *291*, 157–165. [CrossRef]
14. Yu, H.; Ping, L.; Chen, S.; Liu, Y.; Fu, D. The influence of hydrophilic mPEG segment on formation, morphology, and properties of PCL-mPEG microspheres. *Adv. Polym. Technol.* **2017**, *37*, 2281–2287.
15. Zheng, W.; Li, M.; Lin, Y.; Zhan, X. Encapsulation of verapamil and doxorubicin by MPEG-PLA to reverse drug resistance in ovarian cancer. *Biomed. Pharmacother.* **2018**, *108*, 565–573. [CrossRef]
16. Li, X.; Huang, Z. Advance of Preparation and Application of Polylactide and Its Copolymers. *Chin. Polym. Bull.* **1999**, *1*, 24–32.
17. Qing, X.; Yingying, Z.; Xu, C.; Haimei, L.; Evans, D.G.; Wensheng, Y. Nanosheet-based titania microspheres with hollow core-shell structure encapsulating horseradish peroxidase for a mediator-free biosensor. *Biomaterials* **2011**, *32*, 6588–6594.
18. Hou, X.; Hao, Y. *Fabrication of Polystyrene/Detonation Nanographite Composite Microspheres with the Core/Shell Structure via Pickering Emulsion Polymerization*; Hindawi Publishing Corp.: London, UK, 2013.
19. Hiraoka, S. Microspheres Having Core/Shell Structure. U.S. Patent 20100203151A1, 12 August 2010.
20. Keramati, M.; Ghasemi, I.; Karrabi, M.; Azizi, H.; Sabzi, M. Incorporation of surface modified graphene nanoplatelets for development of shape memory PLA nanocomposite. *Fibers Polym.* **2016**, *17*, 1062–1068. [CrossRef]
21. Qi, Y.; Geng, Y.; Ma, B.; Yun, H.; Cui, C. In vitro hemolytic properties' assessment of K2Ti6O13 nanowires. *Proc. Inst. Mech. Eng. Part N J. Nanoeng. Nanosyst.* **2014**, *229*, 201–205. [CrossRef]
22. Henkelman, S.; Rakhorst, G.; Blanton, J.; Oeveren, W.V. Standardization of incubation conditions for hemolysis testing of biomaterials. *Mater. Sci. Eng. C* **2009**, *29*, 1650–1654. [CrossRef]
23. Muthukumarasamyvel, T.; Rajendran, G.; Santhana, P.D.; Kasthuri, J.; Kathiravan, K.; Rajendiran, N. Role of Surface Hydrophobicity of Dicationic Amphiphile-Stabilized Gold Nanoparticles on A549 Lung Cancer Cells. *ACS Omega* **2017**, *2*, 3527–3538. [CrossRef]
24. Qiu, Y.Y.; Chen, Y.; Zhou, W.Y.; Xia, X.L.; Yang, X.J. Biocompatibility and security of calcium sulfate bone substitutes. *Chin. J. Tissue Eng. Res.* **2016**, *20*, 2317–2323.
25. Cao, Y.E.; Wang, H.; Yang, C.; Zhong, R.; Lei, Y.U.; Sun, K.; Liu, J. In vitro studies of PBT Nonwoven Fabrics adsorbent for the removal of low density lipoprotein from hyperlipemia plasma. *Appl. Surf. Sci.* **2011**, *257*, 7521–7528. [CrossRef]
26. Ciochină, A.D.; Bredeţean, O.; Dimitriu, D.C.; Iacob, G. Considerations on in vitro and in vivo magnetic nanoparticles hemocompatibility testing. *Revista Medico-Chirurgicala a Societatii de Medici si Naturalisti din Iasi* **2009**, *113*, 279–285.
27. Ren, B.R.; Liu, J.; Zhang, E.L.; Dong, H. Biocompatibility of a new titanium alloy containing copper. *Chin. J. Tissue Eng. Res.* **2015**, *1*, 5473–5479.
28. Archana, D.; Dutta, J.; Dutta, P.K. Evaluation of chitosan nano dressing for wound healing: Characterization, in vitro and in vivo studies. *Int. J. Biol. Macromol.* **2013**, *57*, 193–203. [CrossRef] [PubMed]
29. Lan, G.; Lu, B.; Wang, T.; Wang, L.; Chen, J.; Yu, K.; Liu, J.; Dai, F.; Wu, D. Chitosan/gelatin composite sponge is an absorbable surgical hemostatic agent. *Colloids Surf. B Biointerfaces* **2015**, *136*, 1026–1034. [CrossRef]

30. Chia-Jen, L.; Chia-Herng, Y.; Yu-Jie, L.; Yu-Yu, L.; Shao-Hsuan, K.; Jer-Yuh, L.; Yieng-How, C. Antitumor activity of acriflavine in lung adenocarcinoma cell line A549. *Anticancer Res.* **2014**, *34*, 6467–6472.
31. Kumar, S.P.; Birundha, K.; Kaveri, K.; Devi, K.T.R. Antioxidant studies of chitosan nanoparticles containing naringenin and their cytotoxicity effects in lung cancer cells. *Int. J. Biol. Macromol.* **2015**, *78*, 87–95. [CrossRef] [PubMed]
32. Ge, Y.; Cheng, R.; Zhou, Y.; Shen, J.; Peng, L.; Xu, X.; Dai, Q.; Liu, P.; Wang, H.; Ma, X. Cryptotanshinone induces cell cycle arrest and apoptosis of multidrug resistant human chronic myeloid leukemia cells by inhibiting the activity of eukaryotic initiation factor 4E. *Mol. Cell. Biochem.* **2012**, *368*, 17–25. [CrossRef] [PubMed]
33. Zhang, H.; Shi, X. Anti-inflammatory and analgesic effects of total flavonoids in flos sophorae immaturus. *Clin. Res. Pract.* **2018**, *11*, 4–6.
34. Wei, Y.; Wang, Y.; Kang, A.; Wang, W.; Ho, S.V.; Gao, J.; Ma, G.; Su, Z. A novel sustained-release formulation of recombinant human growth hormone and its pharmacokinetic, pharmacodynamic and safety profiles. *Mol. Pharm.* **2012**, *9*, 2039. [CrossRef] [PubMed]
35. Yi, W.; Xia, W.Y.; Wei, W.; Ho, S.V.; Feng, Q.; Guang Hui, M.; Guo, S.Z. Microcosmic mechanisms for protein incomplete release and stability of various amphiphilic mPEG-PLA microspheres. *Langmuir* **2012**, *28*, 13984–13992.
36. Zheng, X.L.; Kan, B.; Gou, M.L.; Fu, S.Z.; Zhang, J.; Men, K.; Chen, L.J.; Luo, F.; Zhao, Y.L.; Zhao, X. Preparation of MPEG–PLA nanoparticle for honokiol delivery in vitro. *Int. J. Pharm.* **2010**, *386*, 262–267. [CrossRef] [PubMed]
37. LizhiXiong; ZeqiangHe, Preparation and In-Vitro Properties of mPEG/PLA Microspheres Loaded 5-Fluorouracil for Controlled Release. *J. Macromol. Sci. Part D Rev. Polym. Process.* **2013**, *52*, 268–272.
38. Yuan, L.; Rong, Q.X.; Yoshie, M.; Tsuneji, N. PEG-PLA diblock copolymer micelle-like nanoparticles as all-trans-retinoic acid carrier: In vitro and in vivo characterizations. *Nanotechnology* **2009**, *20*, 055106.
39. Simone, E.A.; Dziubla, T.D.; Francheska, C.G.; Discher, D.E.; Muzykantov, V.R. Effect of polymer amphiphilicity on loading of a therapeutic enzyme into protective filamentous and spherical polymer nanocarriers. *Biomacromolecules* **2007**, *8*, 3914–3921. [CrossRef]
40. Song, X.L.; Bo, L.; Xu, K.; Liu, J.; Wen, J.; Wang, J.; Liu, X.D.; Li, J.; Qi, Y.F. Cytotoxicity of water-soluble mPEG-SH-coated silver nanoparticles in HL-7702 cells. *Cell Biol. Toxicol.* **2012**, *28*, 225–237. [CrossRef] [PubMed]

materials

MDPI

Article

Utilization of Ethylcellulose Microparticles with Rupatadine Fumarate in Designing Orodispersible Minitablets with Taste Masking Effect

Katarzyna Wasilewska [1], Patrycja Ciosek-Skibińska [2], Joanna Lenik [3], Stanko Srčič [4], Anna Basa [5] and Katarzyna Winnicka [1,*]

1 Department of Pharmaceutical Technology, Medical University of Białystok, Mickiewicza 2c, 15-222 Białystok, Poland; katarzyna.wasilewska@umb.edu.pl
2 Chair of Medical Biotechnology, Warsaw University of Technology, Noakowskiego 3, 00-664 Warsaw, Poland; pciosek@ch.pw.edu.pl
3 Department of Analytical Chemistry and Instrumental Analysis, Faculty of Chemistry, Maria Curie-Skłodowska University, M. Curie-Skłodowska Sq. 3, 20-031 Lublin, Poland; j.lenik@poczta.umcs.lublin.pl
4 Department of Pharmaceutical Technology, University of Ljubljana, Aškerčeva c. 7, 1000 Ljubljana, Slovenia; stanko.srcic@ffa.uni-lj.si
5 Department of Physical Chemistry, Faculty of Chemistry, University of Białystok, Ciołkowskiego 1K, 15-245 Białystok, Poland; abasa@uwb.edu.pl
* Correspondence: kwin@umb.edu.pl; Tel.: +48-85-7485616

Received: 5 May 2020; Accepted: 9 June 2020; Published: 15 June 2020

Abstract: Minitablets in orodispersible form constitute a flexible drug delivery tool for paediatric and geriatric population as they eliminate the risk of chocking and do not require drinking water in the application. Due to their direct contact with taste buds, taste sensation is an important factor. Preparing microparticles with taste masking polymers utilizing spray drying is an efficient technique for reducing the bitterness of drugs. Ethylcellulose is a hydrophobic polymer widely used as a taste masking material. Rupatadine fumarate, one of the newest antihistamines, features an intensive bitter taste, hence in designing orodispersible formulations, achieving an acceptable taste is a crucial issue. The main objective of this work was to formulate orodispersible minitablets containing taste masked ethylcellulose-based microparticles with rupatadine fumarate and evaluation of their quality, especially in terms of taste masking efficacy. The accessed data indicated that all obtained minitablets were characterized by beneficial pharmaceutical properties. Three independent methods: in vivo with healthy volunteers, in vitro drug dissolution, and "electronic tongue" confirmed that all designed formulations provided satisfactory taste masking rate and that formulation F15 (prepared with Pearlitol® Flash and Surelease® microparticles with rupatadine fumarate) was characterized by the lowest bitterness score.

Keywords: ethylcellulose; spray drying; microparticles; rupatadine fumarate; orodispersible minitablets; taste masking

1. Introduction

The challenge of modern pharmaceutical technology is designing easy-to-administer drug dosage forms where the dose is sufficiently flexible to enable proper application and dose titration both to paediatric and adult patients. The currently available formulations that might be used in any age population are primarily liquids. However, the barrier of their utilization are difficulties in effective taste masking of bitter active pharmaceutical ingredients (API), the necessity of applying a large volume of medical preparation to adults, as well as low physicochemical and microbiological stability. Therefore, new technological solutions and strategies are being sought, as the appropriate drug formulation should be acceptable for a wide age group, in terms of organoleptic properties (taste, smell, appearance) to ensure regular intake of medicine, even while prolonged therapy. In case of solids, a crucial element that decides if they are swallowed, is their size—the smaller the unit, the easier application [1–4].

The solid drug dosage form, which connects the advantages of liquids (flexibility of dosing, ease of swallowing) with the qualities of solids (taste masking, stability), as well as enables individual dose adjustments for patients of all ages are minitablets (MT). MT are created for those encountering difficulties with application of larger tablets or as a form providing the possibility of dose titration by "multiplication", which allows the use of one product for the entire age population. The appropriate dose is determined by the number of MT administered (e.g., children of different ages will take different number of MT as one dose). MT are characterized by small sizes of one to three millimeters and mass of 5–25 mg. They are produced like traditional tablets, using existing technologies and production lines, as well as standard tableting blends. A promising type of MT are orodispersible minitablets (ODMT), which are characterized by very short disintegration and dissolution rates. ODMT are particularly recommended for patients with swallowing problems, by eliminating the risk of choking to a minimal [5–11].

The major limitation in designing orodispersible formulations is unpleasant taste of API. Insufficient taste masking effect of medicine is the most common reason for refusing of taking the preparation [12]. Taste masking techniques can be divided in two main groups: chemical modifications of API to reduce its solubility or creating a physical barrier between drug molecules and taste receptors. The first group includes primarily conversion into a prodrug (ester, salt), complexation with cyclodextrins, or ion exchange resins. The second mechanism involves designing microparticles or coatings using polymers having limited solubility in the oral cavity environment. A useful method of decreasing unsavory taste of a medicine is obtaining microparticles utilizing taste masking polymers. There are several technologies for preparing microparticles, among which spray drying is one of the most effective and efficient [12–16].

The aim of the following paper was to create ODMT containing microparticles with rupatadine fumarate (RUP) as a model bitter drug. Microparticles were obtained employing the spray-drying, utilizing ethylcellulose as a barrier forming polymer. In our previous work, microparticles prepared using different forms of EC were compared. The microparticles prepared with an aqueous dispersion of EC were found to have better properties in terms of taste masking effectiveness and morphology [17], therefore they were used to formulate ODMT. ODMT were evaluated regarding their morphological structure utilizing scanning electron microscopy (SEM), uniformity of weight and thickness, mechanical properties, drug content, and potential interactions occurring using differential scanning calorymetry. Disintegration time was evaluated in vivo by healthy volunteers on petri dishes and with texture analyzer usage. The crucial test—assessment of taste masking effect was carried out according to three alternative approaches: in vivo, by the drug dissolution and with electronic tongue utilization.

2. Materials and Methods

2.1. Materials

Aquacoat® ECD was donated from FMC BioPolymer, Newark, NJ, USA. Surelease® E-7-19040 was given from Colorcon Inc., Harleysville, PA, USA. Parteck® ODT was received from Merck KGaA, Darmstadt, Germany. SmartEx® QD-50 was handed over from Shin-Etsu Chemical Co., Ltd., Tokyo, Japan. F-Melt C was purchased from Fuji Chemical Industry Co., Ltd., Toyama, Japan. Pearlitol® Flash was a gift from Roquette, Lestrem, France. Magnesium stearate and methylene blue were acquired from POCh, Piekary Śląskie, Poland. RUP was procured from Xi'An Kerui Biotechnology Co., Ltd., Xi'An, China.

2.2. Preparation of ODMT

A traditional tablet presser (Type XP1, Korsch, Berlin, Germany) with 3-mm punches was employed to manufacture ODMT by direct compression. In tableting bulk preparation, optimized spray dried microparticles with RUP prepared with EC aqueous dispersions selected during preliminary studies—urelease® and Aquacoat® ECD as a barrier coatings were utilized [17]. The conditions of the spray drying process were established experimentally: 85 °C, aspirator flow 98%, rate of flow 3.5 mL/min. Efficient barrier for masking the bitterness of drug enclosed in microparticles was obtained utilizing RUP:polymer ratio (0.5:1) with 6% EC concentration and this formulation was used for designing ODMT. ODMT with a mass of 14 mg and amount of microparticles corresponding to 0.5 mg of RUP per single tablet were assumed. The compositions of designed tableting masses utilized in the study are shown in Table 1. Prepared tableting blends were mixed manually for 30 s. To determine relevant conditions of the tableting process, various pressure force grades ranging from 0.6 to 1.2 kN were tested. Tablets with optimal properties that did not have a damaged surface of microparticles were obtained using a 0.9 kN (± 0.1) force. To simplify the formulation of ODMT, multifunctional co-processed mixtures were utilized. Co-processed mixtures are designed by processing several excipients to possess advantages that cannot be achieved by the simple physical mixtures of their components [18–24].

Table 1. Composition of orodispersible minitablets (ODMT) formulations.

Ingredient [%]	Formulation															
	F1	F2	F3	F4	F5	F6	F7	F8	F9	F10	F11	F12	F13	F14	F15	F16
RUP	-	3.5			-	3.5			-	3.5			-	3.5		
SUR MP RUP *[corresponding to 0.5 mg RUP per one tablet]*	-		7.15		-	-	7.15		-	-	7.15		-	-	7.15	
AQ MP RUP *[corresponding to 0.5 mg RUP]*	-			8.65	-	-		8.65	-	-		8.65	-	-		8.65
Parteck® ODT	99	95.5	91.85	90.35	-	-			-	-			-	-		
SmartEx® QD-50	-				99	95.5	91.85	90.35	-	-			-	-		
F-Melt C	-				-	-			99	95.5	91.85	90.35	-	-		
Pearlitol® Flash	-				-	-			-				99	95.5	91.85	90.35
Magnesium stearate	1	1	1	1	1	1	1	1	1	1	1	1	1	1	1	1

2.3. Flow Properties of Powders

The tableting blends (Table 1) were subjected to preformulative quality assessment in accordance with pharmacopoeial requirements [25]. Each study was carried out in triplicate. A tapping apparatus (Electrolab ETD-1020, Mumbai, India) was utilized for the compressibility studies. The bulk and tapped densities were calculated as quotients of the weight of the powder to its volumes occupied before and after tapping and then the powder density index (Index Carr) and the powder flow index (Hausner's ratio) were calculated. To investigate powder flow time, 50 g of the sample was placed in the funnel with the outlet closed and after opening the valve, the flow of the whole sample through the funnel was measured [25].

2.4. Evaluation of Morphology of ODMT

Morphological structure was assessed utilizing scanning electron microscopy (Inspect™S50, FEI Company, Hillsboro, OR, USA). Swatches were placed on adhesive tapes fixed to the surface of a special stand and gold sprayed. Tests were performed at room temperature, using various magnifications.

2.5. Quality Assessment of ODMT

2.5.1. Uniformity of Weight and Thickness

Twenty randomly chosen ODMT were weighted individually, employing analytical balance (Radwag, Radom, Poland) [25]. The thickness was tested with calibrated digital caliper utilization (Beta 1651DGT, Milan, Italy).

2.5.2. Mechanical Properties

Tests were conducted with friability tester (EF-1 W, Electrolab, Mumbai, India) according to pharmacopoeial monograph for a quantity of ODMT corresponding to 6.5 g [25]. The crushing strength of ODMT was tested utilizing a hardness tester (5Y, Pharmaton AG, Thun, Switzerland) and a Texture Analyzer TA.XT. Plus (Stable Microsystems, Godalming, UK) with a steel cylinder of 6 mm diameter and 0.1 mm/s pre-speed [25]. The minimum force (N) needed to crush the ODMT was measured by vertically applying pressure along its diameter. Ten randomly selected tablets from the batch were used for hardness assessment in both methods.

2.5.3. Drug Content

HPLC apparatus (Agilent Technologies 1200) equipped with Waters Spherisorb® 5 μm ODS1 4.6 × 250 mm column (Waters Corporation, Milford, CT, USA) was applied to evaluate RUP content uniformity for individual ODMT. As a mobile phase methanol:phosphate buffer pH = 3.0 (35:65, v/v) was utilized (isocratic elution). Flux was established as 1.0 mL/min and wavelength as 245 nm. Buffer was composed of NaH_2PO_4 and water, adjusted to pH = 3.0 by H_3PO_4. Standard calibration curve was linear in the range of 1–100 μg/mL and the correlation coefficient R^2 was 0.999. The studies were carried out in triplicate [26–28].

2.6. Disintegration Time Assessment

2.6.1. In Vivo

The study was carried out by six probands (Research Ethics Committee at Medical University of Białystok approval number R-I-002/438/2016) in the following stages: oral cavity rinsing with water, placing ODMT in the mouth without chewing until disintegration, splitting out. The time needed for the total disintegration in the mouth was recorded.

2.6.2. Petri Dish

Petri dish having a diameter of 7 cm was filled with 4 mL of phosphate buffer pH = 6.8 imitating natural spit (composing of Na_2HPO_4; KH_2PO_4 and water; adjusted to pH 6.8 by 1-M NaOH) [17] and single ODMT was put in the center. The test was repeated for 6 tablets from each batch. Time for the tablet to completely disintegrate into fine particles was measured.

2.6.3. Texture Analyzer

The study was conducted with a texture analyzer (Stable Microsystems, Godalming, UK) with an ODT disintegration time rig (Figure 1). Single ODMT was attached to the bottom of the probe with double-side adhesive tape. The ODMT was immersed in the medium (4 mL of phosphate buffer pH = 6.8 imitating natural spit) until it comes in contact with the perforated bottom of the container. The test was repeated for 6 tablets from each batch.

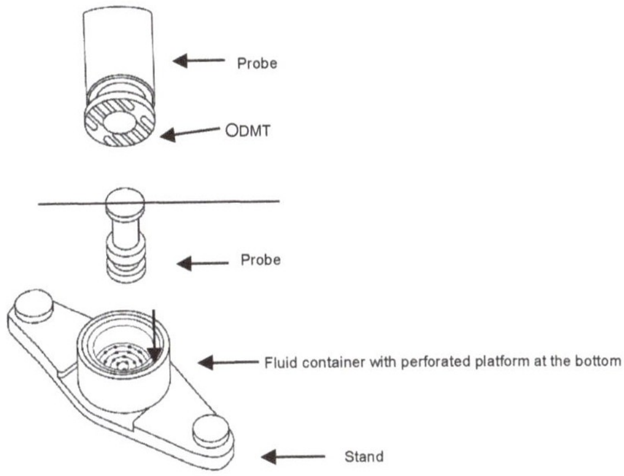

Figure 1. Schematic illustration of disintegration time rig of the texture analyzer.

2.6.4. Wettability

The wettability test was carried out according to Stoltenberg and Breitkreutz [29]. A 96-well plate (Biologix Group Limited, Lenexa, KS, USA) was used and a cellulose filter disc was placed in each well, which was then moistened by adding 20 μL of 0.3% aqueous solution of methylene blue. The time from placing the ODMT on moistened paper to complete coloring of the MT matrix was noted.

2.6.5. Differential Scanning Calorimetry

RUP raw material (API), microparticles placebo (MP AQ placebo, MP SUR placebo), microparticles (MP AQ RUP, MP SUR RUP), ODMT placebo (F1, F5, F9, F13), ODMT with pure RUP (F2, F6, F10, F14) and ODMT with RUP enclosed in SUR MP (F3, F7, F11, F15) and AQ MP (F4, F8, F12, F16) were tested using thermal analyzer system (DSC Mettler Toledo, Greifensee, Switzerland). Swatches were accurately weighed (5 mg), inserted in pans made of aluminum, then warmed up to 300 °C with 10 °C/min rate with 20 mL/min flow of nitrogen.

2.7. Evaluation of Taste Masking Effectiveness

2.7.1. In Vivo

The study was conducted in accordance with the Declaration of Helsinki and the protocol was approved by the Ethics Committee of Medical University of Białystok approval number R-I-002/438/2016. The efficiency of taste masking level was tested by six probands undergoing a test conducted as follows: five ODMTs were placed in the oral cavity for 30 s (the maximum time to dissolve/disintegrate in accordance to FDA guidelines [30]), spitted and mouth were rinsed with water. Sensory evaluation was marked as follows: 0—no bitterness, 1—slightly bitterness, 2—moderately bitterness, 3—significant bitterness. Before the experiment was carried out, the participants were chosen on the basis of sensory sensitivity test, utilizing four main flavors [31].

2.7.2. RUP Dissolution

RUP dissolution was carried out in apparatus II (paddle) (Erweka Dissolution Tester DT 600HH, Heusenstamm, Germany) with phosphate buffer (pH 6.8, 50 mL) imitating natural spit in following conditions: 75 rpm and 37 °C (+/−0.5). The quantity of dissolved RUP was assessed as pointed in 2.5.3.

2.7.3. Electronic Tongue

Reagents and Membrane Materials

Analytical reagent grade chemicals and purified water with 0.07 μS/cm conductivity (Elix Advantage System Mili-Q plus Milipore, Spittal an der Drau, Austria) were used. The membrane consisted of poly(vinylchloride) (PVC) (Tarwinyl, Tarnów, Poland); bis(2-ethylhexyl) sebacate (DOS), and o-nitrophenyl octyl ether (o-NPOE) (Fluka, St. Gallen, Switzerland); potassium tetrakis [3,5-bis(trifluoromethyl)phenyl]-borate (KTFPB), tridodecylmethylammonium chloride (TDMAC), 1-dodecylpyridinium chloride (DDPC), (Sigma—Aldrich, St. Luis, MO, USA), and potassium tetrakis(p-chlorophenyl)borate (KTpCPB) (Fluka, St. Gallen, Switzerland); calix[6]arene-hexaacetic acid hexaethylester (amine ionophore I) (Fluka, St. Gallen, Switzerland), and 3-mercapto-5-/2′-hydroxynaphthyl-azo-triazole (METRIAN) (Department of Drug Chemistry, Medical University of Lublin, Poland).

Membrane Preparation

Each solid contact electrode consists of a conventional body and an exchange Teflon holder in which the two phases are placed. The interior lamina (1) contains PVC with plasticizer in which the Ag/AgCl electrode is inserted and the exterior lamina (2) contains the ion-sensitive component and inner layer components. The exterior lamina is in contact with the tested solutions. The steps of the membrane phase preparation:

(1) weighing inner layer components: 30% (*w/w*) PVC, 70% (*w/w*) of plasticizers, DOS or o-NPOE,
(2) mixing and deaerating of obtained mixture,
(3) filling the Teflon holder with mixture to cover the silver-silver chloride electrode,
(4) gelating inner layer (1) at 373 K for 30 min, cooling of the gelled layer,
(5) weighing outer layer components, 27–33% (*w/w*) of PVC, 64–68% (*w/w*) of plasticizers, 1–5% of electroactive components (2) (Table 2),
(6) dissolving of obtained mixture in THF,
(7) placing drops on the inner layer (1),
(8) gelating outer layer (2) in result of evaporation THF at 293 K; repeating the steps several times.

Table 2. Electrodes forming sensor array of the electronic tongue.

Electrode Number no.	Electrode Type	Ionophore (%, *w/w*)	Lipophilic Salt (%, *w/w*)	Plasticizer (%, *w/w*)	Polymer (%, *w/w*)
1–2	CSF-D	–	KTFPB (1%)	DOS (66%)	PVC (33%)
3–4	CSF-N	–	KTFPB (1%)	o-NPOE (66%)	PVC (33%)
5–6	CSC-D	–	KTpCPB (3%)	DOS (64%)	PVC (33%)
7–8	CSC-N	–	KTpCPB (3%)	o-NPOE (64%)	PVC (33%)
9–10	AM-D	Amine ionophore I (5%)	–	DOS (68%)	PVC (27%)
11–12	MET-N	METRIAN (4%)	–	o-NPOE (66%)	PVC (30%)
13–14	PC-N	–	DDPC (3%)	o-NPOE (64%)	PVC (33%)
15–16	AN-N	–	TDMAC (4%)	o-NPOE (66%)	PVC (30%)

Potentiometric Measurements

EMF electrochemistry interface system (Lawson Labs. Inc., Malvern, PA, USA) and IBM PC were used for potentiometric measurements. The potentiometric sensor array of the system contains 16 solid contact ion-selective electrodes (two sensors of each type) differing active substances and plasticizers each other. The constructed solid contact electrodes were stored for 24 h in RUP solutions before the first measurement. As a standard, Ag/AgCl electrode (Orion 90-02, Thermo Electron Corporation, Beverly, MA, USA) was used. The electrodes calibration curves were performed in 10^{-5}–10^{-3} mol L^{-1} RUP solutions because of the poor solubility of RUP in water. Previously developed measurement protocol for electronic tongue analysis of pharmaceutical samples [32] was applied to test taste masking and API dissolution from designed ODMTs. First, the sensors were immersed in deionized water (50 mL) to obtain 5 min signal stabilization. Then adequate samples were added and released API and excipients influenced electrodes' signals, that were recorded as the changes of potentials (ΔEMF) of particular electrodes in sensor array in a function of time. The release measurements were carried out for solutions of pure API (RUP), all studied ODMTs and respective placebos. The signals of the sensors were registered during 15 min (5 min stabilization, 10 min release), in 5 repetition for each sample type. The sensors were water rinsed and dried between assays.

Data Analysis

Data matrix was composed of data vectors assigned to every investigated formulation (16 variables for every time-point, responding to 16 ΔEMF signals of 16 potentiometric sensors). The data matrixes were processed by chemometric procedures: Principal Component Analysis (PCA) or Partial Least Squares (PLS). These calculations as well as data analysis and presentation were performed in SOLO® software (Eigenvector Research Inc., Manson, WA, USA).

3. Results and Discussion

3.1. Pharmaceutical Evaluation of ODMT

An attempt was made to design and develop ODMT as an innovative drug dosage form, utilizing RUP enclosed in ethylcellulose microparticles to reduce bitterness. In preparation of ODMT, mainly spherical, homogenous, smooth surfaced microparticles based on EC aqueous dispersions were used (Figure 2). The mean size of microparticles made from Surelease® was 3.2 +/− 1.1 µm and from Aquacoat® ECD was 3.6 +/− 1.5 µm

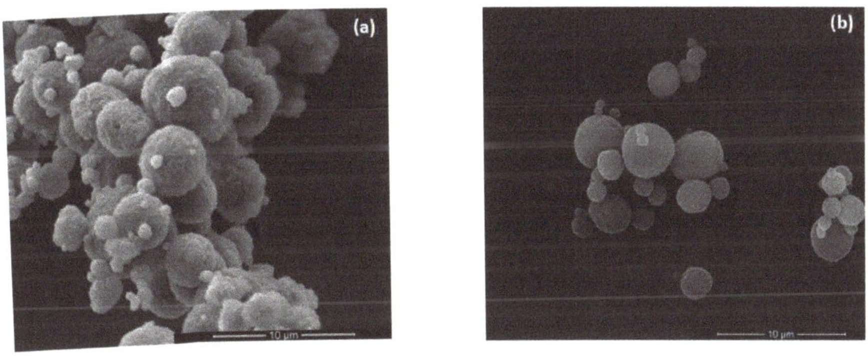

Figure 2. SEM picture of microparticles prepared using: (**a**) Surelease®, (**b**) Aquacoat® ECD under magnification 10,000×.

As a model drug with bitter taste, RUP (a long-acting second generation antihistamine showing anti-allergic and demulcent effect applied both in children and adults) was utilized. RUP is the histamine receptor selective antagonist and receptor for platelet activating factor (PAF), which highlights it from drugs belonging to this group and clarifies its unique mechanism of action. RUP binds to the H_1 receptor permanently and firmly, acting as an inverse agonist, which prolongs duration of its action. It was indicated that RUP is characterized by far greater affinity to H_1 receptor than fexofenadine or levocetirizine. Furthermore, by binding to PAF receptors, RUP causes their blockade, what is clinically relevant to PAF allergic inflammatory processes and bronchial hyperreactivity symptoms. RUP has not only been shown to reduce the amount of erythema (which is characteristic of all antihistamines), but also reduces PAF-induced platelet aggregation. The third component of RUP activity is its additional anti-inflammatory effect consisting in: inhibition of mast cell degranulation and release of histamine and cytokines (e.g., IL-4,5,6,8, TNF α), inhibition of eosinophil and neutrophil chemotaxis, inhibition of expression of adhesive molecules (CD18, CD11b) and transcription factors [33–39]. Commercially it is available in traditional tablets form [40] and due to its unpleasant taste, there is no orodispersible drug dosage forms on the pharmaceutical market.

To reduce bitterness of RUP, ethylcellulose (EC), a hydrophobic polymeric material widespread applied in masking the unpleasant aroma and taste was applied. It belongs to the GRAS (generally regarded as safe) and FDA Inactive Ingredients [41]. Moreover, EC is considered not to carry any health risks, therefore its daily intake has not been explicated by the World Health Organization (WHO) [42]. It is an ethyl ether of cellulose, in the form of a free-flowing, odorless, tasteless, biocompatible, non-allergenic, and nonirritant white to light-tan powder dissolving only in organic media, thus creating a polymeric barrier that allows for temporary isolation of a bitter drug from the oral cavity environment [43–45]. EC is accepted to be utilized in paediatric medicinal products, as well as in non-parenteral formulations authorized in Europe [41,46,47]. EC is available in organic form (e.g., Ethocel®) and as aqueous dispersions (e.g., Surelease®, Aquacoat ECD®). Surelease® contains a 25% of solid EC and dibutyl sebacate and oleic acid as plasticizers. In Aquacoat® ECD there is 27% EC, sodium lauryl sulfate, and cetyl alcohol. The dispersions are accepted for pharmaceutical use in the Europe, United States, and Japan [48,49].

ODMT were prepared by direct compression method, using commercially available mixtures: Parteck® ODT, SmartEx® QD-50, FMelt® C, Pearlitol® Flash. The amount of API was set as 0.5 mg per one ODMT. The dose selection was related to the fact that by multiplication, the dose of 2.5 mg required for children weighting from 10 to 25 kg can be easily achieved. No significant technological problems were observed during tableting process. It was connected with low API content in the tablet masses, so it did not affect the flowing properties of powders. The similar composition of all mixtures based mainly on mannitol caused all the obtained formulations to be characterized by similar physical parameters (Table 3). The best flowability was noted for blends with FMelt® C and Pearlitol® Flash.

Table 3. Characteristics of tableting blends.

Powder Mixture	Density [g/mL]		Flow Properties	
	Bulk	Tapped	Hausner's Ratio	Carr's Index [%]
F1	0.58	0.72	20.45	1.26
F2	0.58	0.71	20.45	1.26
F3	0.56	0.70	20.40	1.27
F4	0.57	0.68	20.20	1.24
F5	0.51	0.65	15.38	1.28
F6	0.52	0.65	15.38	1.28
F7	0.50	0.63	15.39	1.29
F8	0.49	0.62	15.25	1.30
F9	0.56	0.67	13.25	1.16
F10	0.56	0.66	13.24	1.16
F11	0.54	0.62	13.23	1.17
F12	0.52	0.61	13.22	1.15
F13	0.48	0.57	13.10	1.15
F14	0.48	0.58	13.10	1.15
F15	0.45	0.54	13.05	1.19
F16	0.44	0.53	13.13	1.16

Physical parameters of prepared tablets might be described as relatively good as the balance between the mechanical properties (sufficient hardness, friability <1%) and the quick disintegration time was captured. The obtained ODMT were hard enough that they did not crush while handling, and simultaneously, the formulations were characterized by the desired rapid disintegration time (below 30 s) (Table 4). The weight and thickness uniformity of ODMT is essential as it impacts dosing accuracy. The average masses of obtained formulations had values from 12.5 mg to 14.1 mg. Thickness of obtained ODMT was in the range from 1.80 mm to 2.01 mm. The optimal mechanical characteristics and rapid dissolution time are key aspects in orodispersible formulations depending on the conditions applied during the process. Appropriate tensile force is particularly important while tableting microparticles. Too high a value of tensile force could result in cracking microparticles, which has an undesirable effect in the case of taste masking. No microparticles crushing occurred at the applied pressure (Figure 3). The tensile force value 0.9 kN was determined experimentally as optimal. While lower pressure was applied, the tablets were too brittle to handle with, while higher disintegration time was insufficient (>30 s). Friability (in every formulation < 1%) and hardness tests have proven that obtained ODMTs were characterized by mechanical properties adequate enough so as not to be damaged during the manufacturing process or packing. However, the hardness of tablets prepared with microparticles utilization was smaller in comparison to placebo or formulations with pure RUP. RUP loading was in the range from 0.4 to 0.5 mg—the lowest values were marked for F8, F12, F16 and they did not meet pharmacopoeial requirements (<85%) [25].

Table 4. Physicochemical characteristics of prepared ODMT.

Parameter / Formulation	F1	F2	F3	F4	F5	F6	F7	F8	F9	F10	F11	F12	F13	F14	F15	F16
Weight [mg] *	13.4 ± 0.2	13.7 ± 0.3	14.0 ± 0.7	13.6 ± 0.5	14.1 ± 0.3	12.6 ± 0.9	12.9 ± 0.6	12.5 ± 0.7	13.8 ± 0.3	13.4 ± 0.2	13.5 ± 0.4	13.4 ± 0.5	14.1 ± 0.2	13.4 ± 0.5	13.9 ± 0.3	13.70 ± 0.4
Thickness * [mm]	2.01 ± 0.1	2.01 ± 0.1	1.96 ± 0.2	1.94 ± 0.3	1.94 ± 0.1	1.96 ± 0.2	1.82 ± 0.2	1.80 ± 0.3	2.0 ± 0.1	1.98 ± 0.1	1.97 ± 0.3	1.95 ± 0.3	1.97 ± 0.1	1.95 ± 0.1	1.95 ± 0.3	1.93 ± 0.4
Hardness [N] ** *(by hardness tester)*	15.4 ± 3.4	14.4 ± 2.4	8.40 ± 4.5	8.20 ± 4.7	16.8 ± 2.2	16.2 ± 2.5	8.1 ± 2.1	7.8 ± 2.3	15.9 ± 2.5	15.1 ± 2.1	8.1 ± 2.7	7.8 ± 2.9	16.1 ± 1.2	15.8 ± 2.1	8.1 ± 3.7	7.5 ± 4.2
Hardness [N] ** *(by texture analyzer)*	15.1 ± 3.9	14.6 ± 2.3	8.1 ± 4.0	8.0 ± 4.4	16.7 ± 2.2	16.5 ± 2.4	8.1 ± 1.5	7.8 ± 2.5	16.0 ± 2.2	15.0 ± 2.0	8.2 ± 2.5	8.0 ± 2.8	16.1 ± 1.1	15.9 ± 2.3	8.0 ± 3.5	7.6 ± 4.1
Friability [%]	0.1	0.1	0.1	0.2	0.1	0.45	0.1	0.1	0.1	0.1	0.1	0.3	0.1	0.1	0.1	0.2
Drug content * [mg]**	-	0.47 ± 0.1	0.48 ± 0.3	0.43 ± 0.4	-	0.45 ± 0.1	0.43 ± 0.2	0.42 ± 0.3	-	0.47 ± 0.1	0.45 ± 0.3	0.41 ± 0.4	-	0.49 ± 0.1	0.5 ± 0.1	0.40 ± 0.2
% of declared dose	-	94	96	86	-	90	86	84	-	94	90	82	-	98	100	0.80

*—the test was performed for 20 tablets [25]; **—the test was performed for 10 tablets in triplicate; ***—the test was performed for 10 tablets [25].

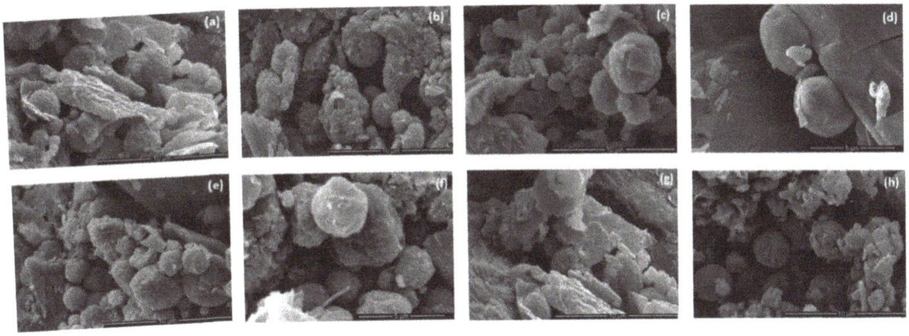

Figure 3. SEM pictures of ODMT cross-sections: (**a**) formulation F3, (**b**) formulation F4, (**c**) formulation F7 under magnification 10,000×, (**d**) formulation F8 under magnification 50,000×, (**e**) formulation F11 under magnification 10,000×, (**f**) formulation F12 under magnification 50,000×, (**g**) formulation F15, (**h**) formulation F16 under magnification 10,000×.

Disintegration time tests, conducted under conditions imitating those prevailing in the oral cavity (2–7 mL) are recommended [50–53], therefore tests in vivo with healthy volunteers, on petri dishes and using a texture analyzer were utilized. Regardless of the method, disintegration time of all ODMT formulations was below 30 s, and most formulations disintegrated even below 15 s. The longest disintegration time was recorded for F13, F14, F15, F16—19–24 s. In all tablets, wetting time below 30 s was noted.

An appropriate selection of pharmaceutical excipients is a key issue in creating drug dosage forms, as the excipients might affect physicochemical properties of API. Differential scanning calorimetry (DSC) is one of the analytical techniques frequently applied to determine drug physical properties, as well as to investigate potential incompatibilities with other components. The procedure provides detailed information about the presence of impurities and energetic properties of substances pointing to the differences in the heat flow generated or absorbed by the sample. To evaluate possible interactions, RUP raw material (API), microparticles placebo (MP AQ placebo, MP SUR placebo), microparticles (MP AQ RUP, MP SUR RUP), ODMT placebo (F1, F5, F9, F13), ODMT with pure RUP (F2, F6, F10, F14) and ODMT with RUP enclosed in SUR MP (F3, F7, F11, F15) and AQ MP (F4, F8, F12, F16) (Figure 4) were assessed. RUP chemical nomenclature is 8-chloro-6,11-dihydro-11-[1-[(5-methyl-3-pyridyl)methyl]-4-piperidylidene]-5H-benzo[5,6]cyclohepta[1,2-b]pyridine fumarate [54]. Its melting point should range from 194 to 201 °C. In the literature, there are no polymorphic forms reported for RUP [54,55]. The thermogram of pure RUP presents endothermic event at 196.44 °C characterized by a sharp pick, corresponding to its melting point. Sample decomposition after melting can be observed. Exothermic event transition is shown at 210.35 °C. Both melting and decomposition was noted in a constricted range of temperatures. No additional thermal events connected with decomposition or loss of surface water were observed. Thermograms of microparticles show that there are no thermal events for AQ MP and SUR MP placebo, which indicates that used aqueous dispersion of EC are in an amorphous state. Converting RUP into microparticle form by the spray drying did not significantly change solid state nature of the drug; however, some changes in its melting point occurred—in case of AQ MP RUP the peak has been shifted to 190.67 °C and for SUR MP RUP to 210.1 °C, which indicates that its melting point decreased about 6 °C or increased about 14 °C in microparticle samples, respectively. This is probably due to the fact that excipients used can slightly change physicochemical properties of API during spray drying. In all ready-made co-processed mixtures, the main ingredient is D-mannitol, whose melting point ranges from 155 °C to 165 °C, what was confirmed in the obtained thermogram. There is also a peak in 87 °C of magnesium stearate. No changes in the position of melting peaks and their specific heats were observed in the thermograms of ODMT. There are no further peaks of RUP as

API dissolves at the mannitol melting point. No distinct interactions between RUP and used excipients were observed.

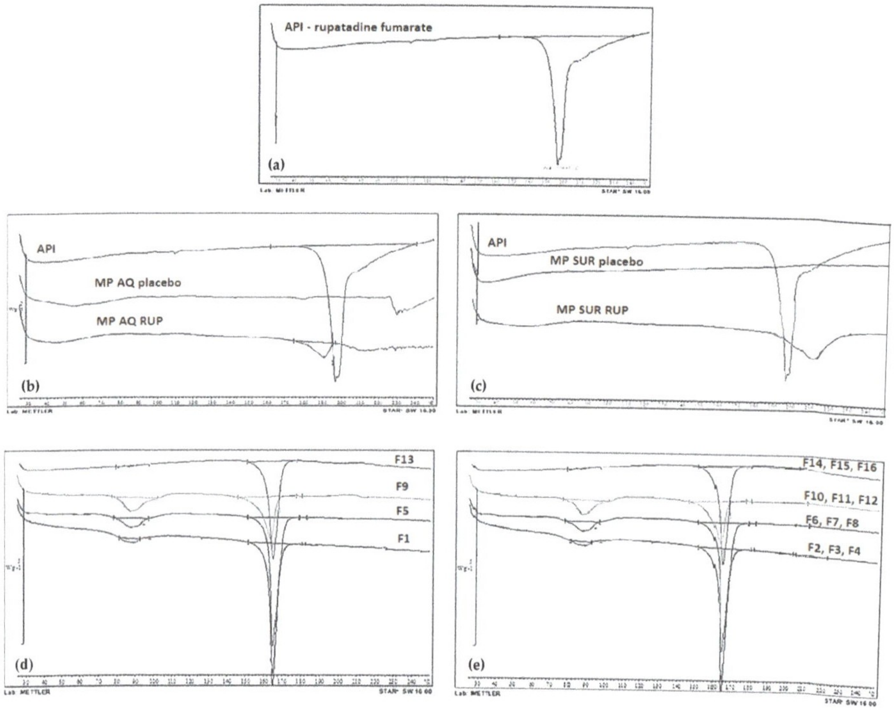

Figure 4. DSC thermograms of RUP - rupatadine fumarate (**a**) microparticles placebo obtained with Aquacoat® or Surelease® (MP AQ placebo MP SUR placebo) and with RUP (MP AQ RUP MP SUR RUP) (**b**,**c**) ODMT placebo (F1, F5, F9, F13) (**d**) ODMT with pure RUP (F2, F6, F10, F14) and with RUP enclosed in microparticles (F3, F4, F7, F8, F11, F12, F15, F16) (**e**).

3.2. *Taste-Masking Efficiency Evaluation*

Evaluation of taste masking effectiveness is a significant issue, as there are no pharmacopoeial and universal methods to assess the taste. To determine taste masking degree, in vivo (human taste panel) and various in vitro methods (e-tongue, drug release) can be utilized. Human taste panel is the most frequently used strategy of taste evaluation as it is widely available; however, it presents a certain challenge. There are high variances in human taste receptors expression and differences in taste perception (e.g., smoking or taking medicines have an impact). As well, children's participation in such a study is considered to be unethical, in turn the results obtained in adults are difficult to extrapolate to the entire population due to the different perception of taste sensations. Nevertheless, the predominant approach of assessing the taste of raw medicines and drug dosage forms is by human volunteers. An alternative approach is electronic tongues utilization. It is an analytical gustatory tool for automatic analysis of drug taste. Its essential element is the sensor array composed of chemical sensors with various selectivity. Potentiometric signals recorded in the tested sample do not provide direct information about the composition of the sample, but create its specific digital chemical image, whose interpretation allows to identify a sample or the content of its individual components, including those responsible for generating the taste. Evaluation of bitter taste can also be correlated to the drug release rate. It seems to be the simplest way to determine taste-masking efficacy based mainly on the quantification of drug concentration [56–59].

3.2.1. In Vivo Taste Evaluation

Initially, six selected healthy volunteers assessed ODMT formulations containing microparticles (F3, F4, F7, F8, F11, F12, F15, F16) as non-bitter or slightly bitter in comparison to those with pure RUP (F2, F6, F10, F14), which were determined as moderately or very bitter (Table 5). It should be also mentioned that mannitol—the main component of obtained ODMT—besides being a sweetening agent, while dissolving in the mouth maintains an impression of cooling, which has a favorable effect on taste sensation during the application [31].

Table 5. Sensory evaluation of designed ODMT formulations, estimated as follows: 0—no bitterness, 1—slightly bitterness, 2—moderately bitterness, 3—significantly bitterness.

Volunteer	Score											
	F2	F3	F4	F6	F7	F8	F10	F11	F12	F14	F15	F16
A	3	0	0	2	0	0	2	0	1	2	0	1
B	3	1	1	3	1	1	3	1	1	3	1	1
C	2	1	0	2	0	0	2	0	1	2	0	1
D	2	0	1	2	0	0	2	0	0	2	0	0
E	3	0	1	3	1	1	3	1	1	3	0	1
F	2	1	0	2	0	0	2	0	0	2	0	0

3.2.2. In Vitro RUP Release

Taste masking was also evaluated by RUP release from obtained formulations. Slowing the release of a drug is associated with better efficacy of masking the taste. ODMT made with microparticles (F3, F4, F7, F8, F11, F12, F15, F16) released RUP significantly slower compared to ODMT with pure RUP (F2, F6, F10, F14), where immediate release of RUP occurred (Figure 5). After one minute of dissolution test, maximum 15% of RUP was released, which indicates satisfactory taste masking effect considering very quick disintegration time (about 20 s) and short residence time in the oral cavity.

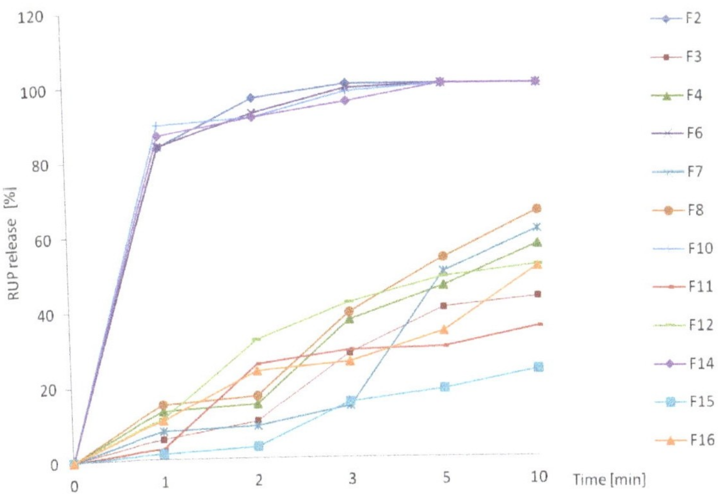

Figure 5. RUP release from designed ODMT performed in paddle apparatus.

3.2.3. Electronic Tongue

To investigate in detail taste masking efficiency for the studied formulations, human panel responses were processed by means of a multivariate technique—PCA. This data analysis method helps to find the most significant information hidden in the multidimensional data structure. As a result, a score plot in principal components (PC) coordinates is obtained, which shows clusters of samples based on their similarity. The more similar multidimensional characteristics are, the closer are the objects in the PC1-PC2 space. PCA scores plot (Figure 6) presents PCA processed data of human panel responses shown in Table 5.

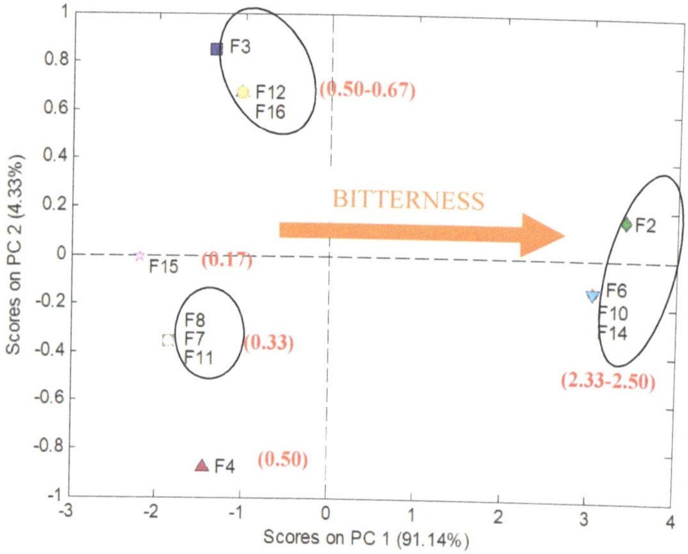

Figure 6. PC1, PCA of human panel responses showing similarity of sensed bitterness for the studied minitablets. Values in brackets show mean values of bitterness score calculated from Table 5.

The formulations form various clusters according to the sensed bitterness. The most bitter formulations: F2, F6, F10, F14, that reached mean score higher than 2, are placed close to each other and are characterized by high value of PC1. F6, F10, and F14 were evaluated identically by all volunteers, therefore they are overlapping, having the same coordinates PC1-PC2. Only one volunteer (a) estimated F2 as very bitter in contrast to F6, F10, F14 scored by him/her as moderately bitter, therefore F2 is similar to F6, F10, F14 PC1-PC2 scores, but not the same. ODMT F15 are placed in the highest distance from F2, F6, F10, and F14 cluster, having the lowest value of PC1, because they were estimated as not bitter by 5 out of 6 volunteers (the lowest mean value of bitterness). All remaining formulations were scored as very slightly bitter (mean values of bitterness from 0.33 to 0.67), and accordingly, they exhibit moderate PC1 values. All these observations perfectly match the dissolution tests (Figure 5), where formulations F2, F6, F10, F14 show high dynamics of RUP release, whereas the slowest release in the first two minutes is observed in the case of F15 minitablets.

Before the measurements of pharmaceutical formulations, an important stage of research was optimization of the sensor array. For this purpose, calibration curves of electrodes towards RUP were determined. As it results from Figure 7, for all electrodes containing various active substances and plasticizers in the membrane, different sensitivity towards RUP was achieved. Sensitivity ranged from about 10 mV decade^{-1} to about 51 mV decade^{-1} of 10^{-5}–10^{-3} mol L^{-1}.

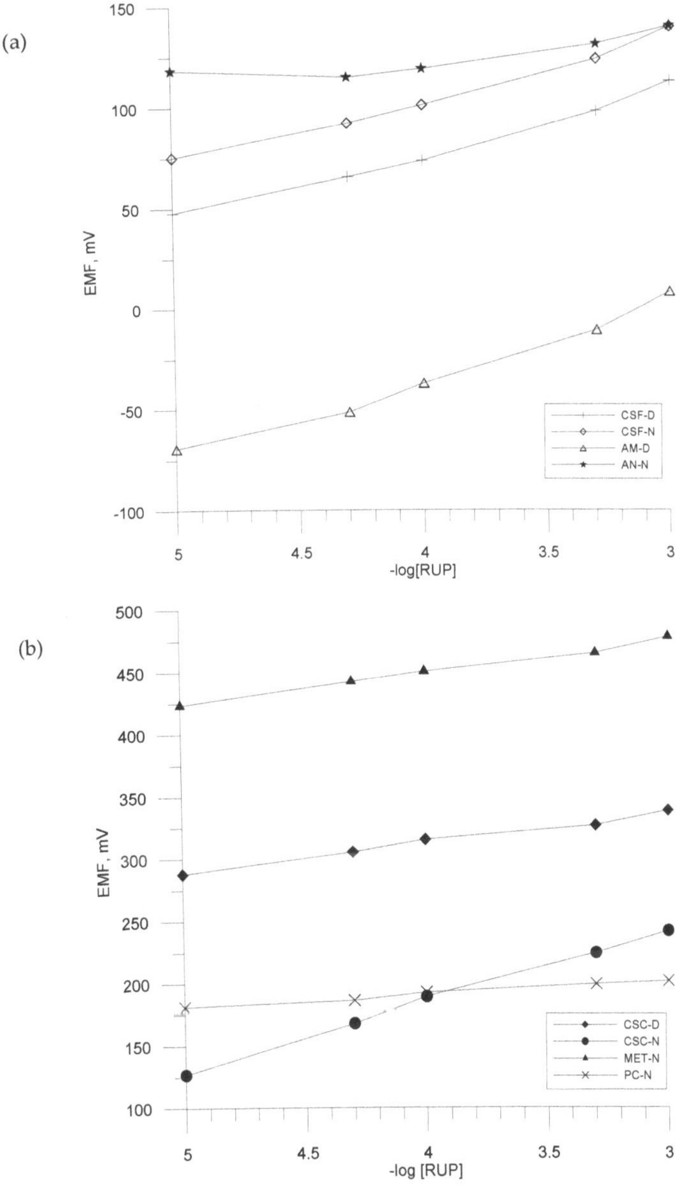

Figure 7. PC2, calibration curves of ion-selective electrodes with CSF, AM, AN (**a**) and CSC, MET, PC (**b**) in 10^{-5}–10^{-3} mol L^{-1} RUP solutions.

The electrodes based on KTFPB (CSF-D, CSF-N) displayed very similar calibration curves, with mean sensitivity 29.2 ± 4.3–32.4 ± 0.89 mV decade^{-1} in the 10^{-5}–10^{-3} mol L^{-1} linear range, good mean correlation coefficient R^2 = 0.9944 (n = 3). The electrodes containing KTpCPB reveal visible differences between each other; the sensor's membrane plasticized with DOS (CSC-D) showed lower mean sensitivity 23.1 ± 4.4 mV decade^{-1} (R^2 = 0.9983) than electrode with o-NPOE. This electrode exhibited linear range with slope close to the near Nernstian 51.5 ± 4.0 mV decade^{-1} and good correlation coefficient R^2 = 0.9993. Slightly lower sensitivity exhibited the electrode with amine ionophore, mean response 41.5 ± 4.6 mV decade^{-1}. Moreover, the lowest slope of characteristics was obtained for electrodes prepared with ammonium and pyridinium ion exchangers. The slope coefficient of linear range of characteristic amounts 10 mV decade^{-1} for PC-N electrode and 6 mV decade^{-1} for AN-N electrode. The sensor based on metrian (MET-N) showed a similar response to electrodes CSF. All electrodes possessed lower or higher sensitivity to ionic RUP molecules (carboxyl groups, protonated nitrogen atom) as a result of the interaction of RUP with the active components of the polymeric membrane. Concluding, the prepared electrodes of electronic tongue sensor array exhibited satisfactory sensitivity towards studied API. According to our previous studies [17,60,61], such sensors are also cross-sensitive, responding to various excipients, which is a necessary condition for electronic tongue study.

3.2.4. Electronic Tongue—Taste Evaluation

The prepared sensor array was applied to check taste masking efficiency of all prepared ODMT. The procedure of measurements are presented in the experimental section. According to it, the responses of every sensor was given as ΔEMF in a function of time. Taste evaluation was performed for signals recorded after two minutes of release and resulting PCA score plot is presented in Figure 8.

All placebos formed a distinct cluster; they are grouped together even though their composition is different. On the opposite side of the plot, pure RUP samples are observed. All studied formulations take place between placebos and API, showing moderate taste between the two, which is correct and was expected. The clusters of formulations are partially overlapping, however similarity between electronic tongue study and human panel evaluation can be noticed. There is a cluster formed by F2, F6, and F14 ODMT in the closest distance to pure RUP, therefore their bitterness is most similar to pure API. Moreover, in proximity of this cluster, F10 can be seen. These four formulations showed the highest bitterness according to human panel (Figure 6) and highest dynamic of RUP release (Figure 8). The closest to RUP samples are F2 minitablets, which were evaluated as the most bitter, having a mean value of bitterness score equal to 2.5. The formulation that was the closest to placebo (not bitter) was F15, having the lowest bitterness score (0.17) and this fact also correlates well with human panel results and dissolution study. However, cluster of F15 is not distinct, it is spread out between other formulations, overlapping, e.g., F3 and F11 minitablets. Generally, samples having similar taste sensed by the human panel are considered similar also in terms of electronic tongue response, e.g., F8 is close to F7, and F12 is close to F16. The correlation is not perfect—the most surprising is the position of F11, in high distance from F7 and F8 minitablets. Nevertheless, the results of electronic tongue study reveal highest efficiency of taste masking for formulation F15 and lowest for formulations F2, F6, F10, and F14, which was confirmed by dissolution tests (Figures 5 and 9) and human panel results.

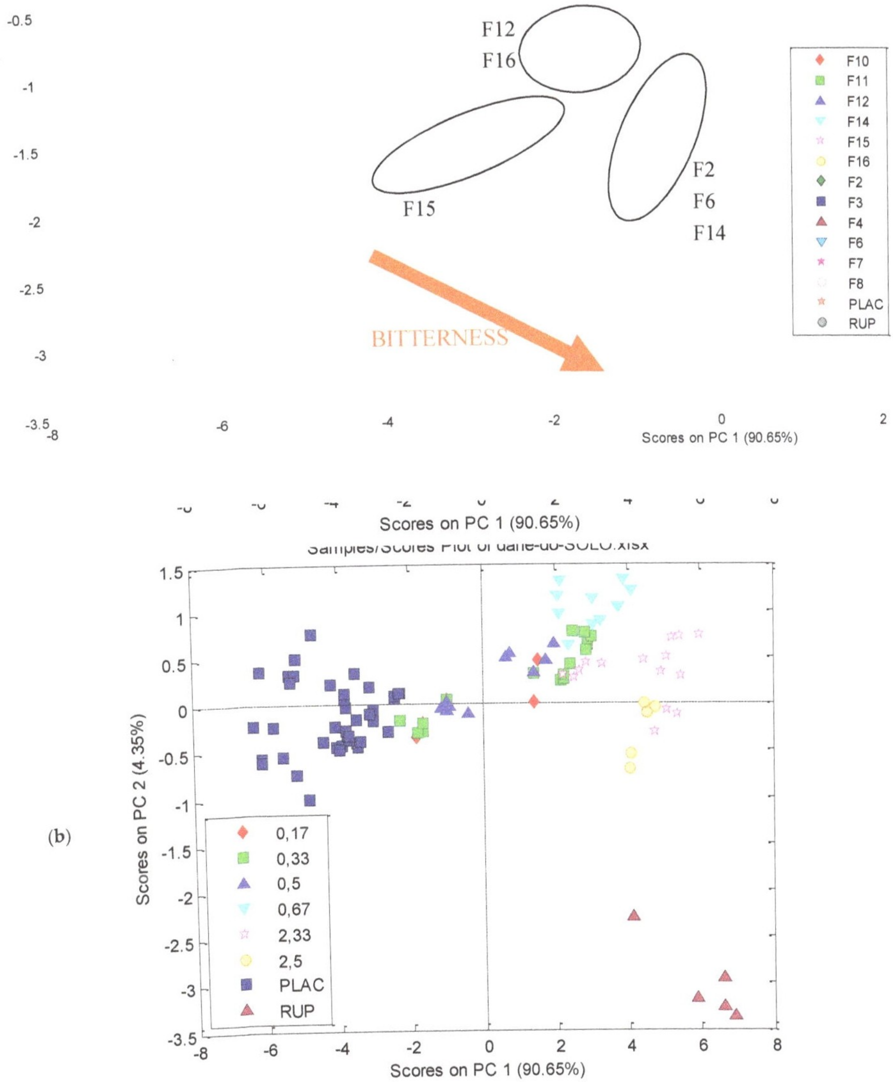

Figure 8. PC3, PCA score plot of electronic tongue responses for all studied formulations (F2-F16), respective placebos (PLAC) and pure RUP. On both plots the same object are presented, therefore they are in the same configuration, but the symbols are given according to: (**a**) formulation type; (**b**) mean values of bitterness score calculated from Table 5.

3.2.5. Electronic Tongue—Prediction of Dissolution Study

Signals of electrodes forming sensor array of electronic tongue were recorded during 10 min of formulation release. These outputs are related to RUP release, because the sensors are sensitive towards this API; however, they are also strongly influenced by increasing concentration of excipients, which are also released, due to cross-sensitivity of sensors. Therefore, to extract information of RUP release from sensor responses, a supervised data analysis technique was applied. First, we attempted to construct a Partial Least Squares (PLS) model. PLS regression is a chemometric procedure combining PCA and multiple regression. This approach leads to the possibility of the prediction of dependent variables from independent variables (measurement data). It is performed by transforming the obtained results into so-called latent components enabling the calculation of the dependent variables [62]. We applied this data analysis technique for various applications of the electronic tongue system [63–65]. In this paper, electronic tongue signals in appropriate time points were used for PLS modeling. All obtained data were divided into the training and test set. Target matrix was constructed based on %RUP release values that were determined by a standard dissolution test. For establishing a PLS model, a training set of data was applied aiming to find a correlation between the sensor array signals in an appropriate time point and % of RUP released in a respective formulation in a respective time point. When model was ready, the electronic tongue system was capable of predicting of amount of RUP that was released from the respective formulation based on electrodes' signals. The values of the RUP release were obtained for all studied formulations for a few time points. The resulting dissolution curves for independent test set data are presented in Figure 9. It must be underlined that what is predicted by electronic tongue system values are estimates, they do not provide accurate values of RUP release. The most evident example of that fact are negative values of RUP release for F3 minitablets. However, the outputs of PLS model show general tendencies discerning the studied formulations according to release dynamics. Two groups of dissolution curves can be observed. According to electronic tongue signals, four kinds of ODMT: F2, F6, F10, and F14, released RUP very fast, whereas all other formulations were characterized by much slower dynamics of its release. This finding correlates well with the standard dissolution test (Figure 5).

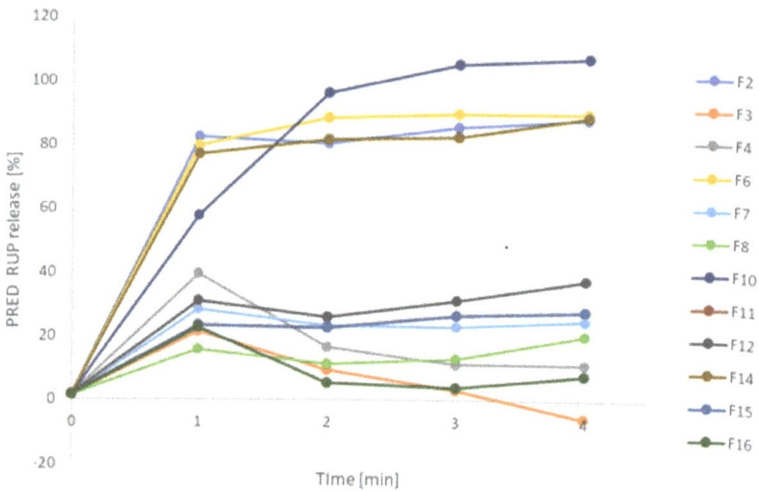

Figure 9. Electronic tongue prediction of RUP release (the results for independent test set).

4. Conclusions

Minitablets are getting increasingly significant among modern solid oral drug dosage forms, enabling the application of medicinal substances for patients of all ages by dose multiplying. Due to their small size they are easy to administer and swallow. Minitablets in orodispersible form improve patient compliance in relation to rapid disintegration in the oral cavity within seconds in the presence of the saliva without requirement of drinking water. Orodispersible minitablets (ODMT) with rupatadine fumarate were successfully prepared by direct compression of commercially available ready to use blends and ethylcellulose microparticles. Designed ODMT were characterized by beneficial physicochemical parameters. The presented study also indicates that MP made of etyhylcellulose with rupatadine fumarare might be efficiently utilized in preparation of ODMT by direct compression technique. The evaluation of taste masking level undertaken by three alternative approaches (e-tongue assessment, volunteers, and the drug release) established that designed ODMT are efficient taste–masking carriers of rupatadine fumarate and are supposed to be a promising alternative for traditional tablets.

Author Contributions: Conceptualization, K.W. (Katarzyna Wasilewska) and K.W. (Katarzyna Winnicka); methodology, K.W. (Katarzyna Wasilewska), P.C.-S., J.L., S.S., A.B., and K.W. (Katarzyna Winnicka); software, P.C.-S. and J.L.; validation, K.W. (Katarzyna Wasilewska), P.C.-S., and J.L.; formal analysis, K.W. (Katarzyna Winnicka); investigation, K.W. (Katarzyna Wasilewska), P.C.-S., J.L., S.S., and A.B.; resources, K.W. (Katarzyna Wasilewska) and K.W. (Katarzyna Winnicka); data curation, K.W. (Katarzyna Wasilewska); writing–original draft preparation, K.W. (Katarzyna Wasilewska), P.C.-S., J.L., and K.W. (Katarzyna Winnicka); writing–review and editing, K.W. (Katarzyna Wasilewska) and K.W. (Katarzyna Winnicka); visualization, K.W. (Katarzyna Wasilewska); supervision, K.W. (Katarzyna Winnicka); project administration, K.W. (Katarzyna Wasilewska) and K.W. (Katarzyna Winnicka); funding acquisition, K.W. (Katarzyna Wasilewska) and K.W. (Katarzyna Winnicka). All authors have read and agreed to the published version of the manuscript.

Funding: This research was funded by Medical University of Bialystok grant SUB/2/DN/19/004/2215. Equipment used in this study was from OP DEP 2007–2013, Priority Axis I.3, contract No. POPW.01.03.00-20-008/09, and the equipment of Center BioNanoTechno—scanning electron microscope (Inspect™S50, FEI Company, Hillsboro, USA)—was partly supported by EU founds via project number POPW.O1.03.00-20-004/11.

Acknowledgments: We are thankful to the following companies: Colorcon Inc., Harleysville, PA, USA; FMC BioPolymer, Newark, NJ, USA; Merck, Darmstadt, Germany; Shin-Etsu Chemical Co., Ltd., Tokyo, Japan; Fuji Chemical Industry Co., Ltd., Toyama, Japan; Roquette, Lestrem, France for kindly donating materials.

Conflicts of Interest: The authors declare no conflicts of interest. The funders had no role in the design of the study; in the collection, analyses, or interpretation of data; in the writing of the manuscript, or in the decision to publish the results.

References

1. Gore, R.; Chugh, P.K.; Tripathi, C.D.; Lhamo, Y.; Gautam, S. Paediatric off-label and unlicensed drug use and its implications. *Curr. Clin. Pharmacol.* **2017**, *12*, 18–25. [CrossRef]
2. McIntyre, J.; Conroy, S.; Avery, A.; Corns, H.; Choonara, I. Unlicensed and of label prescribing of drugs in general practice. *Arch. Dis. Child.* **2000**, *83*, 498–501. [CrossRef] [PubMed]
3. Batchelor, H.K.; Marriott, J.F. Formulations for children: Problems and solutions. *Br. J. Clin. Pharmacol.* **2015**, *79*, 405–418. [CrossRef] [PubMed]
4. Lopez, F.L.; Ernest, T.B.; Tuleu, C.; Gul, M.O. Formulation approaches to paediatric oral drug delivery: Benefits and limitations of current platforms. *Expert Opin. Drug Deliv.* **2015**, *12*, 1727–1740. [CrossRef] [PubMed]
5. Spomer, N.; Klingmann, V.; Stoltenberg, I.; Lerch, C.; Meissner, T.; Breitkreutz, J. Acceptance of uncoated mini-tablets in young children: Results from a prospective exploratory cross-over study. *Arch. Dis. Child* **2012**, *97*, 283–286. [CrossRef] [PubMed]
6. Klingmann, V.; Spomer, N.; Lerch, C.; Stoltenberg, I.; Frömke, C.; Bosse, H.M.; Breitkreutz, J.; Meissner, T. Favorable acceptance of mini-tablets compared with syrup: A randomized controlled trial in infants and preschool children. *J. Pediatr.* **2013**, *163*, 1728–1732. [CrossRef]
7. Klingmann, V.; Seitz, A.; Meissner, T.; Breitkreutz, J.; Moeltner, A.; Bosse, H.M. Acceptability of uncoated mini-tablets in neonates-a randomized controlled trial. *J. Pediatr.* **2015**, *167*, 893–896. [CrossRef]
8. Kluk, A.; Sznitowska, M.; Brandt, A.; Sznurkowska, K.; Plata-Nazar, K.; Mysliwiec, M.; Kaminska, B.; Kotłowska, H. Can preschool-aged children swallow several minitablets at a time? Results from a clinical pilot study. *Int. J. Pharm.* **2015**, *485*, 1–6. [CrossRef]
9. Kumar, K.P.; Teotia, D. A comprehensieve review on pharmaceutical minitablets. *J. Drug Deliv. Ther.* **2018**, *8*, 382–390.
10. Aleksovski, A.; Dreu, R.; Gašperlin, M.; Planinšek, O. Mini-tablets: A contemporary system for oral drug delivery in targeted patient group. *Expert Opin. Drug Deliv.* **2015**, *12*, 65–84. [CrossRef]
11. Guidance for Industry. Orally Disintegrating Tablets. Available online: https://www.fda.gov/media/70877/download (accessed on 20 February 2020).
12. Sharma, S.; Lewis, S. Taste masking technologies: A review. *Int. J. Pharm. Sci.* **2010**, *2*, 1–8.

13. Pandey, S.; Kumar, S.; Prajapati, S.K.; Madhar, N.M. An overview on taste physiology and masking of bitter drugs. *Int. J. Pharm. Bio. Sci.* **2010**, *1*, 1–11.
14. Jyothi, N.V.N.; Sakarkar, S.N.; Kumar, G.Y.; Prasanna, M. Microencapsulation techniques, factors influencing encapsulation efficiency: A review. *J. Microencapsul.* **2010**, *27*, 187–197. [CrossRef] [PubMed]
15. Faisal, W.; Farag, F.; Abdellatif, A.A.H.; Abbas, A. Taste masking approaches for medicines. *Curr. Drug Deliv.* **2018**, *15*, 167–185. [CrossRef] [PubMed]
16. Liu, W.; Chen, X.D.; Selomuyla, C. On the spray drying of uniform functional microparticles. *Particuology* **2015**, *22*, 1–12. [CrossRef]
17. Wasilewska, K.; Szekalska, M.; Ciosek-Skibinska, P.; Lenik, J.; Basa, A.; Jacyna, J.; Markuszewski, M.; Winnicka, K. Ethylcellulose in organic solution or aqueous dispersion form in designing taste-masked microparticles by the spray drying technique with a model bitter drug: Rupatadine fumarate. *Polymers* **2019**, *11*, 17. [CrossRef]
18. Bowles, B.J.; Dziemidowicz, K.; Lopez, F.L.; Orlu, M.; Tuleu, C.; Edwards, A.; Ernest, T.B. Co-processed excipients for dispersible tablets—Part 1: Manufacturability. *AAPS PharmSciTech* **2018**, *19*, 2598–2609. [CrossRef]
19. Prosolv® ODT. Available online: https://www.jrspharma.com/pharma_en/products-services/excipients/hfe/prosolv-odt-g2.php (accessed on 20 February 2020).
20. Parteck® ODT. Available online: http://www.phexcom.cn/uploadfiles/2011126103726995.pdf (accessed on 20 February 2020).
21. Fmelt®. Available online: http://www.fujichemical.co.jp/english/newsletter/newsletter_pharma_0802.html (accessed on 20 February 2020).
22. SmartEx® QD-50. Available online: http://www.metolose.jp/en/pharmaceutical/smartexr.html (accessed on 20 February 2020).
23. Pearlitol® Flash. Available online: https://www.roquette.com/pharma-and-nutraceuticals-coprocessed-mannitol-starch (accessed on 20 February 2020).
24. Ludiflash®. Available online: https://pharmaceutical.basf.com/global/en/drug-formulation/products/ludiflash.html (accessed on 20 February 2020).
25. Council of Europe. *The European Pharmacopoeia*, 9th ed.; Council of Europe: Strasburg, France, 2016.
26. Choudekar, R.L.; Mahajan, M.P.; Sawant, S.D. Validated RP-HPLC method for the estimation of rupatadine fumarate in bulk and tablet dosage form. *Pharma Chem.* **2012**, *4*, 1047–1053.
27. Redasani, V.K.; Kothawade, A.R.; Surana, S.J. Stability indicating RP-HPLC method for simultaneous estimation of rupatadine fumarate and montelukast sodium in bulk and tablet dosage form. *J. Anal. Chem.* **2014**, *69*, 384–389. [CrossRef]
28. Rele, R.V.; Mali, R.N. New validated RP-HPLC method for quantification of rupatadine fumarate impurities in solid dosage form supported by forced degradation studies. *Der Pharm. Lett.* **2016**, *8*, 66–72.
29. Stoltenberg, I.; Breitkreutz, J. Orally disintegrating mini-tablets (ODMTs)—A novel solid oral dosage form for paediatric use. *Eur. J. Pharm. Biopharm.* **2011**, *78*, 462–469. [CrossRef]
30. FDA; CDER. Guidance for Industry—Orally Disintegrating Tablets. 2008. Available online: https://www.fda.gov/downloads/Drugs/Guidances/ucm070578.pdf (accessed on 20 January 2020).
31. Amelian, A.; Szekalska, M.; Ciosek, P.; Basa, A.; Winnicka, K. Characterization and taste masking evaluation of microparticles with cetirizine dihydrochloride and methacrylate-based copolymer obtained by spray drying. *Acta Pharm.* **2017**, *67*, 113–124. [CrossRef] [PubMed]
32. Łabańska, M.; Ciosek-Skibińska, P.; Wróblewski, W. Critical evaluation of laboratory potentiometric electronic tongues for pharmaceutical analysis—An overview. *Sensors* **2019**, *19*, 5376. [CrossRef] [PubMed]
33. Shamizadeh, S.; Brockow, K.; Ring, J. Rupatadine: Efficacy and safety of a non-sedating antihistamine with PAF-antagonist effects. *Allergo J. Int.* **2014**, *23*, 87–95. [CrossRef]
34. Mullol, J.; Gonzalez-Nunez, V.; Bachert, C. Rupatadine: Global safety evaluation in allergic rhinitis and urticaria. *Expert Opin. Drug Saf.* **2016**, *15*, 1439–1448.
35. Rao Sudhakara, M.; Dwarakanatha Reddy, D.; Murthy, P.S.N. Rupatadine: Pharmacological profile and its use in the treatment of allergic rhinitis. *Indian J. Otolaryngol. Head Neck Surg.* **2009**, *61*, 320–332. [CrossRef]
36. Picado, C.S. Rupatadine: Pharmacological profile and its use in the treatment of allergic disorders. *Expert Opin. Pharmacoter.* **2006**, *7*, 1989–2001. [CrossRef]

37. Merlos, M.; Giral, M.; Balsa, D.; Ferrando, R.; Queralt, M.; Puigdemont, A.; García-Rafanell, J.; Forn, J. Rupatadine, a new potent, orally active dual antagonist of histamine and platelet-activating factor (PAF). *J. Pharmacol. Exp. Ther.* **1997**, *280*, 114–121.
38. Chuch, M.K. Efficacy and tolerability of rupatadine at four times the recommended dose against histamine- and platelet-activating factor-induced flare responses and ex vivo platelet aggregation in healthy males. *Br. J. Dermatol.* **2010**, *163*, 1330–1332.
39. Izquierdo, I.; Valero, A.; García, O.; Pérez, I.; Mullol, J.; Van Cauwenberge, P. Clinical efficacy of rupatadine in allergic rhinitis under ARIA criteria: Pooled analysis. *Allergy Clin. Immunol. Int.* **2005**, *1*, 271–277.
40. Rupafin®. Available online: https://www.drugs.com/uk/rupafin-10mg-tablets-leaflet.html (accessed on 20 February 2020).
41. FDA Inactive Ingredients Database. Available online: https://search.fda.gov/search?utf8=%E2%9C%93&affiliate=fda1&query=ethylcellulose&commit=Search (accessed on 20 February 2020).
42. WHO. Available online: https://apps.who.int/iris/bitstream/handle/10665/42601/WHO_TRS_913.pdf (accessed on 20 February 2020).
43. Rowe, R.C.; Sheskey, P.J.; Quinn, M.E. *Handbook of Pharmaceutical Excipients*, 6th ed.; Pharmaceutical Press: London, UK; Chicago, IL, USA; Washington, DC, USA, 2009; pp. 262–267.
44. Ethylcellulose. Available online: https://pubchem.ncbi.nlm.nih.gov/compound/24832091#section=ProbableRoutes-of-Human-Exposure (accessed on 20 April 2020).
45. Wasilewska, K.; Winnicka, K. Ethylcellulose—A pharmaceutical excipient with multidirectional application in drug dosage forms development. *Materials* **2019**, *12*, E3386. [CrossRef] [PubMed]
46. Safety & Toxicity of Excipients for Paediatrics, STEP Database. Available online: http://www.eupfi.org/step-database-info/ (accessed on 26 May 2020).
47. Canadian List of Acceptable Non-Medicinal Ingredients. Available online: https://www.canada.ca/en/health-canada/services/drugs-health-products/natural-non-prescription/applications-submissions/product-licensing/compendium-monographs.html. (accessed on 26 May 2020).
48. Surelease®. Available online: https://www.colorcon.com/products-formulation/all-products/filmcoatings/sustained-release/surelease (accessed on 20 February 2020).
49. Aquacoat® ECD. Available online: http://www.fmcbiopolymer.com/Pharmaceutical/Products/Aquacoat.aspx (accessed on 20 February 2020).
50. Müller, K.; Fingueroa, C.; Martinez, C.; Madel, M.; Obreque, E.; Peña-Neira, A.; Morales-Bozo, I.; Toledo, H.; Lopez-Solis, R.O. Measurement of saliva volume in the mouth of members of a trained sensory panel using a beetroot (Beta vulgaris) extract. *Food Qual. Prefer.* **2010**, *21*, 569–574. [CrossRef]
51. Ali, J.; Zgair, A.; Hameed, G.D.S.; Garnet, M.C.; Roberts, C.J.; Vurley, J.C.; Gershkovich, P. Application of biorelevant saliva-based dissolution for optimization of orally disintegrating formulations of felodipin. *Int. J. Pharm.* **2019**, *30*, 228–236. [CrossRef]
52. Brniak, W.; Jachowicz, R.; Pelka, P. The practical approach to the evaluation of methods used to determinate the disintegration time of orally disintegrating tablets (ODTs). *Saudi Pharm. J.* **2015**, *23*, 437–443. [CrossRef]
53. Rupatadine Fumarate. Available online: https://www.drugbank.ca/salts/DBSALT001922 (accessed on 20 February 2020).
54. Henríquez, L.C.; Redondo, G.M.; Zúñiga, R.V.; Berrocal, G.C. Identification of rupatadine fumarate polymorphic crystalline forms in pharmaceutical raw materials. *AJST* **2018**, *9*, 7482–7487.
55. Henríquez, L.C.; Zúñiga, R.V.; Redondo, G.M.; Berrocal, G.C.; Vargas, G.H. Determination of the impact caused by direct compression on the crystalline state of rupatadine fumarate 10 mg tablets. *Int. J. Pharm. Technol. Biotechnol.* **2019**, *6*, 1–12.
56. Mohamed-Ahmed, A.H.; Soto, J.; Ernest, T.; Tuleu, C. Non-human tools for the evaluation of bitter taste in the design and development of medicines: A systematic review. *Drug Discov. Today* **2016**, *21*, 1170–1180. [CrossRef] [PubMed]
57. Pein, M.; Preis, M.; Eckert, C.; Kiene, F.E. Taste-masking assessment of solid oral dosage forms—A critical review. *Int. J. Pharm.* **2014**, *465*, 239–254. [CrossRef] [PubMed]
58. Clapham, D.; Kirsanov, D.; Legin, A.; Rudnitskaya, A.; Saunders, K. Assessing taste without using humans: Rat brief access aversion model and electronic tongue. *Int. J. Pharm.* **2012**, *435*, 137–139. [CrossRef]
59. Legin, A.; Rudnitskaya, A.; Clapham, D.; Seleznev, B.; Lord, K.; Vlasov, Y. Electronic tongue for pharmaceutical analytics: Quantification of tastes and masking effects. *Anal. Bioanal. Chem.* **2004**, *380*, 36–45. [CrossRef]

60. Wesoły, M.; Cal, K.; Ciosek, P.; Wróblewski, W. Influence of dissolution-modifying excipients in various pharmaceutical formulations on electronic tongue results. *Talanta* **2017**, *162*, 203–209. [CrossRef] [PubMed]
61. Wesoły, M.; Zabadaj, M.; Amelian, A.; Winnicka, K.; Wróblewski, W.; Ciosek, P. Tasting cetirizine-based microspheres with an electronic tongue. *Sens. Actuators B* **2017**, *238*, 1190–1198. [CrossRef]
62. Krishnan, A.; Williams, L.J.; McIntosh, A.R.; Abdi, H. Partial Least Squares (PLS) methods for neuroimaging: A tutorial and review. *NeuroImage* **2011**, *56*, 455–475. [CrossRef] [PubMed]
63. Wesoły, M.; Ciosek-Skibińska, P. Comparison of performance of various data analysis techniques applied for the classification of pharmaceutical samples by electronic tongue. *Sens. Actuators B* **2018**, *267*, 570–580. [CrossRef]
64. Zabadaj, M.; Szuplewska, A.; Kalinowska, D.; Chudy, M.; Ciosek-Skibińska, P. Studying pharmacodynamic effects in cell cultures by chemical fingerprinting—SIA electronic tongue versus 2D fluorescence soft sensor. *Sens. Actuators B* **2018**, *272*, 264–273. [CrossRef]
65. Zabadaj, M.; Ufnalska, I.; Chreptowicz, K.; Mierzejewska, J.; Wróblewski, W.; Ciosek-Skibińska, P. Performance of hybrid electronic tongue and HPLC coupled with chemometric analysis for the monitoring of yeast biotransformations. *Chemometr. Intell. Lab.* **2017**, *167*, 69–77. [CrossRef]

materials

MDPI

Review

Ethylcellulose–A Pharmaceutical Excipient with Multidirectional Application in Drug Dosage Forms Development

Katarzyna Wasilewska and Katarzyna Winnicka *

Department of Pharmaceutical Technology, Medical University of Bialystok, Mickiewicza 2c, 15-222 Bialystok, Poland; katarzyna.wasilewska@umb.edu.pl

* Correspondence: kwin@umb.edu.pl; Tel.: +48-85-7485616

Received: 17 September 2019; Accepted: 15 October 2019; Published: 17 October 2019

Abstract: Polymers constitute the most important group of excipients utilized in modern pharmaceutical technology, playing an essential role in the development of drug dosage forms. Synthetic, semisynthetic, and natural polymeric materials offer opportunities to overcome different formulative challenges and to design novel dosage forms for controlled release or for site-specific drug delivery. They are extensively used to design therapeutic systems, modify drug release, or mask unpleasant drug taste. Cellulose derivatives are characterized by different physicochemical properties, such as swellability, viscosity, biodegradability, pH dependency, or mucoadhesion, which determine their use in industry. One cellulose derivative with widespread application is ethylcellulose. Ethylcellulose is used in pharmaceutical technology as a coating agent, flavoring fixative, binder, filler, film-former, drug carrier, or stabilizer. The aim of this article is to provide a broad overview of ethylcellulose utilization for pharmaceutical purposes, with particular emphasis on its multidirectional role in the development of oral and topical drug dosage forms.

Keywords: ethylcellulose; polymeric material; cellulose derivative; pharmaceutical excipient

1. Introduction

Modern pharmaceutical technology could not exist without polymers, which play an integral role in the advancement of drug delivery (e.g., by providing controlled release of therapeutic agents, masking bitter taste of drugs, or serving as carriers in targeted therapy). They have a wide range of physicochemical characteristics according to their molecular weight and configuration. Therefore, polymers approved for use in medicines are receiving considerable attention as essential excipients in the development of different drug dosage forms. Cellulose derivatives are an important group among the most commonly used polymers in pharmaceutical technology. Cellulose is one of the most abundant polymers in nature. It is produced by photosynthesis and constitutes a basic plant component. It is a linear polymer composed of glucopyranose residues, where the units are connected by 1,4-β-glycosidic bonds. It is a tasteless and odorless component with fibrous structure that is practically insoluble both in cold and hot water. Among the cellulose modifiers, its esters and ethers are of practical importance. Cellulose-based biomaterials are widely utilized as biocompatible templates for designing novel drug delivery systems with a wide range of pharmaceutical applications via different routes and pharmacotherapeutic purposes [1–4].

Ethylcellulose (EC), an ethyl ether of cellulose, is a free-flowing, white to light-tan powder prepared from wood pulp or cotton that is used in pharmaceutical manufacturing industries. The European Pharmacopoeia (Ph. Eur.) and United States Pharmacopoeia (USP) monographs describe EC as partially O-ethylated cellulose [5–7]. EC with ethoxyl substitution molecular formula is $C_{12}H_{23}O_6(C_{12}H_{22}O_5)nC_{12}H_{23}O_5$, where *n* can vary in order to provide a wide variety of molecular weights, which affect its properties. It is insoluble at any pH that occurs in organisms, but it undergoes swelling in the presence of gastric juice [3,4]. There is very limited data detailing possible side effects of EC, hence it is generally regarded as safe (GRAS) and included in the US Food and Drug Administration (FDA) Inactive Ingredients database as well as in the Canadian List of Acceptable Non-medicinal Ingredients to be utilized in oral capsules, suspensions, tablets, topical emulsions, and vaginal or ocular preparations. In contrast to other polymers which are insoluble in the gastrointestinal tract (e.g., nitrocellulose, cellulose acetate, Eudragit RL, or Eudragit RS), EC is characterized as a non-ionic material, having the advantage of being non-reactive. Moreover, safety data on utilizing cellulose acetate and methacrylic acid copolymers in pediatric preparations are limited, while EC is clinically tested and approved for use in pediatric formulations. EC is also allowed for use in non-parenteral medicines licensed in Europe. The polymer is accepted for use as a food additive because it is a non-calorific and metabolically inert substance following oral consumption. As EC is not considered to be a health hazard, the World Health Organisation (WHO) has not specified its acceptable daily intake [4,8,9]. According to the intended use, the commercially available EC can be classified in two categories of products: industrial grade and pharmaceutical grade. For the pharmaceutical grade, the quality standard should meet USP, Ph. Eur., Chinese Pharmacopoeia, and Japanese Pharmacopoeia standards [6,7,10,11].

2. Physicochemical Characteristics of EC

The polymer is obtained by synthesis (etherification) through the substitution of the cellulose hydroxyl groups with ethoxyl ones. The chemical reaction involves cellulose dissolution in sodium hydroxide aqueous solution, which leads to degradation of the cellulose's molecular structure. This results in the formation of an alkali cellulose and exposure of the cellulose hydroxyl group for reaction. Afterwards, ethyl chloride gas is added to the reaction medium, leading to bonding with the alkalized cellulose. EC, sodium chloride, and water are formed (Figure 1) [1–5].

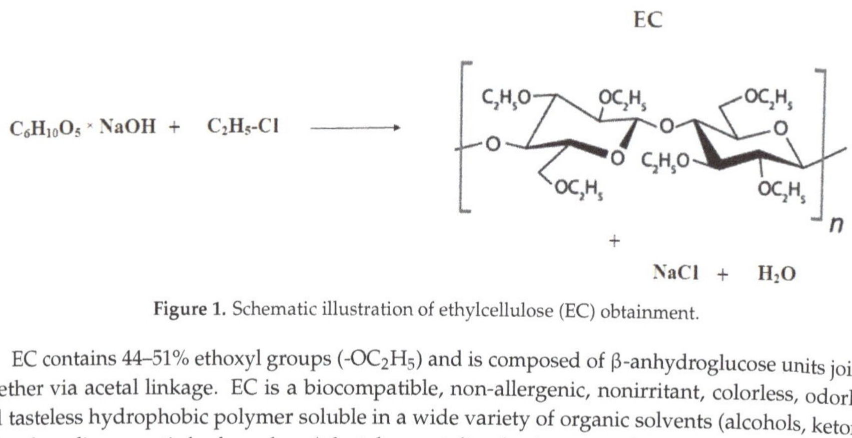

Figure 1. Schematic illustration of ethylcellulose (EC) obtainment.

EC contains 44–51% ethoxyl groups (-OC_2H_5) and is composed of β-anhydroglucose units joined together via acetal linkage. EC is a biocompatible, non-allergenic, nonirritant, colorless, odorless, and tasteless hydrophobic polymer soluble in a wide variety of organic solvents (alcohols, ketones, and polycyclic aromatic hydrocarbons), but does not dissolve in water, glycerin, or propylene glycol. EC that contains no less than 46.5% of ethoxyl groups is freely soluble in chloroform, ethanol, ethyl acetate, methanol, and toluene. EC is compatible with a wide array of excipients and most of the plasticizers used in pharmaceutical formulations, and therefore can be well mixed with many softeners, oils, and waxes (dibutyl phthalate, diethyl phthalate, dibutyl sebacate, triethyl citrate, tributyl citrate, acetylated monoglyceride, butyl and glycol esters of fatty acids, refined mineral oils, oleic acid, stearic acid, stearyl alcohol, castor oil). It shows incompatibilities with paraffin and microcrystalline wax. EC exhibits a high degree of stability within pH 3–11, making it useful both in acidic and alkaline mixtures. EC is also slightly hygroscopic, absorbing very little water from humid air or during immersion. In addition, it can form tough and tensile films and maintains flexibility even at low temperatures. It possesses good thermal stability and low ash content when burning [1,3–5].

The physical characteristic of EC types and their performance depends on the degree of etherification or substitution (ethoxyl content), molecular uniformity, and molecular weight of the cellulosic backbone. Solubility in organic solvents is achieved with degree of substitution (DS) values between 2.2 and 2.6. A range of commercial products is available with a degree of substitution between 2.15 and 2.60, corresponding to a range of ethoxyl contents from 43% to 50%. At low DS values (0.8–1.3), the replacement of some of the hydroxyl groups by ethoxyl groups reduces the hydrogen bonding across the cellulosic chains to such an extent that the material is soluble in water. Further replacement of hydroxyl groups by the less polar and more hydrocarbon ethoxyl groups increases the water resistance. Fully etherified EC is soluble only in non-polar solvents [1–5].

The molecular weight can be regulated by controlled degradation of the alkali cellulose in the presence of air. This can be done either before or during etherification. The molecular weight of commercial grades is usually expressed indirectly as viscosity. The completely etherified material with a degree of substitution of 3 has an ethoxyl content of 54.88%. The viscosities are determined using a capillary viscometer and recorded in millipascal-seconds (mPa·s) or centipoise (cP) synonymous for a 5% *w/w* solution in toluene/ethanol solvent mixture (80:20), being determined at 25 °C and expressed in mPa·s. For each viscosity grade of EC, the pharmacopoeial specification allows variation of 80–120% within the stated nominal viscosity (depending on manufacturer). Several types of EC (e.g., Ethocel™ 4, Ethocel™ 10, and Ethocel™ 45) differ in the length of the polymer chains, the rate of dissolution, and the viscosity of their solutions. The mechanical properties are largely determined by chain length; softening point, hardness, water absorption, and solubility are rather more influenced by the degree of substitution [1–5,12,13].

EC viscosities increase with the increase in the polymeric backbone chain length. The impact of EC molecular weight variation (measured by viscosity) on the drug release was determined by Mehta et al. Metoprolol was selected as a model drug that is highly soluble in water and acetaminophen as one with low aqueous solubility. Drug-layered sugar spheres were coated with EC of different viscosity grades at varying coating weight gains, and their drug release profiles were determined. The study revealed that viscosity variation within the manufacturer's specifications of EC (Ethocel™ 10, 20, and 100 cP) had a minimal effect on release of model drugs. Acetaminophen multiparticulates exhibited slower drug release and longer lag time when compared to the metoprolol beads. The obtained results can be associated with the lower aqueous solubility of acetaminophen compared to metoprolol. Based on the study, it was concluded that all grades of EC were suitable for organic solvent coating of extended release barrier membrane multiparticulates [14]. In another study, sustained-release metformin hydrochloride nanoparticulate systems were developed, and the effect of different viscosity EC grades on their in vitro characteristics was investigated. The sustainability of EC was enhanced by the increase in its apparent viscosity—the EC with higher viscosity grade sustained metformin release more efficiently [15].

3. Pharmaceutical Characteristics of EC

EC has been widely used in the pharmaceutical industry for decades, being utilized in oral and topical pharmaceutical formulations for various purposes. It has the potential to modulate and improve the physiological performance of drug dosage forms because of its hydrophobic nature and swelling capacity. The main goal of EC utilization is the development of drug dosage forms with modified release (MR), as EC ensures drug dissolution in the entire gastrointestinal tract, providing constant drug concentration and eliminating the necessity of taking several doses a day, hence improving pharmacotherapeutic effectiveness. Because it is an inert hydrophobic polymer and possesses properties such as lack of toxicity, stability during storage, and good compressibility, it is suitable for designing sustained-release preparations [1–5,16]. EC provides the formation of hydrophobic coatings, filaments, or backing layers; masks unpleasant medicine tastes; creates matrices and nanostructures for the preparation of bioactive materials or is used as an encapsulation excipient for the preparation of sustained-release microparticles; and serves as moisture protector or binder. It can also be used as a dispersing, stabilizing, and water-retaining agent to prevent drugs from getting wet and to promote the safe storage of drugs. Shell tablets can be obtained by suspending drug in the gastrointestinally insoluble carrier. EC has also been used as a matrix in the preparation of both water-soluble and sparingly water-soluble drugs using solid dispersion technique. It is ubiquitously utilized as a coating material in sustained-release preparations due to its film-forming properties and good mechanical strength. Also, in the coating context, its important feature is insolubility at any physiological pH. However, the polymer exhibits swelling in the presence of gastric juice, making it permeable for water and permitting extended modified drug release [4,17–22]. EC-coated beads have also demonstrated the ability to absorb pressure and hence protect the coating from fracture during compression. Drug release from EC-based film coatings depends on the coating level, drug solubility, and the form in which the polymer is applied in the coating process (e.g., as powder, aqueous dispersion, or organic solution). It is also common to employ blending polymers to get suitable and desired results, as using a single polymer may not give the desired drug release profile. Owing to its hydrophobic properties, EC reduces the penetration of water into the solid polymeric matrix, hence reducing drug release [4,23].

EC can be found in different forms, such as powders with various viscosity grades or aqueous dispersions. Examples of organic solids are Ethocel™ or Aqualon™, and aqueous dispersions include Surelease®, Aquacoat® ECD, Aquarius™ Control ECD, and AshaKote® (Table 1, Figure 2) [12,24–28]. Commercially available types of EC (e.g., Ethocel™ 4, Ethocel™ 10, and Ethocel™ 45) differ in the length of the polymer chains, the rate of dissolution, molecular weights, and hence the viscosity of their solutions (ranging from 3 to 110 mPa·s) [12].

Table 1. Characteristics of commercially available EC *.

Organic EC products
Ethocel™
Available in many varieties that differ in grade and viscosity (e.g., Ethocel™ Standard Grade 4 Premium, 7 Premium, 10 Premium, 20 Premium, 45 Premium, 100 Premium). Ethocel™ are white to light-tan granular powders in physical appearance with bulk density and specific gravity of about 0.4 g/cm^3 and 1.12–1.15 g/cm^3 respectively and glass transition temperatures range between 129 and 133 °C. They dissolve in a wide range of solvents such as aliphatic alcohols, chlorinated solvents, and natural oils. They are practically insoluble in glycerin, propylene glycol, and water. Films made from Ethocel™ are tough, with high tensile strength and high flexibility even at low temperatures. They can be combined with water soluble polymers such as methylcellulose and hypromellose (HPMC) in aqueous coating liquids. They are characterized by thermoplastic nature and ability to soften at 135–160 °C which makes them versatile in pharmaceutical hot melt extrusion processes. They ensure desired drug release properties and improved bioavailability of especially poorly soluble drugs [12].
Aqualon™ ethylcellulose
During preparation, the substitution of ethoxyl groups is controlled to obtain commercially useful range of 48–52% ethoxyl (or 2.3–2.6 ethoxyl groups out of a theoretical maximum of 3.0) per anhydroglucose unit. Over this ethoxyl range, Aqualon™ ethylcellulose is classified into three ethoxyl types: N (low substitution), T (mid substitution), and X (high substitution). The improved compressible grade (Aqualon™ T10) was developed with optimized compactability (high ethoxyl content and low viscosity) and good powder flow. The grades of Aqualon™ ethylcellulose are compliant with the monograph requirements of the United States Pharmacopoeia (USP) and the European Pharmacopoeia (Ph. Eur.) [24].
Aqueous dispersion of ethylcellulose

Table 1. *Cont.*

Surelease®
Surelease® is a family of fully formulated, aqueous dispersion products which constitute plasticized aqueous dispersions of EC with 25% (*w/w*) solid content available in four types: E-7-19029, E-7-19030, E7-19040, E-7-19050. They are plasticized (depending on type) with dibutyl sebacate (3.5%-E-7-19029, E-7-19030) and oleic acid (1.9%-E7-19040). Surelease® is produced in the following steps: EC is blended with plasticizer, then extruded and melted. The molten plasticized EC is then directly emulsified in ammoniated water in a high-shear mixing device under pressure. Ammonium oleate is formed in situ in order to stabilize and form the dispersion of plasticized EC particles. Then, purified water is added to achieve the final solids content. Applications of Surelease® include beads and particles coating, matrix granulation (the dispersion can be used as a binder in wet granulation for the production of free-flowing granules, which can subsequently be compressed into tablets), taste-masking coating, or nutritional enteric coating [25].
Aquacoat® ECD
It contains primarily EC with a surfactant and a stabilizer from the emulsion stage (sodium lauryl sulfate (SLS) and cetyl alcohol (CA)). Depending on type it consists of EC (24.5–29.5%), CA (1.7–3.3%), and SLS (0.9–1.7%). Traces of dimethylopolysiloxane to enhance foaming during distillation may also be present. EC is dissolved in a water-immiscible organic solvent with CA addition as a dispersion stabilizer. The solution is then emulsified into an aqueous SLS solution. The resulting crude emulsion is homogenized to yield a submicron emulsion which is then distilled to remove the organic solvent and water to yield a solid dispersion. EC is present in the dispersion as spherical particles in the size range of 0.1–0.3 μm. It exists as a milky white liquid with the characteristic odor of EC. Product does not contain plasticizer. Recommended plasticizers include dibutyl sebacate, acetylated monoglycerides, triacetin or glyceryl triacetate, acetyltriethyl citrate, and triethyl citrate. It is used for the aqueous film coating of solid dosage forms to extend drug release, taste mask, or to protect against moisture [26].

* The table presents examples of selected, most commonly used, products.

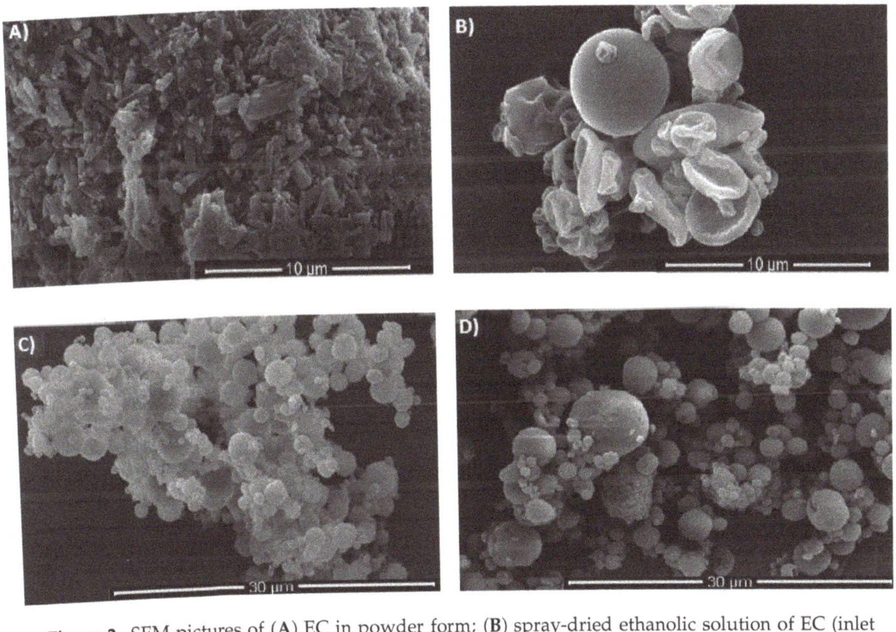

Figure 2. SEM pictures of (**A**) EC in powder form; (**B**) spray-dried ethanolic solution of EC (inlet temperature 65 °C, aspirator flow 98%, feed flow 3.5 mL/min); (**C**) spray-dried Surelease®; and (**D**) spray-dried Aquacoat™ ECD-30 (author's pictures under magnification 10,000× (**A**,**B**); 5000× (**C**,**D**)).

4. Applicability of EC in Pharmaceutical Formulations

4.1. EC as An MR Coating Material in Oral Delivery Systems

Because of EC's potential to modify drug release, it is widely studied as a coating agent. EC as release retardant was employed in a once-a-day sustained-release system of tacrolimus. The pellets were coated with EC using a fluid bed granulator. Drug release was markedly impeded by the outer EC-based coating layer, displaying about 60% drug release after 8 h, regardless of the acidity of the medium. It was assessed that there were no statistical differences between the obtained pellets and the marketed sustained-release capsules (Advagraf®) [29,30]. Shah et al. developed a multiunit formulation for the colon-targeted delivery of metronidazole using EC and Eudragit® S 100 as coating polymers to prevent initial drug release in the gastric region (Figure 3). Cores of mini tablets containing drug were prepared using suitable swelling agents to provide pH-sensitive pulsatile drug delivery [31].

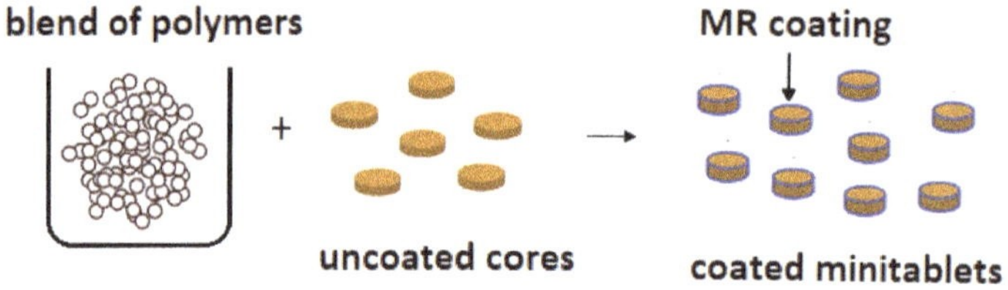

Figure 3. Schematic illustration of coated modified release (MR) minitablets.

A double-layer coated colon-specific drug delivery system was developed by Kim et al. The system consisted of a chitosan-based polymeric subcoating of the core tablet (containing citric acid for microclimate acidification), followed by an enteric EC coating. The system showed drug release in a controlled manner by inhibiting drug release in the stomach and intestine, but releasing the drug gradually in the colon [32]. The example of utilizing EC as a release-retarding coating is a commercially available preparation called Micro-K® in hard gelatin capsules, containing small crystalline particles of potassium chloride (KCl). Each crystal of KCl is microencapsulated by a patented process with an insoluble EC polymeric coating which functions as a semi-permeable membrane ensuring controlled release of K^+ and Cl^- ions. Fluids pass through the membranes and gradually dissolve KCl within the microcapsules. As a result, drug slowly diffuses outward through the membrane over an 8–10 h period [33,34]. Another example is Theo-24®—the first commercial product for 24-h theophylline therapy launched on the market. The technology utilized in the formulation uses a chemical timing complex (protected by patent) to produce very small theophylline-coated beads that provide dependable zero-order controlled drug release. Tiny spheres of a sugar and starch blend form the core of the beads. The core is first coated with theophylline and then with a timing complex utilizing EC. The resulting beads are put into capsules. When the capsule dissolves in the gastrointestinal tract, the insoluble coating on the bead slowly erodes. The drug, which is highly soluble, moves through the coating. In the core, the starch swells and pushes the drug out while the dissolving sugar also helps carry the drug through the chemical timing complex, which results in a constant release [33,35].

Matrix tablets of hydrochlorothiazide were coated with an insoluble barrier membrane using aqueous EC coating (Surelease®) and HPMC-based Opadry® as a pore former, at 85:15 *w/w* ratio. The combination of barrier membrane and hydrophilic matrix system was utilized as a strategy to modulate drug diffusion from hydrophilic matrices and to reduce the overall variability in release [36]. The aims of another study were to control the release of a water-soluble theophylline from mini matrices made of HPMC by applying an EC film coating (Surelease®) and to assess coating load on release rates. At low coating weight gains, tablets released the entire drug within 0.5 h, while at high coating weight gains only a very small amount (<5%) of drug was released after 12 h [17].

EC has found its application as a coating material in commercially available modern MR solid dosage forms (Diffucaps, DiffCORE™, SODAS®, or Geomatrix® systems). It is also used in medicines dedicated to patients suffering from ADHD: Adzenys XR®-ODT and Cotempla® XR-ODT [16,33,37–42]. Diffucaps are a multiparticulate system, where drug profiles are created by layering a drug onto a neutral core (e.g., sugar spheres, crystals, or granules) followed by the application of a rate-controlling functional EC membrane. Such a system is applied in Metadate CD® capsules (Figure 4) [16,37].

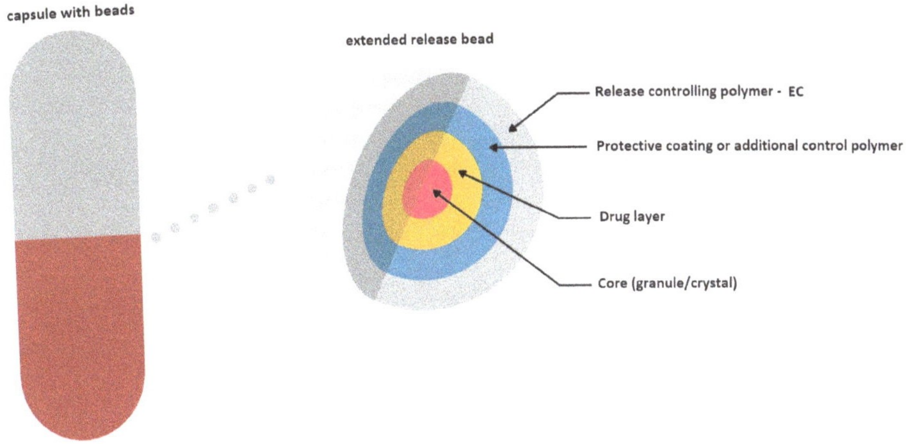

Figure 4. Schematic illustration of Diffucaps system in a Metadate CD® capsule. Reprinted with permission from Reference [16]. Copyright 2019 MDPI.

Another example of an EC-coating-based oral system is DiffCORE™ (Lamictal® XR) (Figure 5). DiffCORE™ is a technology developed to achieve extended release by delivering drug from a tablet core through one or several apertures in an impermeable coat made of EC. The technology combines the use of apertures that are mechanically drilled into functional film-coated tablets (on both faces of the tablet's structure) with a polymer coating that controls the mechanism of diffusion. This combination is designed to ensure the dissolution rate of drug over a period of approximately 12–15 h [16,38].

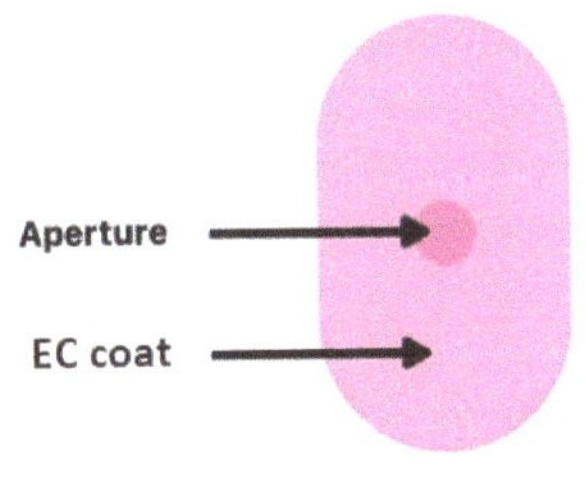

Figure 5. Schematic illustration of DiffCORE™ system. Reprinted with permission from Reference [16]. Copyright 2019 MDPI.

Adzenys® XR-ODT and Cotempla® XR-ODT are examples of orally disintegrating tablets which dissolve quickly in the mouth, maintaining extended drug release along gastrointestinal tract. The technology utilized in the medicines uses two different types of particles (in different ratios depending on the formulation): immediate release and extended release. Two different polymer coatings are applied to the extended release beads: interior polymer coating as diffusion barrier (EC) and pH-dependent exterior polymer coating (methacrylic acid) (Figure 6) [16,41,42].

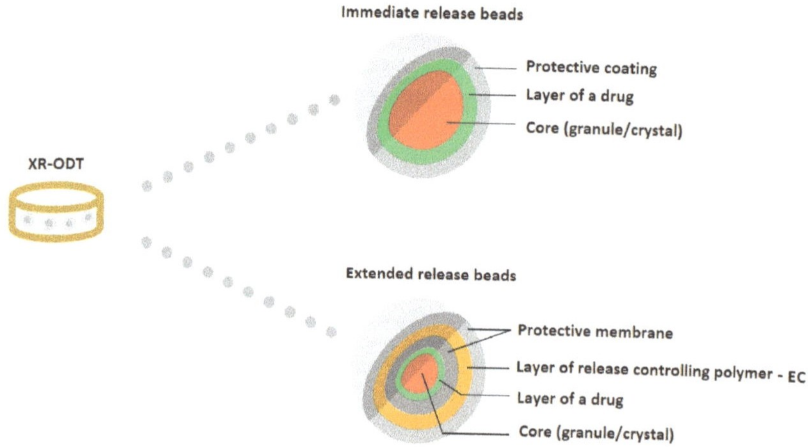

Figure 6. Schematic illustration of XR-ODT system. Reprinted with permission from Reference [16]. Copyright 2019 MDPI.

Inderal® LA, long-acting capsules with propranolol, uses polymer-coated controlled diffusion technology to achieve 12 h release of drug for hypertension treatment. Inderal® LA consists of small spheroids placed in a gelatin capsule. Each spheroid containing propranolol and a microcrystalline cellulose is coated with a porous membrane made of a mixture of EC, HPMC, and plasticizer [33,43]. Dilacor XR® is a one-per-day oral formulation of diltiazem provided in capsule. The capsules contain degradable, controlled-release tablets designed based on Geomatrix® technology to release diltiazem over a 24 h period. Geomatrix™ is a patented controlled-release system incorporated in the tablets. Each capsule contains multiple extended-release diltiazem tablets, consisting of two inactive surfaces sandwiched around the drug core (made of HPMC and hydrogenated castor oil) swellable in an aqueous medium. The inactive surfaces are composed of a methylcellulose–EC combination. The drug is released as a result of the swelling of the core, which acts as a drug reservoir. Controlled release of diltiazem begins within 1 h, with maximum plasma concentrations being achieved 4–6 h after administration. The inactive surfaces hydrate at a rate much slower than the core, thus regulating the drug release and assuring constant 24 h medicine delivery (Figure 7) [33,44].

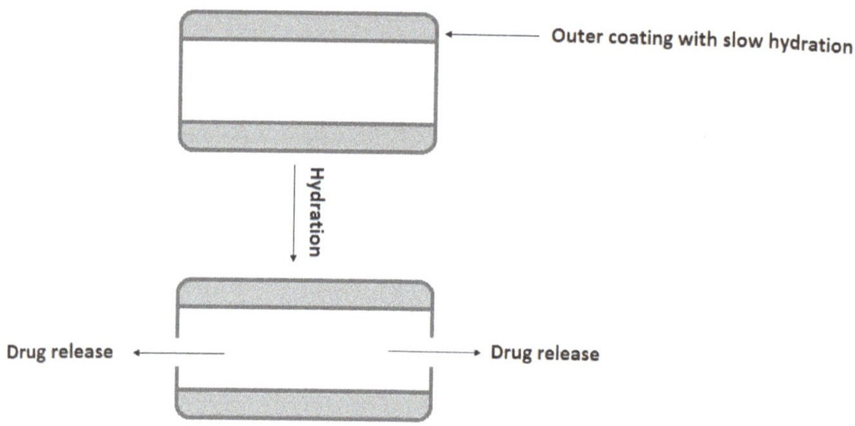

Figure 7. Schematic illustration of drug release from Geomatrix™ system modified from Wan et al. [33].

An interesting example of applying EC as a coating is SODAS® (Spheroidal Oral Drug Absorption System), utilized in Cardizem® CD. The medicine is formulated as capsules and consists of two populations of sustained-release beads that differ only by the thickness of the polymer (EC) surrounding them. The EC membrane contains added water-soluble polymers which dissolve and create pores in the membrane. The polymer beads release 40% of the total diltiazem amount in the first 12 h (Figure 8) [33,39,40,45].

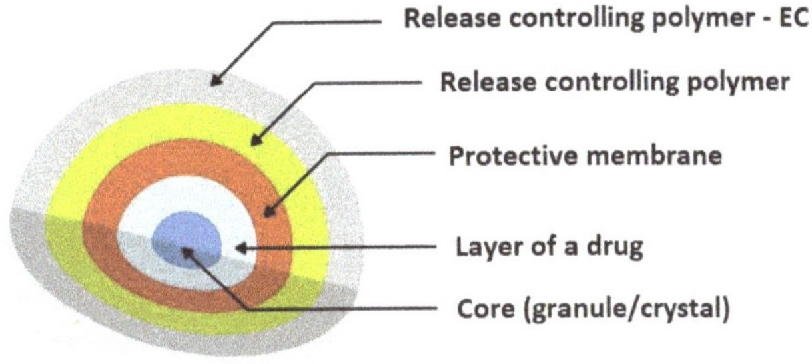

Figure 8. Scheme illustration of SODAS® delivery system modified from Elan drug technologies [39]. Reprinted with permission from Reference [16]. Copyright 2019 MDPI.

Core elements of drugs coated with a water-insoluble polymer such as EC offer reduced dissolution profiles, and as a consequence provide a taste-masking effect [46,47]. EC was used as taste-masking agent for quinine, utilizing a spray drying method. The obtained results showed that the quinine dissolution rate was altered, the bitter taste of quinine was successfully masked, and its intestinal absorption was simultaneously controlled [48]. EC for taste-masking purposes was tested in organic (Ethocel™) and aqueous form (Surelease®). To determine whether the designed particles effectively masked the bitter taste of rupatadine fumarate, three independent methods were used: a human taste panel, an in vitro release test conducted in conditions mimicking the oral cavity environment, and an electronic taste sensing system (electronic tongue). It was clearly confirmed that particles formulated from aqueous dispersions of EC provided a very effective taste-masking barrier [49]. Evaluation of the use of Surelease® as a barrier membrane coating with pediatric precedence of use on the taste-masking of immediate-release acetaminophen granules was also conducted. The dissolution profiles of obtained granules were successfully modified using a Surelease® and Opadry® blend as a barrier membrane, providing slow initial drug release [50]. Another study concerned the development of orally disintegrating caffeine citrate tablets utilizing hot melt extrusion technology (EC was used as a polymeric taste-masking carrier) [51]. EC was also applied as a taste-masking and release-slowing agent to develop a gabapentin nanosponge-based controlled release dry suspension for pediatric use, using the suspension layering technique [52].

Utilizing of EC as release modifier in marketed oral formulations is presented in Table 2.

Table 2. Examples of commercially available oral solid dosage forms with EC as release modifier.

Product (Manufacturer)	Drug	Dosage Form	EC Role	Reference
Aciphex® Sprinkle™ (Eisai Management Co., Ltd.)	rabeprazole	granules in capsule	coating, delayed release	[53]
Adzenys XR-ODT™ (Neos Therapeutics)	amphetamine	extended release orally disintegrating tablet with beads	coating, extended release	[41]
Advagraf XL™	tacrolimus	capsules	extended release	[30]
Cotempla XR-ODT® (Neos Therapeutics)	methylphenidate	extended release orally disintegrating tablet with beads	coating, extended release	[42]
Cardizem CD® (Abbot Laboratories)	diltiazem	tablets in capsule	coating, extended release	[45]
Dilacor XR® (Actavis Pharma, Inc.)	diltiazem	tablets in capsule	extended release	[44]
Dilatrate-SR (Epic Pharma, LLC)	diltiazem	beads in capsule	extended release	[54]
Durlaza (New Haven Pharmaceuticals, Inc.)	acetylsalicylic acid	capsules	extended release	[55]
Enjuvia (Teva Pharmaceuticals USA, Inc.)	estrogens	coated tablets	extended release	[56]
Entocort EC *(Astra Zeneca)*	budesonide	coated granules in capsules	sustained release	[57]
Inderal® LA (Wyeth Pharmaceuticals, Inc.)	propranolol	granules in capsules	coating, extended release	[43]
Levbid® (Mylan Pharmaceuticals Inc.)	hyoscyamine	tablets	extended release	[58]
Metadate CD® (UCB Manufacturing, Inc.)	methylphenidate	granules in capsule	coating, extended release	[37]
Micro-K® (KV Pharmaceutical)	potassium	microcapsules in capsules	coating, extended release	[34]
Lamictal® XR (GSK)	lamotrigine	tablets	coating, extended release	[38]
Orfiril Long (Desitin Arzneimittel GmbH)	natrii valproas	minitablets	extended release	[59]
Palladone *(PF Laboratories Inc.)*	hydromorphone	capsules	extended release	[60]
Pentasa® (Ferring GmbH)	mesalamine	granules	prolonged release	[61]
Qsymia® (Vivus, Inc.)	phentermine, topiramate	capsules	extended release	[62]
Tegretol® XL (Novartis)	carbamazepine	tablets	matrix, extended release	[63]
Theo-24® (UCB Pharma, Inc.)	theophylline	beads in capsule	coating, extended release	[35]
Trokendi XR™ (Catalent Pharma Solutions)	topiramate	capsules	extended release	[64]

4.2. EC as a Sustained Release Material in Topical Delivery Systems

The utilization of EC to obtain a sustained release profile is also exploited in preparations for external use (e.g., transungual, ocular, vaginal, or transdermal). Successful topical therapy depends on effective drug release and penetration into the objective area, which can be achieved using an adequately developed drug dosage form [1–5,65]. EC can be applied in transungual delivery systems as a release modifier and a biocompatible agent ensuring optimal viscosity of the formulation. An isotretinoin nail lacquer intended for the treatment of nail psoriasis was developed, and its penetration efficiency across the nail plate was accessed. Lacquer with EC (at 6% concentration) possessed improved handling characteristics and enhanced drug distribution across the nail [66]. Another study documented an antifungal nail lacquer with miconazole developed with 0.5% EC which extended drug release up to 48 h [67]. Šveikauskaitė et al. studied naftifine hydrochloride release from experimental nail lacquer formulations obtained from EC and Eudragit. By using microcalorimetry they revealed possible interactions between naftifine and EC [68].

Eye administration is a demanding issue in pharmaceutical technology, and to enhance ocular bioavailability, sustained-release drug dosage forms such as hydrogels, inserts, contact lenses, or minitablets are designed. Among the materials employed in developing ophthalmic formulations, EC can be utilized as a polymer extending drug release [69] (Table 3).

Reports have also documented the application of EC as matrix material for transdermal patches for systemic delivery, showing desired permeation enhancement and flexibility [1–5]. Interestingly, EC has also been indicated in the formulation of a transdermal spray with clotrimazole to improve drug transport through the skin up to 12 h and to promote its in vivo antifungal activity [70].

Table 3 presents literature examples of EC utilization in ocular and transdermal drug delivery systems.

Table 3. Examples of ocular and transdermal delivery systems based on EC.

Drug Dosage Form	Active Substance	Used Excipients	EC Role	Reference
		Ocular formulations		
ocular inserts	brimonidine	sodium alginate, EC	hydrophobic, sustained release coating	[71]
ocular inserts	ciprofloxacin	gelatin, EC	rate-controlling, increasing residence time membrane	[72]
ocular insert	acyclovir	sodium alginate, EC	rate-controlling membrane	[73]
in situ ophthalmic hydrogel	besifloxacin	sodium alginate, xanthan gum, EC	increase the pre-corneal residence time	[74]
minitablet	ciprofloxacin	HPMC, sodium carboxymethyl cellulose, EC, hydroxyethyl cellulose, carbopol	sustained release	[75]
		Transdermal formulations		
transdermal patch	amlodipine	EC	sustained release	[76]
transdermal patch	flurbiprofen	EC, propylene glycol, dibutyl phthalate	constant drug release	[77]
transdermal delivery system	topiramate	EC, povidone, Eudragit L 100, carbopol	extended release	[78]
transdermal patch	dexibuprofen	EC, povidone, di-N-butyl phthalate	matrix formation	[79]
transdermal patch	centchroman	EC, polydimethylsiloxane, propylene glycol, Di-n-butyl-phthalate	Film-forming polymer	[80]

4.3. EC-Based Solid Dispersions

In preparation of MR dosage forms, the concepts of solid dispersions and drug incorporation in an insoluble EC carrier have been also explored [81–85]. Tsunashima et al. aimed to prepare MR formulation of tacrolimus to achieve both an extended release profile and improved drug solubility. Extended release solid dispersions of tacrolimus were successfully prepared via the solvent evaporation method using EC and HPMC as polymeric materials [82]. Sadeghi et al. compared characteristics of EC matrices prepared from solid dispersion systems with those prepared from a physical mixture of drug and polymer. Sodium diclofenac was used as a model drug. It was shown that matrices prepared from physical mixtures were harder than those prepared from solid dispersion systems, and their release rates were considerably faster. In the solid dispersion system, drug particles were incorporated in an EC matrix, which caused a great delay in drug diffusion through the polymer and made diffusion a rate-retarding process in the drug release mechanism [83]. Evaluation of the release mechanism of a poorly water-soluble drug (indomethacin) from extended-release solid dispersion systems with EC and HPMC was performed by Ohara et al. The dissolution behavior of indomethacin depended on the structures of EC–HPMC matrices and showed pH dependency—the drug dissolution rate was slower in an acidic environment than in a neutral one. The obtained results revealed that hydrophobic interaction between indomethacin and EC occurred under lower pH and strongly delayed the dissolution rate of the drug [84]. Sustained-release solid dispersions with EC and Eudragit RSPO were designed to control the release of verapamil hydrochloride. Solid dispersions obtained by a simple solvent evaporation technique and physical mixture formulations were compressed to tablets. The in vitro drug release study revealed that solid dispersions containing a combination of EC and Eudragit RSPO extended the release rate for 20 h compared to the physical mixtures at the same ratio, and that the release of verapamil from tablets containing solid dispersion could be effectively controlled [85].

4.4. EC-Based Micro- and Nanocarriers

A common approach to modifying drug release is preparing polymeric micro- or nanocarriers which contain drug enclosed in the polymer shell or incorporated in the polymer matrix (Figure 9). Micro- and nanoparticles have gained significant interest for their use in various drug formulations not only to sustain drug release, but also in order to improve bioavailability, decrease side effects, or increase drug stability [86]. An extended-release oral flexible tablet (ER-OFT) formulation was developed using carbamazepine as model drug for pediatric and geriatric compliance. Microparticles of carbamazepine were prepared using EC as a matrix polymer and HPMC as a hydrophilic pore former utilizing a high-shear granulator fitted with an atomizing spray system. Granulation of carbamazepine and EC with ethanolic binder solution resulted in obtaining microparticles with 16 h extended release [87]. EC microparticles can be formulated using a variety of techniques: interfacial and in situ polymerization, coacervation phase separation, spray drying, spray congealing, or by rotary fluidization bed granulator method [88].

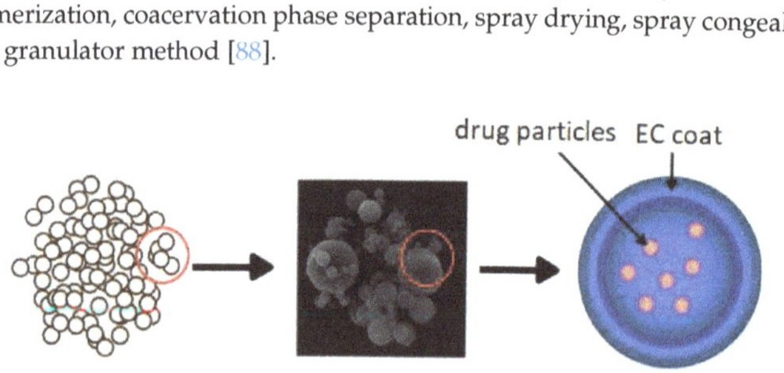

Figure 9. Schematic illustration of microparticles as a drug delivery system. Reprinted with permission from Reference [16]. Copyright 2019 MDPI.

As the extreme acidic environment of the stomach contributes to poor success in the treatment of *Helicobacter pylori* infections, Pan-In et al. encapsulated clarithromycin into particles made of pH-resistant polymer (i.e., EC). Nanoparticles were prepared via a simple anti-solvent particle induction method [89]. Similarly, EC nanocomposites containing rifampicin obtained by supercritical anti-solvent process prolonged drug release and increased its bioavailability [90]. Interestingly, EC was also tested as a carrier for protecting the L-alanyl-L-glutamine peptide and simultaneously providing its prolonged release [91].

Shankar et al. formulated EC microparticles for the vaginal delivery of metronidazole by thermal change method [92]. EC was also utilized in the development of nanoparticles for the topical delivery of corticosteroids [93] or to modulate the release and reduce the ulcerogenicity of piroxicam after oral administration [93]. A reduction of 66% in mean ulcer index was observed, indicating that the obtained particles had a significant potential of offsetting deleterious side effects common in piroxicam use [94]. Similarly, as etoricoxib has many side effects when taken orally, EC nanoparticles produced by a nanoprecipitation technique were designed. Etoricoxib-loaded EC nanoparticles for local drug delivery in arthritis provided sustained drug release, thereby improving patient compliance [95]. Interestingly, nanocapsules of beclomethasone were prepared for pulmonary delivery using EC as a protective polymer. They were characterized by delayed drug photodegradation, prolonged drug release without burst effect, and insignificant cytotoxic effect [96].

EC can also be utilized to prepare sustained-release oral liquids, such as syrup with hydrocodone and chlorpheniramine (Tussionex®) or suspension with mirabegron based on ion-exchange technology [97–99].

Maulavi et al. tested EC-based-microparticles-laden hydrogel contact lenses to provide the sustained ocular delivery of timolol. Prototype poly(hydroxyethyl methacrylate) hydrogel contact lenses improved drug bioavailability due to the increase in ocular residence time [100]. Timolol-encapsulated EC nanoparticles lenses for controlled sustained ocular drug delivery were also evaluated in vivo using rabbits. Nanoparticles were prepared by a double emulsion method and incorporated into acrylate hydrogel, then implanted in hydrogel contact lenses. They exhibited extended timolol release and an intra-ocular pressure lowering effect for 192 h in rabbits without significant ocular complications [101].

Application of EC microparticles containing naproxen to textile materials was studied by Arici et al. Microparticles were prepared by spray-drying using an aqueous EC dispersion and applied to orthopedic support materials. Since these materials are used for traumatic irritations, over-forced ligaments and tendons, and specific problems of various joints, it was concluded that they can serve as potential drug delivery systems [102].

Another approach to obtaining microcarriers is microsponge development. Microsponges are novel drug delivery systems, which ensure effective sustained drug release. They possess unique properties including self-sterilizability, due to the minimal pore size which does not allow bacteria to enter and contaminate the formulation. Microsponges can entrap drugs up to three times their weight due to their very high porosity. EC is utilized as a foundation material for microsponges and for microsponge engineering due to its nonirritating, nontoxic, and nonallergic nature [103–105]. Bothiraja et al. focused on the development of an EC microsponge gel as a topical carrier for the controlled release and cutaneous drug deposition of eberconazole. The microsponges were prepared using the quasi emulsion solvent diffusion method. An in vivo skin deposition study demonstrated higher retention in the stratum corneum layer as compared with commercial cream with eberconazole [106]. Jelvehgari et al. developed a microsponge delivery system to facilitate the topical delivery of benzyl peroxide. As benzyl peroxide has many side effects when applied per se in gel form, such a delivery system to the skin could reduce those effects while reducing its percutaneous absorption [107]. EC was also utilized to prepare microsponges to overcome problems connected with the poor water solubility, low stability, and high volatility of tea tree oil. Microsponges loaded with tea tree oil were prepared using EC and polyvinyl alcohol and incorporated into carbopol gel [108]. Another study describes the development of the topical delivery of oxybenzone—one of the most widely used chemical filters found in commercial sunscreens—using a microsponge-loaded gel. As the ingredient can cause dermatitis and skin irritation, oxybenzone-loaded microsponges were successfully formulated by a quasi emulsion solvent diffusion method with EC utilization. It was shown that controlled release of oxybenzone from the microsponge structures and barrier effect of gel resulted in prolonged retention with reduced permeation, reduced irritation, and enhanced sun protection factor [109]. The aim of another study was to develop a sustained-release delivery system of tacrolimus, formulated using a microsponge base, employing EC and xanthan gum by modified double emulsification techniques [110]. EC microsponges can be designed for ocular delivery; acetazolamide formulated in a microsponges–Pluronic F-127 in situ gel was characterized by improved therapeutic efficacy and reduced systemic side effects of oral acetazolamide. The obtained formulation showed higher therapeutic efficacy compared to free drug in gel [111].

EC has also been studied as a material for nanofibers, which possess a broad range of new applications in pharmaceutical technology [112]. Over the past years, electrospinning has been attractive as a simple, reproducible, versatile, and cost-effective technique for nanofiber production. Electrospun nanofibers have been extensively used for different biomedical applications, including wound dressing, tissue engineering, and drug delivery. The fiber morphology can be manipulated by changing the solvent ratio, resulting in a decreased fiber diameter. Membranes made of EC nanofibers possess good physical properties and are characterized by beneficial air permeability [113,114]. Drug delivery rate is affected by the polymer type, fiber diameter, and drug concentration in the fiber [114,115]. EC as a filament-forming matrix was tested in ketoprofen nanofiber delivery systems designed by triaxial electrospinning [116]. EC was also utilized for creating a 5-fluorouracil loaded core for the fabrication of electrospun fibers [117] and in preparing water-stable composite nanofibers loaded with indomethacin with a sustained, diffusion-controlled release profile [118]. The inclusion of EC as a matrix former improved mechanical properties of the entire delivery system [118]. Another study employed blend fibers of poly(N-vinylcaprolactam) and EC with the aim of developing thermoresponsive sustained release formulations fabricated by twin-jet electrospinning containing ketoprofen. The obtained fibers were largely smooth and homogeneous, and the addition of a drug did not affect their morphology [115]. Additionally, Liu et al. reported that nanofibers obtained by electrospinning using blends of EC and gelatin were characterized by fine morphology and possessed improved thermal stability [119].

4.5. EC in Mucoadhesive Delivery Systems

In the development of dosage forms applied on mucous membranes, the selection of suitable polymers with adhesive properties is a crucial issue. Polymers that are bioadhesive and do not dissolve before releasing the incorporated drug are highly appreciated for sustain drug release. As a water-insoluble polymer, EC is often used as backing membrane for its film-forming property, low water permeability, drug impermeability, and moderate flexibility. It possesses bioadhesive properties, however lesser than Carbopol and chitosan [120]. Bagul et al. evaluated the in vitro mucoadhesive strength of various polymers and reported the following ascending order for force of adhesion expressed in Newtons (N): gelatin (1.42) < gum dammar (1.47) < gum copal (1.52) < ethyl cellulose (1.60) < sodium alginate (1.71) < xanthan gum (1.81) < chitosan (1.91) < HPMC (2.25) < carbopol (2.40) [120].

EC forms a hydrophobic network when the mixture comes into contact with water, resulting in sustained drug release. The utilization of EC as a single polymer or in conjunction with other adjuvants in the preparation of film-like drug carriers is well documented. Drug-loaded EC films are characterized by good adhesion, mechanical strength, and sustained release profile. They provide a flexible diffusion barrier and its properties can be changed by the amount of pore-forming agent, film thickness, and EC molecular weight [1,3–5]. Abruzzo et al. designed buccal films for propranolol administration. A polymeric layer was prepared by casting and drying with polyvinylpyrrolidone or polyvinylalcohol and the addition of gelatin or chitosan. EC was employed to formulate a second layer was applied onto the primary one in order to obtain prolonged drug delivery, increase adhesion, and mask the drug's bitter taste [121]. A bilayer mucoadhesive buccal film containing a combination of ornidazole and dexamethasone was prepared using solvent casting to treat oral ulcers, with EC utilized as a backing layer. The formulation showed favorable swelling characteristics, and both drugs were released at 95% after 4 h [122]. EC was also applied for the preparation of allantoin-loaded films for the management of dry mouth syndrome. The findings revealed that the produced films were functional, mucoadhesive, flexible, and stable, with the potential for treating various intraoral diseases [123]. EC-based mucoadhesive buccal films containing fluticasone formulated by solvent casting technique showed acceptable physicochemical properties, homogenous drug distribution, adequate mucoadhesion time, moderate swelling, and sustained drug release up to 12 h [124]. The objective of another study was to develop a two-layered buccal mucoadhesive system consisting of a highly water-soluble drug (i.e., risedronate). Varied concentrations of chitosan, HPMC, and EC acting as an impermeable backing membrane ensuring sustained release were tested. The obtained systems showed good swelling and mucoadhesive characteristics, with 90–100% drug release within 8–12 h [125]. EC, as a biocompatible backing layer, was also utilized in the development of mucoadhesive bi-layered strips used in dental treatment for the controlled delivery of lidocaine [126]. EC dissolved in N-methyl pyrrolidone was successfully used as a polymeric matrix for the in situ forming gel (with doxycycline, metronidazole, and benzyl peroxide) as dosage form applied in periodontal pocket in periodontitis treatment. It turned out that increasing the amount of EC increased the viscosity of system while still exhibiting Newtonian flow and simultaneously decreasing the release of drug [127]. Another prolonged-release mucoadhesive gel containing metronidazole for periodontal application was developed basing on a mixture of glycerylmonooleate and EC. EC reduced the initial metronidazole release and significantly sustained its release profile [128]. Pathak et al. developed a mucoadhesive coating for fluconazole tablets for oral thrush treatment. EC coating reduced fluconazole permeation through the buccal mucosa, thereby ensuring high local drug concentration and faster provision of the fluconazole minimum inhibitory concentration in the oral cavity [129].

5. Conclusions

EC is a water-insoluble cellulose derivative with attractive features such as biocompatibility, gastroresistance, and degradation to non-toxic and readily excreted products. It is extensively utilized in pharmaceutical technology. It is generally recognized as safe to use, even in pediatric therapy. EC is easy to process alone or with the addition of other excipients (e.g., plasticizers or other polymers). Considering its multifunctional properties, EC is being widely exploited in oral and topical drug dosage forms. EC's main advantage is its ability to modify the release of the drug, which allows for the creation of controlled delivery systems—the unique tailored carriers for many pharmaceuticals. EC-based materials are widely utilized as matrices for designing novel drug delivery systems with a wide range of applications.

Author Contributions: Conceptualization, K.W. (Katarzyna Wasilewska) and K.W. (Katarzyna Winnicka); Formal Analysis, K.W. (Katarzyna Winnicka); Resources, K.W. (Katarzyna Wasilewska); Data Curation, K.W. (Katarzyna Wasilewska); Writing—Original Draft Preparation, K.W. (Katarzyna Wasilewska); Writing—Review and Editing, K.W. (Katarzyna Wasilewska) and K.W. (Katarzyna Winnicka); Visualization, K.W. (Katarzyna Wasilewska); Supervision, K.W. (Katarzyna Winnicka); Project Administration, K.W. (Katarzyna Wasilewska); Funding Acquisition, K.W. (Katarzyna Wasilewska) and K.W. (Katarzyna Winnicka).

Funding: This research was funded by Medical University of Bialystok grant numbers N/ST/MN/18/002/2215 and SUB/2/DN/19/004/2215.

Acknowledgments: We are grateful to the Editors for the permission to reprint selected figures from our previous publication (Reference number 15): Trofimiuk, M.; Wasilewska, K.; Winnicka, K. How to modify drug release in paediatric dosage forms? Novel technologies and modern approaches with regard to children's population. *Int. J. Mol. Sci.* **2019**, 29. We are thankful to Anna Basa (Institute of Chemistry, University of Białystok, Poland) for preparing SEM pictures (Inspect™S50, FEI Company, Hillsboro, OR, USA).

Conflicts of Interest: The authors declare no conflicts of interest.

References

1. Qiu, Y.; Lee, P.I. Rational design of oral modified-release drug delivery systems. In *Developing Solid Oral Dosage Forms. Pharmaceutical Theory and Practice*, 2nd ed.; Qui, Y., Zhang, G.G.Z., Mantri, R.V., Chen, Y., Yu, L., Eds.; Academic Press: Cambridge, MA, USA, 2017; pp. 1127–1160.
2. FDA. Available online: https://www.fda.gov/downloads/Drugs/.../UCM291010.pdf (accessed on 20 July 2019).
3. Rathbone, M.J.; Hadgraft, J.; Roberts, M.S.; Lane, M.E. *Modified-Release Drug Delivery Technology*, 2nd ed.; Informa Healthcare: London, UK, 2008; Volume 2.
4. Rowe, R.C.; Sheskey, P.J.; Quinn, M.E. *Handbook of Pharmaceutical Excipients*, 6th ed.; Pharmaceutical Press: London, UK; Chicago, IL, USA; Washington, DC, USA, 2009; pp. 262–267.
5. Ethyl Cellulose. Available online: https://pubchem.ncbi.nlm.nih.gov/compound/24832091#section=Probable-Routes-of-Human-Exposure (accessed on 21 July 2019).
6. *The European Pharmacopoeia*, 9th ed.; Council of Europe: Strasburg, France, 2016.
7. *The United States Pharmacopeia and National Formulary (USP41-NF 36)*; Pharmacopeia Convention: Rockville, MD, USA, 2018; Volume 2.
8. FDA Inactive Ingredients Database. Available online: https://search.fda.gov/search?utf8=%E2%9C%93&affiliate=fda1&query=ethylcellulose&commit=Search (accessed on 20 July 2019).
9. Canadian List of Acceptable Non-Medicinal Ingredients. Available online: https://www.canada.ca/en/health-canada/services/drugs-health-products/natural-non-prescription/applications-submissions/product-licensing/compendium-monographs.html (accessed on 20 July 2019).
10. *The Pharmacopoeia of the People's Republic of China 2015 Edition*; The Stationery Office: London, UK, 2015.
11. *The Japanese Pharmacopoeia, Seventeenth Edition (JP17)*; The Ministry of Health, Labour and Welfare: Tokyo, Japan, 2016.
12. Ethocel, Colorcon. Available online: https://www.pharma.dupont.com/pharmaceutical-products/ethocel.html (accessed on 20 July 2019).
13. Brydson, J.A. *Plastics Materials*, 6th ed.; Butterworth-Heinemann: Oxford, UK, 1995.
14. Mehta, R.; Teckoe, J.; Schoener, C.; Workentine, S.; Ferrizzi, D.; Rajabi-Siahoboomi, A. Investigation into effect of ethylcellulose viscosity variation on the drug release of metoprolol tartrate and acetaminophen extended release multiparticulates—Part I. *AAPS PharmSciTech* **2016**, *17*, 1366–1375. [CrossRef]
15. Lokhande, A.B.; Ravindra, S.M.; Jitendra, D.K.; Naik, B. Influence of different viscosity grade ethylcellulose polymers on encapsulation and in vitro release study of drug loaded nanoparticles metformin. *J. Pharm. Res.* **2013**, *5*, 414–420.
16. Trofimiuk, M.; Wasilewska, K.; Winnicka, K. How to modify drug release in paediatric dosage forms? Novel technologies and modern approaches with regard to children's population. *Int. J. Mol. Sci.* **2019**, *20*, 3200. [CrossRef]
17. Hamedelniel, E.I.; Bajdik, J.; Pintye-Hódi, K. Optimization of preparation of matrix pellets containing ethylcellulose. *Chem. Eng. Process.* **2010**, *49*, 120–124. [CrossRef]
18. Mallipeddi, R.; Saripella, K.K.; Neau, S.H. Use of fine particle ethylcellulose as the diluent in the production of pellets by extrusion-spheronization. *SPJ.* **2014**, *22*, 360–371. [CrossRef]
19. Grund, J.; Koerber, M.; Walther, M.; Bodmeier, R. The effect of polymer properties on direct compression and drug release from water-insoluble controlled release matrix tablets. *Int. J. Pharm.* **2014**, *469*, 94–101. [CrossRef]
20. Cifuentes, C.; Aguilar-de-Leyva, A.; Rajabi-Siahboomi, A.R.; Caraballo, I. Critical points in ethylcellulose matrices: Influence of the polymer, drug and filler properties. *Acta Pharm.* **2013**, *63*, 115–129. [CrossRef]
21. Bruschi, M. *Strategies to Modify Drug Release from Pharmaceutics Systems*, 1st ed.; Woodhead Publishing: Cambridge, UK, 2015; pp. 87–119.
22. Mehta, R.Y.; Missaghi, S.; Tiwari, S.B.; Rajabi-Siahboomi, A.R. Application of ethylcellulose coating to hydrophilic matrices: A strategy to modulate drug release profile and reduce drug release variability. *AAPS Pharm. Sci. Tech.* **2014**, *15*, 1049–1059. [CrossRef]
23. Pawar, A.; Deepak, V.B.; Vineeta, V.K.; Vilasrao, J.K. Advances in pharmaceutical coatings. *Int. J. Chem. Tech. Res.* **2010**, *1*, 733–737.
24. Aqualon, Ashland. Available online: https://www.ashland.com/industries/pharmaceutical/oral-soliddose/aqualon-ethylcellulose (accessed on 20 July 2019).

25. Surelease, Colorcon. Available online: https://www.colorcon.com/products-formulation/all-products/film-coatings/sustained-release/surelease (accessed on 20 July 2019).
26. Aquacoat, ECD. Available online: http://www.fmcbiopolymer.com/Pharmaceutical/Products/Aquacoat.aspx (accessed on 20 July 2019).
27. Aquarius, Ashland. Available online: https://www.ashland.com/file_source/Ashland/links/PTR-112_Aquarius_Control_ECD_reconstitution_instructions.pdf (accessed on 20 July 2019).
28. AshaKote, Ashaland. Available online: http://www.ashacel.com/AshaKote.PDF (accessed on 20 July 2019).
29. Shin, T.H.; Ho, M.J.; Kim, S.R.; Im, S.H.; Kim, C.H.; Lee, S.; Kang, M.J.; Choi, Y.W. Formulation and in vivo pharmacokinetic evaluation of ethyl cellulose-coated sustained release multiple-unit system of tacrolimus. *Int. J. Biol. Macromol.* **2018**, *1*, 544–550. [CrossRef]
30. Full Prescribing Information Advagraf®. Available online: https://www.ema.europa.eu/en/documents/product-information/advagraf-epar-product-information_en.pdf (accessed on 20 July 2019).
31. Shah, N.; Sharma, O.P.; Mehta, T.; Amin, A. Design of experiment approach for formulating multi-unit colon-targeted drug delivery system: In vitro and in vivo studies. *Drug Dev. Ind. Pharm.* **2016**, *42*, 825–835. [CrossRef]
32. Kim, M.S.; Yeom, D.W.; Kim, S.R.; Yoon, H.Y.; Kim, C.H.; Son, H.Y.; Kim, J.H.; Lee, S.; Choi, Y.W. Development of a chitosan based double layer-coated tablet as a platform for colon-specific drug delivery. *Drug Des. Devel. Ther.* **2017**, *11*, 45–57. [CrossRef]
33. Wan, H.; Park, K. *Oral Controlled Release Formulation Design and Delivery. Theory to Practice*; Wiley: Hoboken, NJ, USA, 2010.
34. Full Prescribing Information Micro-K™. Available online: https://www.rxlist.com/micro-k-drug.htm (accessed on 20 July 2019).
35. Full Prescribing Information Theo-24®. Available online: https://www.drugs.com/dosage/theo-24.html (accessed on 20 July 2019).
36. Mohamed, F.A.; Roberts, M.; Seton, L.; Ford, J.L.; Levina, M.; Rajabi-Siahboomi, A.R. Film-coated matrix mini-tablets for the extended release of a water-soluble drug. *Drug Dev. Ind. Pharm.* **2015**, *41*, 623–630. [CrossRef]
37. Full Prescribing Information Metadate CD®. Available online: https://www.accessdata.fda.gov/drugsatfda_docs/label/2009/021259s021lbl.pdf (accessed on 20 July 2019).
38. Full Prescribing Information Lamictal®XR. Available online: https://www.fda.gov/downloads/drugs/developmentapprovalprocess/developmentresources/ucm215664.pdf (accessed on 20 July 2019).
39. Venkata, P.D.B. Spheroidal oral drug absorption system (SODAS). *J. Glob. Pharma Technol.* **2011**, *3*, 1–5.
40. Elan Drug Technologies. Spheroidal Drug Absorption System (SODAS®). Available online: http://www.elandrugtechnologies.com/oral_controlled_release/sodas (accessed on 20 July 2019).
41. Full Prescribing Information Adzenys XR-ODT®. Available online: https://www.accessdata.fda.gov/drugsatfda_docs/label/2017/204326s002lbl.pdf (accessed on 20 July 2019).
42. Full Prescribing Information Cotempla XR-ODT®. Available online: https://www.accessdata.fda.gov/drugsatfda_docs/label/2017/205489s000lbl.pdf (accessed on 20 July 2019).
43. Full Prescribing Information Inderal® LA. Available online: https://www.accessdata.fda.gov/drugsatfda_docs/label/2011/016418s080,016762s017,017683s008lbl.pdf (accessed on 20 July 2019).
44. Full Prescribing Information Dilacor-XR®. Available online: https://www.drugs.com/pro/dilacor-xr.htmlDilcaor (accessed on 20 July 2019).
45. Full Prescribing Information Cardizem® CD. Available online: https://www.accessdata.fda.gov/drugsatfda_docs/label/2010/020062s040lbl.pdf (accessed on 20 July 2019).
46. Amelian, A.; Winnicka, K. Polymers in pharmaceutical taste masking applications. *Polimery* **2017**, *6*, 417–496. [CrossRef]
47. Faisal, W.; Farag, F.; Abdellatif, A.A.H.; Abbas, A. Taste masking approaches for medicines. *Curr. Drug Deliv.* **2018**, *15*, 167–185. [CrossRef]
48. Taki, M.; Tagami, T.; Ozeki, T. Preparation of polymer-blended quinine nanocomposite particles by spray drying and assessment of their instrumental bitterness-masking effect using a taste sensor. *Drug. Dev. Ind. Pharm.* **2017**, *43*, 715–722. [CrossRef]

49. Wasilewska, K.; Szekalska, M.; Ciosek-Skibinska, P.; Lenik, J.; Jacyna, J.; Markuszewski, M.; Winnicka, K. Ethylcellulose in organic solution or aqueous dispersion form in designing taste-masked microparticles by the spray drying technique with a model bitter drug: Rupatadine fumarate. *Polymers* **2019**, *11*, 522. [CrossRef]
50. Ito, D.; Tackoe, J.; Rajabi-Siahboomo, A. Investigation of Taste Masking Performance of an Aqueous Ethylcellulose Dispersion (Surelease®) on Acetaminophen Granules. Poster (AAPS 2013). Available online: https://www.colorcon.com/products-formulation/all-products/download/311/552/34?method=view (accessed on 20 July 2019).
51. Pimparade, M.B.; Morott, J.T.; Park, J.B.; Kulkarni, V.I.; Majumdar, S.; Murthy, S.N.; Lian, Z.; Pinto, E.; Bi, V.; Durig, T.; et al. Development of taste masked caffeine citrate formulations utilizing hot melt extrusion technology and in vitro-in vivo evaluations. *Int. J. Pharm.* **2015**, *20*, 167–176. [CrossRef]
52. Rao, M.R.; Bhingole, R.C. Nanosponge-based pediatric-controlled release dry suspension of gabapentin for reconstitution. *Drug. Dev. Ind. Pharm.* **2015**, *41*, 2029–2036. [CrossRef]
53. Full Prescribing Information Aciphex®Sprinkle™. Available online: https://www.accessdata.fda.gov/drugsatfda_docs/label/2014/020973s035204736s005lbl.pdf (accessed on 20 July 2019).
54. Full Prescribing Information Dilatrate-SR. Available online: https://www.drugs.com/pro/dilatrate-sr.html (accessed on 20 July 2019).
55. Full Prescribing Information Durlaza. Available online: https://www.drugs.com/pro/durlaza.html (accessed on 20 July 2019).
56. Full Prescribing Information Enjuvia. Available online: https://www.drugs.com/pro/enjuvia.html (accessed on 20 July 2019).
57. Full Prescribing Information Entocort EC. Available online: https://www.drugs.com/pro/entocort-ec.html (accessed on 20 July 2019).
58. Full Prescribing Information Levbid®. Available online: https://www.rxlist.com/levbid-extended-release-drug.htm#indications (accessed on 20 July 2019).
59. Orfiril Long Minitablets Information. Available online: https://ec.europa.eu/health/documents/community-register/2018/20180531140837/anx_140837_en.pdf (accessed on 20 July 2019).
60. Full Prescribing Information Palladone. Available online: https://www.drugs.com/pro/palladone.html (accessed on 20 July 2019).
61. Full Prescribing Information Pentasa. Available online: https://www.accessdata.fda.gov/drugsatfda_docs/label/2018/020049s031lbl.pdf (accessed on 20 July 2019).
62. Full Prescribing Information Qsymia. Available online: https://www.drugs.com/qsymia.html (accessed on 20 July 2019).
63. Full Prescribing Information Tegretol®-XR. Available online: https://www.accessdata.fda.gov/drugsatfda_docs/label/2009/016608s101,018281s048lbl.pdf (accessed on 20 July 2019).
64. Full Prescribing Information Trokendi XR. Available online: https://www.drugs.com/pro/trokendi-xr.html (accessed on 20 July 2019).
65. Kathe, K.; Kathpalia, H. Film forming systems for topical and transdermal drug delivery. *Asian J. Pharm. Sci.* **2017**, *12*, 487–497. [CrossRef]
66. Joshi, M.; Sharma, V.; Pathak, K. Matrix based system of isotretinoin as nail lacquer to enhance transungal delivery across human nail plate. *Int. J. Pharm.* **2015**, *478*, 268–277. [CrossRef] [PubMed]
67. Vipin, K.V.; Sarath, C.C.; Ann, R.A.; Premaletha, K.; Muriakose, M.R. Formulation and evaluation of an antifungal nail lacquer for onychomycosis. *Br. Biomed. Bull.* **2014**, *2*, 242–248.
68. Šveikauskaitė, I.; Briedis, V. Effect of film-forming polymers on release of naftifine hydrochloride from nail lacquers. *Int. J. Polym. Sci.* **2017**, *1*, 7. [CrossRef]
69. Baranowski, P.; Karolewicz, B.; Gajda, M.; Pluta, J. Ophthalmic drug dosage forms: Characterisation and research methods. *Sci. World J.* **2014**, *2014*, 1–14. [CrossRef]
70. Paradkar, M.; Thakkar, V.; Soni, T.; Gandhi, T.; Gohel, M. Formulation and evaluation of clotrimazole transdermal spray. *Drug Dev. Ind. Pharm.* **2015**, *41*, 1718–1725. [CrossRef]
71. Goudanavar, P.; Ambhore, N.; Hiremath, D.; Udupi, R. Comparative evaluation of polymer combination in the design of brimonidine tartrate ocular inserts. *Indian Drugs* **2012**, *49*, 30–35.
72. Mundada, A.S.; Shrikhande, B.K. Design and evaluation of soluble ocular drug insert for controlled release of ciprofloxacin hydrochloride. *Drug Dev. Ind. Pharm.* **2006**, *32*, 443–448. [CrossRef]

73. Khan, S.; Ali, A.; Singhavi, D.; Yeole, P. Preparation of ophthalmic insert of acyclovir using ethylcellulose rate-controlling membrane. *J. Appl. Ther. Res.* **2008**, *6*, 19–24.
74. Kala, S.; Gurudiwn, P.; Juyal, D. Formulation and evaluation of besifloxacin loaded in situ gel for ophthalmic delivery. *UKJPB.* **2018**, *6*, 36–40. [CrossRef]
75. Mortazavi, S.A.; Jaffariazar, Z.; Damercheli, E. Formulation and in-vitro evaluation of ocular ciprofloxacin-containing minitablets prepared with different combinations of carbopol 974P and various cellulose derivatives. *IJPR.* **2010**, *9*, 107–114. [PubMed]
76. John, L.; Kumar, A.; Samuel, S. Formulation and evaluation of amlodipine transdermal patches using ethyl cellulose. *Int. Res. J. Pharm.* **2013**, *4*, 84–88. [CrossRef]
77. Idrees, A.; Rahman, N.U.; Javaid, Z.; Kashif, M.; Aslam, I.; Abbas, K.; Hussain, T. In vitro evaluation of transdermal patches of flurbiprofen with ethyl cellulose. *Acta Pol. Pharm.* **2014**, *71*, 287–295. [PubMed]
78. Cherukuri, S.; Ba, U.R. Formulation and evaluation of transdermal drug delivery of topiramate. *Int. J. Pharm. Investig.* **2017**, *7*, 10–17. [PubMed]
79. Akhlaq, M.; Arshad, M.S.; Mudassir, A.M.; Hussain, A.; Kucuk, I.; Haj-Ahmad, R.; Rasekh, M.; Ahmad, Z. Formulation and evaluation of anti-rheumatic dexibuprofen transdermal patches: A quality-by-design approach. *J. Drug. Targ.* **2016**, *24*, 603–612. [CrossRef]
80. Gupta, V.; Singh, S.; Srivarstava, M.; Ahmad, H.; Pachauri, S.D.; Khandelwal, K.; Dwivedi, P.; Dwivedi, A.K. Effect of polydimethylsiloxane and ethylcellulose on in vitro permeation of centchroman from its transdermal patches. *Drug Deliv.* **2016**, *23*, 113–122. [CrossRef]
81. Huang, Y.; Dai, W.G. Fundamental aspects of solid dispersion technology for poorly soluble drugs. *Acta Pharm. Sin. B* **2014**, *4*, 18–25. [CrossRef]
82. Tsunashima, D.; Yamashita, K.; Ogawara, K.; Sako, K.; Higaki, K. Preparation of extended release solid dispersion formulations of tacrolimus using ethylcellulose and hydroxypropylmethylcellulose by solvent evaporation metod. *J. Pharm. Pharmacol.* **2016**, *68*, 316–323. [CrossRef]
83. Sadeghi, F.; Garekani, H.A.; Sadeghi, R. Comparison of ethylcellulose matrix characteristics prepared by solid dispersion technique or physical mixing. *Daru J. Fac. Pharm.* **2012**, *11*, 27–34.
84. Ohara, T.; Kitamura, S.; Kitagawa, T.; Terada, K. Dissolution mechanism of poorly water-soluble drug from extended release solid dispersion system with ethylcellulose and hydroxypropylmethylcellulose. *Int. J. Pharm.* **2005**, *302*, 95–102. [CrossRef]
85. Samba, M.; Kumawat, M.; Sambamoorthly, U. Formulation of sustain release solid dispersion of verapamil hydrochloride using ethyl cellulose and Eudragit RSPO. *Int. J. Pharm. Pharm. Sci.* **2011**, *3*, 116–122.
86. Singh, M.N.; Hemant, K.S.Y.; Ram, M.; Shivakumar, H.G. Microencapsulation: A promising technique for controlled drug delivery. *Res. Pharm. Sci.* **2010**, *5*, 65–77. [PubMed]
87. Chandrasekaran, P.; Kandasamy, R. Solid oral flexible formulations for pediatric and geriatric patients: Age-appropriate formulation platforms. *Indian J. Pharm. Sci.* **2018**, *80*, 14–25. [CrossRef]
88. Murtaza, G. Ethylcellulose microparticles: A review. *Acta Pol. Pharm.* **2012**, *69*, 11–22.
89. Pan-In, P.; Banlunara, W.; Chaichanawongsaroj, N.; Wanichwecharungruang, S. Ethyl cellulose nanoparticles: Clarithomycin encapsulation and eradication of *H. pylori*. *Carbohydr. Polym.* **2014**, *109*, 22–27. [CrossRef]
90. Djerafi, R.; Swanepoel, A.; Crampon, C.; Kalombo, L.; Labuschagne, P.; Badens, E.; Masmoudi, Y. Supercritical antisolvent co-precipitation of rifampicin and ethyl cellulose. *Eur. J. Pharm. Sci.* **2017**, *102*, 161–171. [CrossRef]
91. Villa Nova, M.; Gonçalves Mde, C.; Nogueira, A.C.; Herculano Lda, S.; Medina, A.N.; Bazotte, R.B.; Bruschi, M.L. Formulation and characterization of ethylcellulose microparticles containing L-alanyl-L-glutamine. *Drug Dev. Ind. Pharm.* **2014**, *40*, 1308–1317. [CrossRef]
92. Shankar, N.B.; Kumar, R.P.; Kumar, N.U.; Brata, B.B. Development and characterization of bioadhesive gel of microencapsulated metronidazole for vaginal use. *Iran J. Pharm. Res.* **2010**, *9*, 209–219.
93. Balzus, B.; Sahle, F.F.; Hönzke, S.; Gerecke, C.; Schumacher, F.; Hedtrich, S.; Kleuser, B.; Bodmeier, R. Formulation and ex vivo evaluation of polymeric nanoparticles for controlled delivery of corticosteroids to the skin and the corneal epithelium. *Eur. J. Pharm. Biopharm.* **2017**, *115*, 122–130. [CrossRef]
94. El-Habashy, S.E.; Allam, A.N.; El-Kamel, A.H. Ethyl cellulose nanoparticles as a platform to decrease ulcerogenic potential of piroxicam: Formulation and in vitro/in vivo evaluation. *Int. J. Nanomed.* **2016**, *11*, 2369–2380.

95. Begum, N.; Sailaja, A.K. Formulation and evaluation of Cox-2 inhibitor (etoricoxib) loaded ethyl cellulose nanoparticles for topical drug delivery. *Nano. Biomed. Eng.* **2018**, *10*, 1–9.
96. Chassot, J.M.; Ribas, D.; Silveira, E.F.; Grünspan, L.D.; Pires, C.C.; Farago, P.V.; Braganhol, E.; Tasso, L.; Cruz, L. Beclomethasone dipropionate-loaded polymeric nanocapsules: Development, in vitro cytotoxicity, and in vivo evaluation of acute lung injury. *J. Nanosci. Nanotechnol.* **2015**, *15*, 855–864. [CrossRef] [PubMed]
97. Tussionex®Drug Information: Description, User Reviews, Drug Side Effects, Interactions—Prescribing Information. Available online: https://www.rxlist.com/tussionex-drug.htm (accessed on 20 July 2019).
98. Kasashima, Y.; Uchida, S.; Yoshihara, K.; Yasuji, T.; Sako, K.; Namiki, N. Oral sustained-release suspension based on a lauryl sulfate salt/complex. *Int. J. Pharm.* **2016**, *30*, 677–683. [CrossRef] [PubMed]
99. Singh, I.; Rehni, A.K.; Kalra, R.; Joshi, G.; Kumar, M.; Aboul-Enein, H.Y. Ion exchange resins: Drug delivery and therapeutic applications. *FABAD J. Pharm. Sci.* **2007**, *32*, 91–100.
100. Maulvi, F.A.; Soni, T.G.; Shah, D.O. Extended release of timolol from ethyl cellulose microparticles laden hydrogel contact lenses. *Open Pharm. Sci.* **2015**, *2*, 1–12. [CrossRef]
101. Maulvi, F.A.; Lakdawala, D.H.; Shaikh, A.A.; Desai, A.R.; Choksi, H.H.; Vaidya, R.J.; Ranch, K.M.; Koli, A.R.; Vyas, B.A.; Shah, D.O. In vitro and in vivo evaluation of novel implantation technology in hydrogel contact lenses for controlled drug delivery. *J. Control Release* **2016**, *226*, 47–56. [CrossRef]
102. Arici, M.; Topbas, O.; Karavana, S.Y.; Ertan, G.; Sariisik, M.; Ozturk, C. Preparation of naproxen-ethyl cellulose microparticles by spray drying technique and their application to textile materials. *J. Microencapsul.* **2014**, *31*, 654–666. [CrossRef]
103. Patel, S.; Patel, H.; Seth, A. Microsponge drug delivery system: An overview. *JGPT.* **2010**, *2*, 1–9.
104. Aloorkar, N.; Kulkarni, A.; Ingale, D.; Patil, R. Microsponges as innovative drug delivery systems. *Int. J. Pharm. Sci. Nanotechnol.* **2012**, *5*, 1597–1606.
105. Kappor, D.; Patel, M.; Vyas, R.; Lad, C.; Tyagi, B. A review on microsponge drug delivery system. *JDDT.* **2014**, *4*, 29–35. [CrossRef]
106. Bothiraja, C.; Gholap, A.D.; Shaikh, K.S.; Pawar, A.P. Investigation of ethyl cellulose microsponge gel for topical delivery of eberconazole nitrate for fungal therapy. *Ther. Deliv.* **2014**, *5*, 781–794. [CrossRef] [PubMed]
107. Jelvehgari, M.; Siahi-Shadbad, M.R.; Azarmi, S.; Martin, G.P.; Nokhodchi, A. The microsponge delivery system of benzoyl peroxide: Preparation, characterization and release studies. *Int. J. Pharm.* **2006**, *308*, 124–132. [CrossRef] [PubMed]
108. Yadav, E.; Rao, R.; Kumar, S.; Mahant, S.; Vohra, P. Microsponge based gel of tea tree oil for dermatological microbial infections. *Nat. Prod. J.* **2018**, *8*, 1. [CrossRef]
109. Pawar, A.P.; Gholap, A.P.; Kuchekar, A.B.; Bothiraja, C.; Mali, A.J. Formulation and evaluation of optimized oxybenzone microsponge gel for topical delivery. *J. Drug. Deliv.* **2015**, 1–9. [CrossRef]
110. Zaman, M.; Qureshi, S.; Sultana, K.; Hanif, M.; Mahmood, A.; Shaheryar, Z.A.; Gulzar, F.; Barkat, K.; Abdel-Daim, M.M. Application of quasi-emulsification and modified double emulsification techniques for formulation of tacrolimus microsponges. *Int. J. Nanomed.* **2018**, *13*, 4537–4548. [CrossRef]
111. Obiedallah, M.M.; Abdel-Mageed, A.M.; Elfaham, T.H. Ocular administration of acetazolamide microsponges in situ gel formulations. *SPJ* **2018**, *6*, 909–920. [CrossRef]
112. Bharathala, S.; Sharma, P. Biomedical applications of nanoparticles. Recent Trends and Future Perspectives. In *Nanotechnology in Modern Animal Biotechnology*; Maurya, P.K., Singh, S., Eds.; Springer Singapore: Singapore, 2019; pp. 113–132.
113. Liu, M.; Zhang, Y.; Sun, S.; Khan, A.R.; Ji, J.; Yang, M.; Zhai, G. Recent advances in electrospun for drug delivery purpose. *J. Drug Target.* **2019**, *27*, 270–282. [CrossRef]
114. Goyal, R.; Macri, L.K.; Kaplan, H.M.; Kohn, J. Nanoparticles and nanofibers for topical drug delivery. *J. Control Release* **2016**, *28*, 77–92. [CrossRef]
115. Li, H.; Liu, K.; Williams, G.R.; Wu, J.; Wu, J.; Wang, H.; Niu, S.; Zhu, L.M. Dual temperature and pH responsive nanofiber formulations prepared by electrospinning. *Colloids Surf. B* **2018**, *171*, 142–149. [CrossRef]
116. You, D.G.; Li, X.Y.; Wang, X.; Yang, J.H.; Bligh, S.A.; Williams, G.R. Nanofibers fabricated using triaxial electrospinning as zero order drug delivery systems. *ACS Appl. Mater. Interfaces* **2015**, *7*, 18891–18897. [CrossRef]
117. Illangakoon, U.E.; Yu, D.G.; Ahmad, B.S.; Chatterton, N.P.; Williams, D.R. 5-Fluorouracil loaded Eudragit fibers prepared by electrospinning. *Int. J. Pharm.* **2015**, *495*, 895–902. [CrossRef] [PubMed]

118. Lu, H.; Wang, Q.; Li, G.; Qiu, Y.; Wei, Q. Electrospun water-stable zein/ethyl cellulose composite nanofiber and its drug release properties. *Mater. Sci. Eng. C Matter. Biol. Appl.* **2017**, *74*, 86–93. [CrossRef] [PubMed]
119. Liu, Y.; Deng, L.; Zhang, C.; Feng, F.; Zhang, H. Tunable physical properties of ethylcellulose/gelatin composite nanofibers by electrospinning. *J. Agric. Food Chem.* **2018**, *28*, 1907–1915. [CrossRef] [PubMed]
120. Bagul, U.; Gujar, K.; Dhat, S.; Aphale, S.; Bhavsar, M. In vitro study of mucoadhesive strength of polymers for mucoadhesive drug delivery systems. *Int. J. Curr. Pharm. Res.* **2009**, *1*, 42–46.
121. Abruzzo, A.; Nicoletta, F.P.; Dalena, F.; Cerchiara, T.; Luppi, B.; Bigucci, F. Bilayered buccal films as child-appropriate dosage form for systemic administration of propranolol. *Int. J. Pharm.* **2017**, *5*, 257–265. [CrossRef] [PubMed]
122. Zhang, C.; Liu, Y.; Li, W.; Gao, P.; Xiang, D.; Ren, X.; Liu, D. Mucoadhesive buccal film containing ornidazole and dexamethasone for oral ulcers: In vitro and in vivo studies. *Pharm. Dev. Technol.* **2019**, *24*, 118–126. [CrossRef] [PubMed]
123. Laffleur, F.; Krouská, J.; Tkacz, J.; Pekař, M.; Aghai, F.; Netsomboon, K. Buccal adhesive films with moisturizer-the next level for dry mouth syndrome? *Int. J. Pharm.* **2018**, *550*, 309–315. [CrossRef] [PubMed]
124. Ammar, H.O.; Ghorab, M.M.; Mahmoud, A.A.; Shahin, H.I. Design and in vitro/in vivo evaluation of ultra-thin mucoadhesive buccal film containing fluticasone propionate. *AAPS PharmSciTech.* **2017**, *18*, 93–103. [CrossRef]
125. Mukherjee, D.; Bharath, S. Design and characterization of double layered mucoadhesive system containing bisphosphonate derivative. *ISRN Pharm.* **2013**, *19*, 604690. [CrossRef]
126. Roh, J.; Han, M.; Kim, K.N.; Kim, K.M. The in vitro and in vivo effects of a fast-dissolving mucoadhesive bi-layered strip as topical anesthetics. *Dent. Mater. J.* **2016**, *35*, 601–605. [CrossRef]
127. Phaechamud, T.; Mahadlek, J. Solvent exchange-induced in situ forming gel comprising ethyl cellulose-antimicrobial drugs. *Int. J. Pharm.* **2015**, *494*, 381–392. [CrossRef] [PubMed]
128. Sallam, A.S.; Hamudi, F.F.; Khalil, E.A. Effect of ethylcellulose and propylene glycol on the controlled-release performance of glyceryl monooleate–mertronidazole periodontal gel. *Pharm. Dev. Technol.* **2015**, *20*, 159–168. [CrossRef] [PubMed]
129. Pathak, K.; Sharma, V.; Akhtar, N.; Rastogi, P. Localization of fluconazole in oral cavity by preferential coating of buccoadhesive tablet for treatment of oral thrush. *Int. J. Pharm. Invest.* **2016**, *6*, 106–115. [CrossRef] [PubMed]

materials

MDPI

Article

Versatile Layer-By-Layer Highly Stable Multilayer Films: Study of the Loading and Release of FITC-Labeled Short Peptide in the Drug Delivery Field

Kun Nie [1], Xiang Yu [1,*], Navnita Kumar [2] and Yihe Zhang [1,*]

1 Beijing Key Laboratory of Materials Utilization of Nonmetallic Minerals and Solid Wastes, National Laboratory of Mineral Materials, School of Materials Science and Technology, China University of Geosciences (Beijing), Beijing 100083, China; nk@cugb.edu.cn

2 Department of Chemistry and Biochemistry, University of California, Los Angeles, CA 90095, USA; navnitakumar77@gmail.com

* Correspondence: yuxiang@cugb.edu.cn (X.Y.); zyh@cugb.edu.cn (Y.Z.); Tel.: +86-189-1180-7048 (X.Y.)

Received: 7 March 2019; Accepted: 11 April 2019; Published: 12 April 2019

Abstract: A viable short FITC-peptide immobilization is the most essential step in the fabrication of multilayer films based on FITC-peptide. These functional multilayer films have potential applications in drug delivery, medical therapy, and so forth. These FITC-peptides films needed to be handled with a lot of care and precision due to their sensitive nature. In this study, a general immobilization method is reported for the purpose of stabilizing various kinds of peptides at the interfacial regions. Utilizing Mesoporous silica nanoparticles can help in the preservation of these FITC-peptides by embedding themselves into these covalently cross-linked multilayers. This basic outlook of the multilayer films is potent enough and could be reused as a positive substrate. The spatio-temporal retention property of peptides can be modulated by varying the number of capping layers. The release speed of guest molecules such as tyrosine within FITC-peptide or/and adamantane (Ad)-in short peptides could also be fine-tuned by the specific arrangements of the multilayers of mesoporous silica nanoparticles (MSNs) and hyaluronic acid- cyclodextrin (HA-CD) multilayer films.

Keywords: mesoporous silica; layer-by-layer; FITC-peptide; hyaluronic acid; multilayer film; host-guest interaction

1. Introduction

Layer-by-layer self-assembly technology is a useful method to prepare multilayer films with a thickness which can be controlled even down to the nanoscale [1–5], and is among one of the most widely used techniques to weave the organic-inorganic multilayer films together [6–10]. Mesoporous silica nanoparticles (MSNs) have been widely used in adsorption and drug delivery due to their unique pore structure, large specific surface area and good biocompatibility [11]. Some of these can directly act as catalysts as well because the ordered mesoporous material can accelerate the diffusion rate of the product. The selectivity is 100% and the conversion rate is up to 90%. Due to the flexibility of the structure and the narrow pore distribution, the doping of metal oxide and other complexes to these ordered mesoporous material matrix makes them better catalysts. Biological macromolecules, such as enzymes, proteins, nucleic acids, etc., have molecular sizes less than 10 nm, while viruses have sizes around 30 nm. However, the ordered mesoporous materials have quite a range of pore sizes between 2 and 50 nm. Due to the non-toxicity of these MSNs, they play an important role in the decomposition and fixation of enzymes, proteins and other substances. The mesoporous silica used for a particular function needs to be of a particle size, otherwise it will have an adverse effect [11–13]. A lot of research has been done in the area of multilayer films conjoining polymer and inorganic nanoparticles [14–16], and this can be attributed to the variable functionalities from the organic and inorganic parts of multilayer films [17–20]. The multilayer films which incorporate different inorganic parts can generally be prepared either by forming a covalent cross-link or by utilizing the non-covalent interactions between the polymer and the inorganic parts [21–25]. According to the literature, under ultraviolet irradiation, poly (allylamine hydrochloride) (PAH) and 4,4′-diazostilbene-2,2′-disulfonic acid disodium salt (DAS) can form covalent cross-linked layers in situ [26]. Non-covalent multilayer films have been obtained by using different kinds of supramolecular interactions [27–29], electrostatic interactions [30], hydrophobic interactions [29], host–guest interactions [30–32] and so on and so forth.

The knowledge of diffusion mechanism of fluorescent agent in these multilayers is very important and can provide a deeper perspective into the molecular interaction in organisms, soft material systems, and various advanced functional films [33–35]. Due to their good absorptivity, water solubility, and fluorescence quantum yield, Fluorescein derivatives, especially Fluorescein Isothiocyanate (FITC) have become the most popular fluorescent reagents in this field. Owing to their wide fluorescence emission range, their good performance on photo-bleaching and fluorescence quenching on conjugation to biopolymers, the scope of the application can be widened for FITC-based dyes and their conjugates [36].

Hyaluronic acid (HA) consists of repeatable disaccharides of the β-glucuronic acid and N-acetyl-d-glucosamine [37]. Hyaluronic gel is found in synovial fluid and is the main component of a glycosaminoglycan superstructure complex which is associated with different polysaccharides such as the chondroitin sulfate [38]. Hyaluronic acid can anchor to the surface of the cell by attaching itself to the cell surface receptors. HA is an upcoming drug delivery molecule for soft tissue repair and regeneration [39]. Studies to control the host-guest interaction within supramolecular structures are gradually rising because of their potential application as stimuli-responsive hydrogels, nanoparticles, smart biosensor devices and so on [40,41].

Amino-betacyclodextrin (CD) is a member of cyclic oligosaccharides and consists of the lipophilic central cavity and a hydrophilic outer surface [42,43]. Amino-betacyclodextrin has a large structure with a many hydrogen donors and acceptor groups. However, these groups are impermeable to the lipophilic films, making them difficult to use as a drug delivery vehicle. To enhance aqueous solubility of those poorly soluble drugs and to also raise their bioavailability, cyclodextrin has been utilized as a complexing agent [44]. CD can be grafted onto HA by chemical synthesis, and then the properties of both the moieties in the complex can be utilized. The grafting rate can be measured by ^{1}H NMR [45]. In addition, there is supramolecular hydrophobic host-guest interaction between adamantine (Ad) and cyclodextrin (CD) [46]. Peptides with Ad can be used as targets to study the effect of host-guest interactions in molecular diffusion. At present, there are many reported articles on the host-guest inclusion phenomena that delays the release of peptides [47].

In this paper, we have explored a new method to construct multilayer film materials that effectively preserves different kinds of short peptides. Mesoporous silica serves as a short peptide-container and functional multilayers were embedded in these via electrostatic interactions. Because of the exquisite film structure, the magnitude of the supramolecular interactions within the multilayer films can be modulated. The release curves of the short peptides vary with the number of laminated multilayers. The supramolecular force-delayed molecular release was taken as an example to further demonstrate the functions of the reported peptides immobilization strategy. This technique can be utilized to prepare multilayers which can further be used to control the fabrication of these versatile multilayer films.

2. Materials and Methods

2.1. Materials and Instruments

The following chemicals were used without any further treatment: Sodium hyaluronate (HA, 95%), amino-betacyclodextrin, fluorescein isothiocyanate (FITC, purity > 90), phosphate buffer solution (PBS) and peptides (FITC-RGD & FITC-RGD-Ad) modified with fluorescein isothiocyanate were provided by Sinopharm Chemical Reagent Beijing Co., Ltd (Beijing, China). Mono-(6-Amino-6-deoxy)-Beta-cyclodextrin was bought from Shandong Binzhou Zhiyuan Biotechnology Co., Ltd (Shandong, China). Peptides (FITC-$(SGGYGGS)_4$ & FITC-$(SGGSGGS)_4$) were bought from LifeTein LLC (Beijing, China). Cetyltrimethylammonium bromide surfactant ($CH_3(CH_2)_{15}N(CH_3)_3Br$, CTAB), tetraethoxysilane (TEOS) and Poly(allylamine hydrochloride) (PAH, Mw = 15,000) were purchased from Sigma-Aldrich (St. Louis, MO, USA). 4,4′-diazostilbene-2,2′-disulfonic acid disodium salt (DAS), N-Hydroxy succinimide (NHS) and 1-(3-Dimethylaminopropyl)-3-ethylcarbodiimide hydrochloride (EDC) were purchased from TCI (Tokyo Chemical Industry, Tokyo, Japan). Only deionized water was used for all the syntheses.

UV-vis spectra were obtained using a Hitachi U-3900 spectrophotometer (Hitachi, Tokyo, Japan). The Surface morphologies of multilayer films were characterized using transmission electron microscope (TEM) (TEM, Tecnai T12, Field Electron and Ion Company, FEI, Hillsboro, OR, USA). TEM experiments were carried out on a Titan S/TEM (Field Electron and Ion Company, FEI, Hillsboro, OR, USA) microscope. ^{1}H NMR spectra were obtained on Bruker Avance III 400MHz WB (Bruker, Baden, Switzerland). The Brunauer-Emmett-Teller(BET)-Barret-Joyner-Halenda(BJH) BET-BJH data were obtained from the Autosorb-iQ2 (Quantachrome, Boynton Beach, FL, USA).

2.2. *The Preparation of HA-β-CD Gels*

HA-CD gel was prepared according to literature [48]. In a 100 mL Erlenmeyer flask, 0.3 g of sodium hyaluronate and 50 mL of morpholinoethanesulfonic buffer (MES) were added and rapid stirred for 6 h. Subsequently, 0.285 g of NHS and 0.342 g of EDC were added into the Erlenmeyer flask. Then the mixture was stirred for another hour. To the mixture, 0.1686 g of amino cyclodextrin was added, followed by stirring for 24 h. The detailed experimental procedure and the sample codes are listed in Table S1. ^{1}H NMR spectroscopy was used to determine the degree of grafting of HA-CD. The graft ratio was calculated by using the integral area of the ^{1}H NMR spectrum. As shown in Figure S1, the peak at 5.10 ppm represents the proton of amino (NH) (5.10 ppm) group and its integral area was 0.11. Therefore, the integral area of each hydrogen atom of the aminocyclodextrin was 0.11/7. Similarly, the peak at 2.05 ppm represents the proton of CH_3 (2.05 ppm) group and its integral area was 1.00. Therefore, the integral area of each hydrogen atom of hyaluronic acid was 1/3. Grafting ratio was obtained by dividing the integral area of each hydrogen atom of amino cyclodextrin by the integral area of each hydrogen atom of hyaluronic acid. So, according to the ^{1}H NMR (400 MHz, D_2O) data, the grafting rate of HA-CD was 4.71%.

2.3. *The Synthesis of Mesoporous Silica Nanoparticles (MSNs)*

MSNs were synthetized according to the literature [49]. In brief, an aqueous solution containing of 2.0 g CTAB, 7.0 mL NaOH (2 mol L^{-1}) and H_2O (480 g) was heated for 30 min at 80 degrees with rapid stirring. To this solution, 8.31 g ethyl orthosilicate was added. The white solid appeared within two minutes. The product was further stirred at 80 degrees for two more hours, followed by centrifugation, washing with a lot of ultra-pure water, and drying in vacuum oven for 12 h. The CTAB extraction was carried at 60-degrees Celsius. This white powder was added to a mixture of ethanol (120 mL) and concentrated hydrochloric acid (1.0 mL) and stirred for 8 h at 60 degrees Celsius. The product was then centrifuged and washed with a lot of water and ethanol, followed by drying under vacuum. The diameter of meso-SiO_2 is about 100 nm as seen in TEM images shown in Figure 1c. The surface area of the MSN is 646.715 m^2 g^{-1} and the pore diameter of the MSN was about 3.4 nm, which was characterized via the BET instrument.

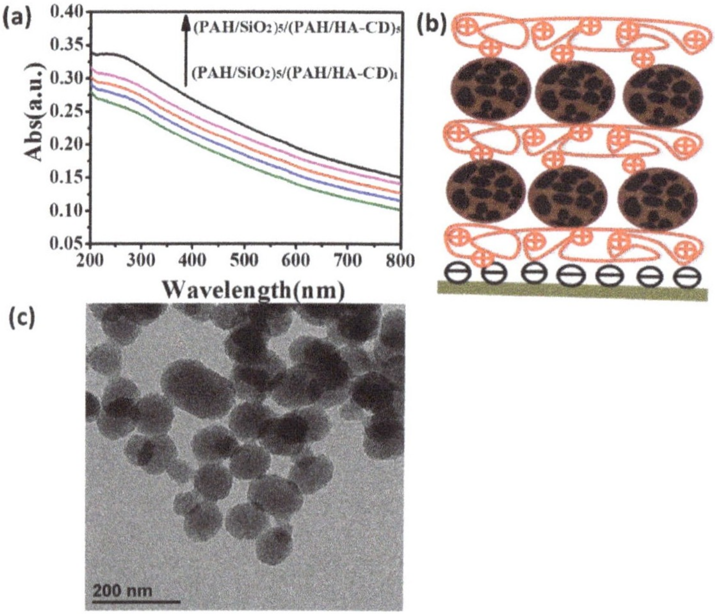

Figure 1. (**a**) Assembly process followed by UV-vis spectrum of $(SiO_2/PAH)_5/(PAH/HA\text{-}CD)_5/(PAH/DAS)_5$ multilayers; (**b**) Cartoon showing the process of layer-by-layer self-assembly; (**c**) TEM image of MSNs.

2.4. Layer-By-Layer Assembly Multilayer

Quartz slides cleaned by piranha solution (concentrated H_2SO_4/H_2O_2 (v:v = 7:3)) for 10 h were immersed in PAH (aq., MW = 15,000, 1 mg ml^{-1}, pH = 9) for 25 min in a typical layer-by-layer cycle. They were rinsed in ultrapure water and then dried with nitrogen. After that, the slides were dipped into the SiO_2 solution for 25 min and rinsed and dried as before. The cycle was continued until desirable thickness was obtained and the related coating film will be designated as $(PAH/SiO_2)_n$. Generally, the LBL film was assembled in pH = 9 solution. The substrate was immersed in PAH (aq., MW = 15,000, 1 mg mL^{-1}, pH = 9) for 25 min, rinsed in ultrapure water, and then dried with nitrogen. Subsequently, these slides were soaked into the HA-CD (1 mg mL^{-1}) solution for 25 min, rinsed and dried as before so that (PAH/HA-CD) LBL films could be deposited onto the surface of the $(PAH/SiO_2)_n$ films. The film should be fixed on the quartz substrate to avoid bending during handling so that the good quality of multilayers on the quartz substrate can be guaranteed.

2.5. Peptide Loading and Release

1 mg of FITC-labeled peptide solution was dissolved in 10 mL of PBS buffer solution (pH = 7.4) in a 20 mL glass beaker at room temperature. The quartz sheets with the multilayers on the surface were then immersed in the peptide solution for 24 h. The quartz sheets were then removed and the surface was immediately rinsed with deionized water and blown dry with nitrogen. Finally, these quartz sheets were used for release experiments in PBS buffer solution.

3. Results and Discussions

The surfaces of PAH and DAS are positively charged. PAH and DAS can form covalent cross-linked layers in situ under ultraviolet irradiation. The versatile multilayer films reported here were built by integrating MSN, which act as the storage space of polypeptide of FITC-peptide-Ad and the covalently cross-linking multilayers to increase the stability of the whole system of the multilayer films

(Scheme 1). A multilayer film of $(PAH/SiO_2)_5/(PAH/HA\text{-}CD)_5$ was prepared as illustrated in Figure 1a. The formation of this multilayer was monitored by UV-vis spectroscopy. The cartoon as shown in Figure 1b, displays the process of layer by layer self-assembly between MSN and PAH. The MSN used in the experiment were synthesized in our laboratory. The diameters for these MSNs vary from small to 100 nm as shown in Figure 1c. The thickness of the cross-section of multilayer films prepared on the silica substrate was about 140 nm with surface roughness of around 35 nm (Figure S2). As compared to uncross-linked $(PAH/DAS)_5$ films, the stability of the $(PAH/SiO_2)_5$ multilayer films is not great and could not be enhanced, and the absorbance of the basic solution remains 32.26% of the original absorbance after the immersion as shown by the UV-vis spectra (Figure 2a,b).

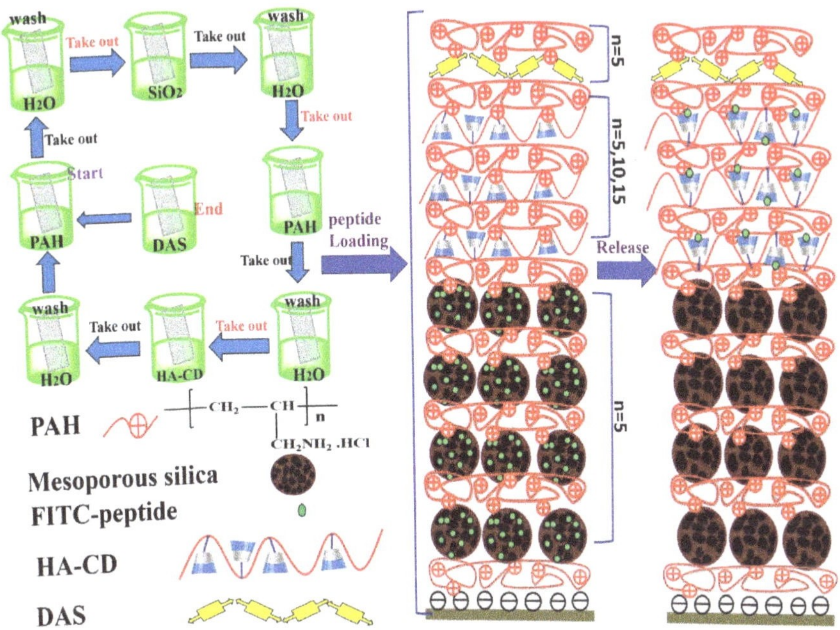

Scheme 1. Description of the loading and release using multilayer film: polymer layers and the mesoporous silica reservoir both play an important role in fabricating functional and stable multilayer films

UV irradiation was used to enhance the stability of the multilayer films. $(PAH/DAS)_5$ multilayer films were subsequently prepared by a self-assembly method on the top of the $(PAH/SiO_2)_5/(PAH/HA\text{-}CD)_5$ multilayers, and cross-link of $(PAH/DAS)_5$ multilayer films was analyzed using UV spectra. The decrease of the absorbance peak at 340 nm indicates that the DAS decomposes and forms covalent cross-linkages within the $(PAH/DAS)_5$ films. The stability of these films can be verified by soaking the cross-linked multilayer films into the basic solution. After the treatment with basic solution for 2 h, the peak absorbance of the multilayer films increased by 94.12% (Figure 2a).

The result shows that the covalently cross-linked $(PAH/DAS)_5$ acts as an outermost film which contributes in the enhancement of the stability of the multilayer films within the MSNs. Based on these results, the presence of $(PAH/DAS)_5$ multilayer films can enhance the stability of the uncross-linked multilayer films. It was supposed that the nano-net effect resulted in the ability of $(PAH/DAS)_5$ multilayer films to improve the stability of the films by incorporating silica underneath. In addition, DAS could enter the multilayer films and also enhance the stability of the multilayer films through covalent cross-linking bonding. This effective and facile strategy can play an important role in practical purposes because of its ability in stabilizing different kinds of multilayer films.

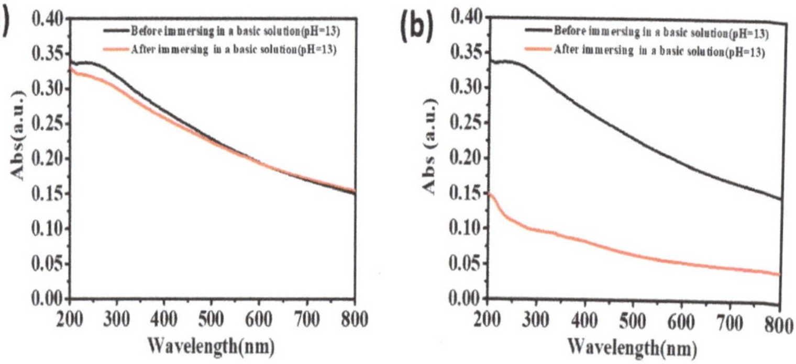

Figure 2. (**a**) UV-vis spectrum of the cross-linked $(SiO_2/PAH)_5(PAH/HA\text{-}CD)_5(PAH/DAS)_5$ film before (black) and after (red) immersion in a basic solution (pH = 13); (**b**) UV-vis spectrum of the uncross-linked $(SiO_2/PAH)_5/(PAH/HA\text{-}CD)_5/(PAH/DAS)_5$ film before (black) and after (red) immersion in a basic solution (pH = 13).

In consideration of the above facts, a model multilayer film for the release of FITC-labeled peptides can be employed as a model drug. The host-guest interaction between CD and Ad existed on the three-dimensional scale. The cyclodextrin molecule is hydrophilic from outside and hydrophobic from inside. It indicates that the CD possesses the external hydrophilic surface and the hydrophobic central cavity.

Figure 3 shows the short peptides release profiles from cross-linked multilayer films. Figure 3a clearly shows that the drug release rate of FITC-RGD is faster than the drug release rate of FITC-RGD-Ad. The release curve clearly shows the release time of FITC-RGD from $(SiO_2/PAH)_5/(PAH/HA\text{-}CD)_5/(PAH/DAS)_5$ is about 180 min (in the red line) (Figure 3a); however, the release time of FITC-RGD-Ad from $(SiO_2/PAH)_5/(PAH/HA\text{-}CD)_5/(PAH/DAS)_5$ is about 360 min (in the black line) (Figure 3a). As shown in Figure 3b, the release time of FITC-RGD and FITC-RGD-Ad is about 360 min and 480 min, respectively. As shown in Figure S3, ten layers of multilayers were used for experiments to prove that the molecular size had little effect on the release rate. As shown in Figure S3, the release time of both FITC-RGD and FITC-RGD-Ad is about 480 min. As illustrated in Figure 3c, it takes about 480 min for the release of FITC-RGD, and 720 min for FITC-RGD-Ad from $(SiO_2/PAH)_5/(PAH/HA\text{-}CD)_5/(PAH/DAS)_5$ multilayer film. Moreover, it shows that the release profiles are very smooth in Figure 3. Despite the number of multilayers in the film being same, the release speed and release time is different. The delayed effect of the films was caused by the host-guest supramolecular interactions between cyclodextrin and adamantane and also the release process increased with number of the multilayers. In addition, the release time of different kinds of multilayer is summarized in Figure 3d. In the experiment, the controlled variable method was used to study the release of peptide. Figure 4a is the same as Figure 3a. Compared with the release profiles in Figures 3 and 4a, Figure 4b clearly shows that the peptide release speed of FITC-RGD is faster than the release speed of FITC-RGD-Ad. The release time of FITC-RGD and FITC-RGD-Ad is about 210 min and 400 min, respectively. As seen in Figure 4c, the release time of FITC-RGD and FITC-RGD-Ad is 350 min and 470 min, respectively. Delayed effect of the multilayer films resulted from the host-guest supramolecular interactions, which show an increase in the release process with an increasing number of multilayers.

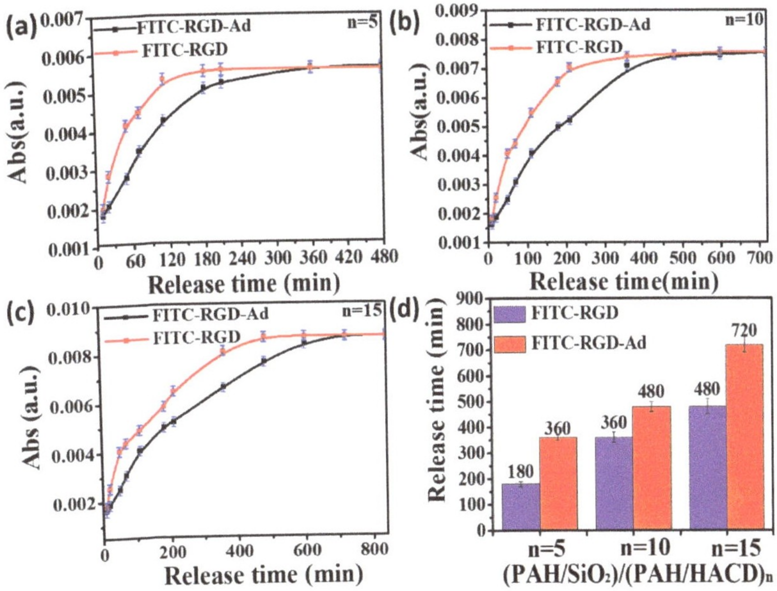

Figure 3. Release profiles of FITC-RGD-Ad and FITC-RGD from the cross-linked (**a**) $(SiO_2/PAH)_5/(PAH/HA\text{-}CD)_5/(PAH/DAS)_5$; (**b**) $(SiO_2/PAH)_5/(PAH/HA\text{-}CD)_{10}/(PAH/DAS)_5$ and (**c**) $(SiO_2/PAH)_5/(PAH/HA\text{-}CD)_{15}/(PAH/DAS)_5$ multilayer films. (**d**) the release time of FITC-RGD-Ad and FITC-RGD from different kinds of multilayer films.

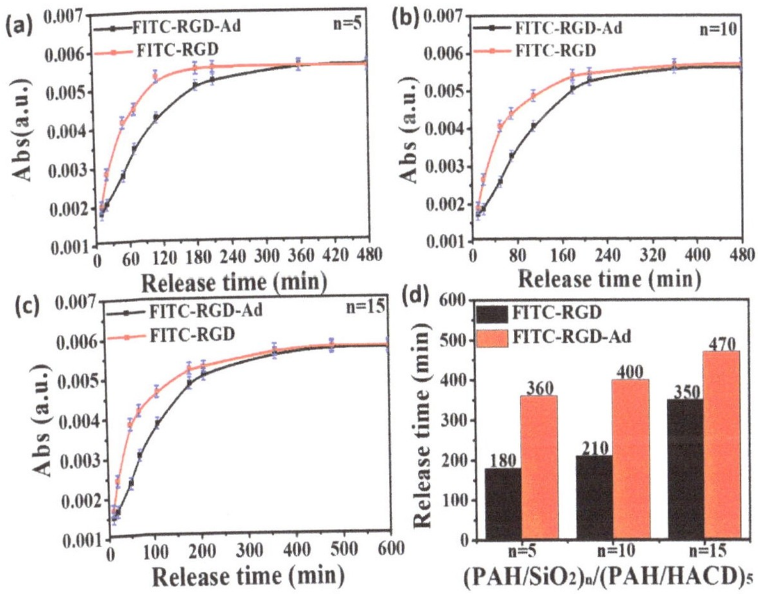

Figure 4. Release profiles of FITC-RGD-Ad and FITC-RGD from the cross-linked (**a**) $(SiO_2/PAH)_5/(PAH/HA\text{-}CD)_5/(PAH/DAS)_5$. (**b**) $(SiO_2/PAH)_{10}/(PAH/HA\text{-}CD)_5/(PAH/DAS)_5$ and (**c**) $(SiO_2/PAH)_{15}/(PAH/HA\text{-}CD)_5/(PAH/DAS)_5$ multilayer films. (**d**) The release time of FITC-RGD-Ad and FITC-RGD from different kinds of multilayer films.

It is well known that tyrosine (Y) is a hydrophobic amino acid. We infer that host-guest interactions between CD and Y will delay the release speed of Y-based peptide. For proof of concept, different kinds of multilayers were immersed into different solution of peptide, as shown in Figure 5.

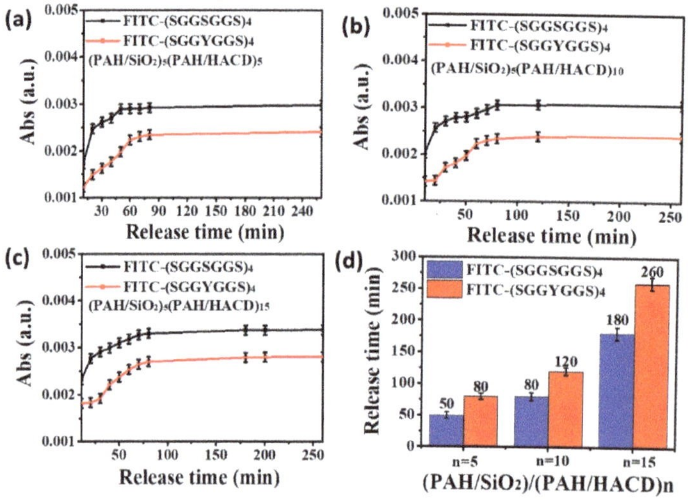

Figure 5. Release profiles of FITC-(SGGSGGS)$_4$ or FITC-(SGGYGGS)$_4$ from the cross-linked (**a**) (SiO_2/PAH)$_5$/(PAH/HA-CD)$_5$/(PAH/DAS)$_5$ (**b**) (SiO_2/PAH)$_5$/(PAH/HA-CD)$_{10}$/(PAH/DAS)$_5$ and (**c**) (SiO_2/PAH)$_5$/(PAH/HA-CD)$_{15}$/(PAH/DAS)$_5$ multilayer films. (**d**) The release time of FITC-RGD-Ad and FITC-RGD from different kinds of multilayer films.

As illustrated in Figure 5a, the release time of FITC-(SGGYGGS)$_4$ from (SiO_2/PAH)$_5$(PAH/HA-CD)$_5$(PAH/DAS)$_5$ is about 50 min. This is the time taken to obtain saturated release of fluorescent probe. For FITC-(SGGSGGS)$_4$ from (SiO_2/PAH)$_5$(PAH/HA-CD)$_5$(PAH/DAS)$_5$ this release time is 80 mins. As shown in Figure 5b, the release time of FITC-(SGGYGGS)$_4$ from (SiO_2/PAH)$_5$(PAH/HA-CD)$_{10}$(PAH/DAS)$_5$ is about 80 min and around 120 min for FITC-(SGGYGGS)$_4$ from (SiO_2/PAH)$_5$(PAH/HA-CD)$_{10}$(PAH/DAS)$_5$. Figure 5c shows that it takes around 180 min for the release of FITC-(SGGSGGS)$_4$ and 260 min for FITC-(SGGYGGS)$_4$ from (SiO_2/PAH)$_5$(PAH/HA-CD)$_{15}$(PAH/DAS)$_5$ multilayer films. The more the number of the multilayers in the film, the greater the host-guest interactions. By changing variables and adjusting parameters, it is proved that upon coming in contact with supramolecular forces, host-guest interactions between HA-CD and FITC-(SGGYGGS)$_4$ delayed the release speed of fluorescent molecules as shown in Figure 5.

4. Conclusions

In this study, a method has been developed which can raise the stability of multilayer films made up of polymer and inorganic nanoparticles. The covalently cross-linked super stratum was formed on outermost layer of the multilayer films so that the structure of multilayers can be stabilized. The super stratum acts as a nano-net which prevents the diffusion of the nanoparticles from the substrate. It is worth mentioning that the super stratum possesses permeability for small molecules like FITC-RGD, FITC-RGD-Ad, FITC-(SGGYGGS)$_4$ or FITC-(SGGSGGS)$_4$. The difference in permeability of the super stratum towards nanoparticles and small molecules is utilized to synthesize a functional system for drug delivery incorporating mesoporous silica as the molecule reservoir. The construction of functional multilayer films that incorporate inorganic nanoparticles has potential promotional value benefitting from the mild synthesis conditions of the super stratum.

Supplementary Materials: The following are available online at http://www.mdpi.com/1996-1944/12/8/1206/s1.

Author Contributions: K.N., X.Y., N.K. and Y.Z. conceived and designed the experiments; K.N. conducted the experiments and analyzed the data; K.N. wrote the draft, and K.N., N.K. and X.Y. improved it.

Funding: This research was funded by [National Natural Science Foundation of China] grant number [21303169, 21673209].

Acknowledgments: This work was supported by the NSFC (21303169, 21673209) and the National Scholarship Fund of China Scholarship Council (CSC). I am very grateful to Jeffrey I. Zink and Navnita Kumar for helping me to polish and revise the paper. Navnita Kumar is a postdoctoral faculty in the Department of Chemistry and Biochemistry at University of California, Los Angeles (UCLA).

Conflicts of Interest: The authors declare no conflict of interest.

References

1. Manna, U.; Bharani, S.; Patil, S. Layer-by-Layer Self-Assembly of Modified Hyaluronic Acid/Chitosan Based on Hydrogen Bonding. *Biomacromolecules* **2009**, *10*, 2632–2639. [CrossRef]
2. Campbell, M.G.; Liu, Q.; Sanders, A.; Evans, J.S.; Smalyukh, I.I. Preparation of nanocomposite plasmonic films made from cellulose nanocrystals or mesoporous silica decorated with unidirectionally aligned gold nanorods. *Materials* **2014**, *7*, 3021–3033. [CrossRef]
3. Nie, K.; An, Q.; Tao, S.Y.; Zhang, Z.P.; Luan, X.L.; Zhang, Q.; Zhang, Y.H. Layer-by-layer reduced graphene oxide (rGO)/gold nanosheets (AuNSs) hybrid films: significantly enhanced photothermal transition effect compared with rGO or AuNSs films. *RSC Adv.* **2015**, *5*, 57389–57394. [CrossRef]
4. Anirudhan, T.S.; Vasantha, C.S.; Sasidharan, A.V. Layer-by-layer assembly of hyaluronic acid/carboxymethylchitosan polyelectrolytes on the surface of aminated mesoporous silica for the oral delivery of 5-fluorouracil. *Eur. Polym. J.* **2017**, *93*, 572–589. [CrossRef]
5. Wang, D.G.; Sheridan, M.V.; Shan, B.; Famum, B.H.; Marquard, S.L.; Sherman, B.D.; Eberhart, M.S.; Nayak, A.; Dares, C.J.; Das, A.K.; Bullock, R.M.; Meyer, T.J. Layer-by-Layer Molecular Assemblies for Dye-Sensitized Photoelectrosynthesis Cells Prepared by Atomic Layer Deposition. *J. Am. Chem. Soc.* **2017**, *139*, 14518–14525. [CrossRef]
6. Kang, E.H.; Bu, T.J.; Jin, P.C.; Sun, J.Q.; Yang, Y.Q.; Shen, J.C. Layer-by-layer deposited organic/inorganic hybrid multilayer films containing noncentrosymmetrically orientated azobenzene chromophores. *Langmuir* **2007**, *23*, 7594–7601. [CrossRef]
7. Wang, H.J.; Ishihara, S.; Ariga, K.; Yamauchi, Y. All-Metal Layer-by-Layer Films: Bimetallic Alternate Layers with Accessible Mesopores for Enhanced Electrocatalysis. *J. Am. Chem. Soc.* **2012**, *134*, 10819–10821. [CrossRef]
8. Nie, K.; An, Q.; Zhang, Y.H. A functional protein retention and release multilayer with high stability. *Nanoscale* **2016**, *8*, 8791–8797. [CrossRef]
9. Xu, Q.W.; Li, X.; Jin, Y.Y.; Sun, L.; Ding, X.X.; Liang, L.; Wang, L.; Nan, K.H.; Ji, J.; Chen, H.; Wang, B.L. Bacterial self-defense antibiotics release from organic-inorganic hybrid multilayer films for long-term anti-adhesion and biofilm inhibition properties. *Nanoscale* **2017**, *9*, 19245–19254. [CrossRef]
10. Slowing, I.I.; Vivero-Escoto, J.L.; Wu, C.W.; Lin, V.S.Y. Mesoporous silica nanoparticles as controlled release drug delivery and gene transfection carriers. *Adv. Drug Deliv. Rev.* **2008**, *60*, 1278–1288. [CrossRef]
11. Lau, H.H.; Murney, R.; Yakovlev, N.L.; Novoselova, M.V.; Lim, S.H.; Roy, N.; Singh, H.; Sukhorukov, G.B.; Haigh, B.; Kiryukhin, M.V. Protein-tannic acid multilayer films: A multifunctional material for microencapsulation of food-derived bioactives. *J. Colloid Interface Sci.* **2017**, *505*, 332–340. [CrossRef]
12. Nguyen, T.T.T.; Belbekhouche, S.; Dubot, P.; Carbonnier, B.; Grande, D. From the functionalization of polyelectrolytes to the development of a versatile approach to the synthesis of polyelectrolyte multilayer films with enhanced stability. *J. Mater. Chem. A* **2017**, *5*, 24472–24483. [CrossRef]
13. Jin, W.; Shi, X.Y.; Caruso, F. High activity enzyme microcrystal multilayer films. *J. Am. Chem. Soc.* **2001**, *123*, 8121–8122. [CrossRef]
14. Kanazawa, A.; Ikeda, T.; Abe, J. Supramolecular polar thin films built by surfactant liquid crystals: Polarization-tunable multilayer self-assemblies with in-plane ferroelectric ordering of ion-based dipoles. *J. Am. Chem. Soc.* **2001**, *123*, 1748–1754. [CrossRef]

15. Ma, R.Z.; Sasaki, T.; Bando, Y. Layer-by-layer assembled multilayer films of titanate nanotubes, Ag- or Au-loaded nanotubes, and nanotubes/nanosheets with polycations. *J. Am. Chem. Soc.* **2004**, *126*, 10382–10388. [CrossRef]
16. Ribeiro, T.; Baleizao, C.; Farinha, J.P.S. Functional films from silica/polymer nanoparticles. *Materials* **2014**, *7*, 3881–3900. [CrossRef]
17. Rest, C.; Kandanelli, R.; Fernandez, G. Strategies to create hierarchical self-assembled structures via cooperative non-covalent interactions. *Chem. Soc. Rev.* **2015**, *44*, 2543–2572. [CrossRef]
18. Sakata, S.; Inoue, Y.; Ishihara, K. Molecular Interaction Forces Generated during Protein Adsorption to Well-Defined Polymer Brush Surfaces. *Langmuir* **2015**, *31*, 3108–3114. [CrossRef]
19. Zhang, D.W.; Tian, J.; Chen, L.; Zhang, L.; Li, Z.T. Dimerization of Conjugated Radical Cations: An Emerging Non-Covalent Interaction for Self-Assembly. *Chem. Asian J.* **2015**, *10*, 56–68. [CrossRef]
20. Ma, X.X.; Mei, L.F.; Liu, H.K.; Liao, L.B.; Liu, Y.Q.; Nie, K.; Li, Z.H. Structure and fluorescent properties of Ba3Sc(PO4)(3):Sm3+ red-orange phosphor for n-UV w-LEDs. *Chem. Phys. Lett.* **2016**, *653*, 212–215. [CrossRef]
21. Steiner, C.; Gebhardt, J.; Ammon, M.; Yang, Z.C.; Heidenreich, A.; Hammer, N.; Gorling, A.; Kivala, M.; Maier, S. Hierarchical on-surface synthesis and electronic structure of carbonyl-functionalized one- and two-dimensional covalent nanoarchitectures. *Nat. Commun.* **2017**, *8*, 14765. [CrossRef]
22. Zhang, X.S.; Jiang, C.; Cheng, M.J.; Zhou, Y.; Zhu, X.Q.; Nie, J.; Zhang, Y.J.; An, Q.; Shi, F. Facile Method for the Fabrication of Robust Polyelectrolyte Multilayers by Post-Photo-Cross-Linking of Azido Groups. *Langmuir* **2012**, *28*, 7096–7100. [CrossRef]
23. Nguyen, H.D.; Dang, D.T.; van Dongen, J.L.J.; Brunsveld, L. Protein Dimerization Induced by Supramolecular Interactions with Cucurbit[8]uril. *Angew. Chem. Int. Edit.* **2010**, *49*, 895–898. [CrossRef]
24. Boraste, D.R.; Chakraborty, G.; Ray, A.K.; Shankarling, G.S.; Pal, H. Supramolecular host-guest interaction of antibiotic drug ciprofloxacin with cucurbit[7]uril macrocycle: Modulations in photophysical properties and enhanced photostability. *J. Photoch. Photobio. Chem.* **2018**, *358*, 26–37. [CrossRef]
25. Conesa-Egea, J.; Redondo, C.D.; Martinez, J.I.; Gomez-Garcia, C.J.; Castillo, O.; Zamora, F.; Amo-Ochoa, P. Supramolecular Interactions Modulating Electrical Conductivity and Nanoprocessing of Copper-Iodine Double-Chain Coordination Polymers. *Inorg. Chem.* **2018**, *57*, 7568–7577. [CrossRef]
26. Lee, H.Y.; Park, S.H.; Kim, J.H.; Kim, M.S. Temperature-responsive hydrogels via the electrostatic interaction of amphiphilic diblock copolymers with pendant-ion groups. *Polymer Chem.* **2017**, *8*, 6606–6616. [CrossRef]
27. Derbenev, I.N.; Filippov, A.V.; Stace, A.J.; Besley, E. Electrostatic interactions between charged dielectric particles in an electrolyte solution: constant potential boundary conditions. *Soft Matter* **2018**, *14*, 5480–5487. [CrossRef]
28. Mishra, A.K.; Weissman, H.; Krieg, E.; Votaw, K.A.; McCullagh, M.; Rybtchinski, B.; Lewis, F.D. Self-Assembly of Perylenediimide-Single-Strand-DNA Conjugates: Employing Hydrophobic Interactions and DNA Base-Pairing To Create a Diverse Structural Space. *Chem. Eur. J.* **2017**, *23*, 10328–10337. [CrossRef]
29. Moon, S.; Park, S.O.; Ahn, Y.H.; Kim, H.; Shin, E.; Hong, S.; Lee, Y.; Kwak, S.K.; Park, Y. Distinct hydrophobic-hydrophilic dual interactions occurring in the clathrate hydrates of 3,3-dimethyl-1-butanol with help gases. *Chem. Eng. J.* **2018**, *348*, 583–591. [CrossRef]
30. Shumilova, T.A.; Ruffer, T.; Lang, H.; Kataev, E.A. Straightforward Design of Fluorescent Receptors for Sulfate: Study of Non-Covalent Interactions Contributing to Host-Guest Formation. *Chem. Eur. J.* **2018**, *24*, 1500–1504. [CrossRef]
31. Xue, F.C.; Wang, Y.Q.; Zhang, Q.X.; Han, S.L.; Zhang, F.Z.; Jin, T.T.; Li, C.W.; Hu, H.Y.; Zhang, J.X. Self-assembly of affinity-controlled nanoparticles via host-guest interactions for drug delivery. *Nanoscale* **2018**, *10*, 12364–12377. [CrossRef]
32. Furchner, A.; Kroning, A.; Rauch, S.; Uhlmann, P.; Eichhorn, K.J.; Hinrichs, K. Molecular Interactions and Hydration States of Ultrathin Functional Films at the Solid-Liquid Interface. *Anal. Chem.* **2017**, *89*, 3240–3244. [CrossRef]
33. Palao, E.; Sola-Llano, R.; Tabero, A.; Manzano, H.; Agarrabeitia, A.R.; Villanueva, A.; Lopez-Arbeloa, I.; Martinez-Martinez, V.; Ortiz, M.J. Acetylacetonate BODIPY-Biscyclometalated Iridium(III) Complexes: Effective Strategy towards Smarter Fluorescent Photosensitizer Agents. *Chem. Eur. J.* **2017**, *23*, 10139–10147. [CrossRef]

34. Schwenck, J.; Maier, F.C.; Kneilling, M.; Wiehr, S.; Fuchs, K. Non-invasive In Vivo Fluorescence Optical Imaging of Inflammatory MMP Activity Using an Activatable Fluorescent Imaging Agent. *Jove-J. Vis. Exp.* **2017**, *123*, e55180. [CrossRef]
35. Luan, L.; Lin, Z.J.; Wu, G.H.; Huang, X.L.; Cai, Z.M.; Chen, X. Encoding electrochemiluminescence using Ru(bpy)(3)$^{(2+)}$ and fluorescein isothiocyanate co-doped silica nanoparticles. *Chem. Commun.* **2011**, *47*, 3963–3965. [CrossRef]
36. Tang, A.M.; Mei, B.; Wang, W.J.; Hu, W.L.; Li, F.; Zhou, J.; Yang, Q.; Cui, H.; Wu, M.; Liang, G.L. FITC-quencher based caspase 3-activatable nanoprobes for effectively sensing caspase 3 in vitro and in cells. *Nanoscale* **2013**, *5*, 8963–8967. [CrossRef]
37. Tay, A.; Sohrabi, A.; Poole, K.; Seidlits, S.; Di Carlo, D. A 3D Magnetic Hyaluronic Acid Hydrogel for Magnetomechanical Neuromodulation of Primary Dorsal Root Ganglion Neurons. *Adv. Mater.* **2018**, *30*, 1800927. [CrossRef]
38. Souchek, J.J.; Wojtynek, N.E.; Payne, W.M.; Holmes, M.B.; Dutta, S.; Qi, B.W.; Datta, K.; LaGrange, C.A.; Mohs, A.M. Hyaluronic acid formulation of near infrared fluorophores optimizes surgical imaging in a prostate tumor xenograft. *Acta Biomater.* **2018**, *75*, 323–333. [CrossRef]
39. Hunt, J.A.; Joshi, H.N.; Stella, V.J.; Topp, E.M. Diffusion and Drug Release in Polymer-Films Prepared from Ester Derivatives of Hyaluronic-Acid. *J. Control. Release* **1990**, *12*, 159–169. [CrossRef]
40. Pitarresi, G.; Palumbo, F.S.; Albanese, A.; Fiorica, C.; Picone, P.; Giammona, G. Self-assembled amphiphilic hyaluronic acid graft copolymers for targeted release of antitumoral drug. *J. Drug Target* **2010**, *18*, 264–276. [CrossRef]
41. Ramamurthy, V.; Jockusch, S.; Pore, M. Supramolecular Photochemistry in Solution and on Surfaces: Encapsulation and Dynamics of Guest Molecules and Communication between Encapsulated and Free Molecules. *Langmuir* **2015**, *31*, 5554–5570. [CrossRef]
42. Mondal, P.; Rath, S.P. Efficient Host-Guest Complexation of a Bisporphyrin Host with Electron Deficient Guests: Synthesis, Structure, and Photoinduced Electron Transfer. *Isr. J. Chem.* **2016**, *56*, 144–155. [CrossRef]
43. Kanagaraj, K.; Pitchumani, K. The Aminocyclodextrin/Pd(OAc)(2) Complex as an Efficient Catalyst for the Mizoroki-Heck Cross-Coupling Reaction. *Chem. Eur. J.* **2013**, *19*, 14425–14431. [CrossRef]
44. Kaur, N.; Garg, T.; Goyal, A.K.; Rath, G. Formulation, optimization and evaluation of curcumin-β-cyclodextrin-loaded sponge for effective drug delivery in thermal burns chemotherapy. *Drug Deliv.* **2016**, *23*, 2245–2254. [CrossRef]
45. Nakahata, M.; Takashima, Y.; Yamaguchi, H.; Harada, A. Redox-responsive self-healing materials formed from host-guest polymers. *Nat. Commun.* **2011**, *2*, 511. [CrossRef]
46. Wei, S.J.; Chu, H.M.; Xu, L.S.; Wang, Z.Z.; Huang, Q. Macrocyclic drug conjugates of metronidazole-cyclodextrin for colon-targeted delivery. *J. Control. Release* **2017**, *259*, E120–E121. [CrossRef]
47. Parlati, S.; Gobetto, R.; Barolo, C.; Arrais, A.; Buscaino, R.; Medana, C.; Savarino, P. Preparation and application of a beta-cyclodextrin-disperse/reactive dye complex. *J. Incl. Phenom. Macrocycl. Chem.* **2007**, *57*, 463–470. [CrossRef]
48. Nie, K.; An, Q.; Zink, J.I.; Yu, X.; Zhang, Y.H. Layer by Layer Mesoporous Silica-Hyaluronic Acid-Cyclodextrin Bifunctional "Lamination": Study of the Application of Fluorescent Probe and Host-Guest Interactions in the Drug Delivery Field. *Materials* **2018**, *11*, 1745. [CrossRef]
49. Trebosc, J.; Wiench, J.W.; Huh, S.; Lin, V.S.Y.; Pruski, M. Solid-state NMR study of MCM-41-type mesoporous silica nanoparticles. *J. Am. Chem. Soc.* **2005**, *127*, 3057–3068. [CrossRef]

materials

Article

Sirolimus-Eluting Electrospun-Produced Matrices as Coatings for Vascular Stents: Dependence of Drug Release on Matrix Structure and Composition of the External Environment

Zhanna K. Nazarkina [1,*,†], Boris P. Chelobanov [1,†], Vera S. Chernonosova [1,2], Irina V. Romanova [1], Andrey A. Karpenko [2] and Pavel P. Laktionov [1,2]

1 Institute of Chemical Biology and Fundamental Medicine, Siberian Branch, Russian Academy of Sciences, Novosibirsk 630090, Russia; boris.p.chelobanov@gmail.com (B.P.C.); vera_mal@niboch.nsc.ru (V.S.C.); irin-romanova@yandex.ru (I.V.R.); lakt@niboch.nsc.ru (P.P.L.)

2 Meshalkin National Medical Research Center, Ministry of Health of the Russian Federation, Novosibirsk 630055, Russia; andreikarpenko@rambler.ru

* Correspondence: zha_naz@niboch.nsc.ru

† The first two authors should be regarded as joint first authors.

Received: 19 May 2020; Accepted: 5 June 2020; Published: 12 June 2020

Abstract: Although a number of drug-eluting coatings for vascular stents (VSs) have been developed and are in commercial use, more efficient stent coatings and drug delivery systems are needed. Sirolimus (SRL) is a clinically important drug with antiproliferative and immunosuppressive activities that is widely used for coating stents. Here, we characterized SRL-enriched matrices, intended for coating vascular stents, that were produced by electrospinning (ES) on a drum collector from a solution of polycaprolactone (PCL) and human serum albumin (HSA), 1,1,1,3,3,3-hexafluoroisopropanol (HFIP), dimethyl sulfoxide (DMSO), and SRL. The release of tritium-labeled SRL (^{3}H-SRL) from matrices in phosphate-buffered saline (PBS) or human blood plasma (BP) was studied. The introduction of DMSO in the ES blend decreased SRL release. The use of BP significantly accelerated SRL release through binding with serum biomolecules. The exchange of PBS or BP after every time point also increased SRL release. The maximum SRL release in BP was observed at 3 days. The matrices produced from the ES solution with DMSO and HSA released no more than 80% SRL after 27 days in BP, even under medium exchange conditions. Therefore, PCL-based matrices containing HSA, SRL, and DMSO can be used for coating VSs with prolonged SRL delivery.

Keywords: sirolimus; electrospinning; drug release; polycaprolactone; 3D matrix; drug-eluting stents

1. Introduction

Stent implantation is one of the most common procedures for the reconstitution of blood flow through atherosclerosis-occluded arteries. The placement of the stent initiates a variety of reactions, including endothelial injury, crushing of the plaque, and stretching and lacerations of the arterial wall, which can lead to stent-induced restenosis [1]. These processes stimulate an inflammatory response—the predominant cause of neointimal proliferation and in-stent restenosis [2] accompanied by the migration and proliferation of smooth muscle cells (SMCs) and fibroblasts from the arterial wall [3]. It was suggested that neointima growth after application of angioplasty drug-eluting stents (DESs) could be prevented by reducing the growth and migration of SMCs and other cell types. Sirolimus (SRL)-eluting stents, which were approved in Europe in 2002 and in the United States in 2003 [4], are the most popular and have been shown to be the most effective among the relevant analogs in reducing coronary restenosis, re-intervention rates, and other adverse cardiac events in patients with coronary artery disease [5,6].

SRL (rapamycin) is a polyketide macrolide compound obtained by a strain of *Streptomyces hygroscopicus* [7]. In mammals, the molecular target of rapamycin, mTOR, is a cell cycle-specific kinase involved in complex intracellular signaling pathways that regulate cell growth and proliferation, immunity, angiogenesis, fibrogenesis, and metabolism [8]. After entering into the cell, SRL interacts with the immunophilin FKBP-12. This complex interacts with mTOR kinase and inhibits its activity [7]. Thus, SRL blocks the activation of p70S6 kinase, which leads to cell cycle arrest at the G1 to S phase.

SRL is a potent inhibitor of the antigen-induced T- and B-cells' proliferation and antibody production [7], as well as the proliferation of a variety of cell types of nonlymphoid origin [9], including the inhibition of human coronary artery SMC proliferation [10]. Complete inhibition of PDGF-induced DNA synthesis in SMCs was achieved at 10 ng/mL SRL, and more than 50% inhibition was observed at an SRL concentration of 0.1 ng/mL. The inhibitory properties of SRL could be observed from a threshold concentration of 0.01 ng/mL. SRL is able to inhibit both quiescent and actively cycling cells [10].

SRL has a broad therapeutic window. Preclinical animal trials have shown that SRL demonstrates biological activity in doses from 18 to 1200 μg, without showing toxicity to the vessel wall [4]. Despite the promising pharmacological activities of SRL, however, its clinical application is complicated by poor aqueous solubility, first-pass metabolism, and nonspecific distribution in off-target sites [8]. SRL has some adverse, dose-dependent effects, including immunosuppression; the inhibition of bone growth; increased cholesterol, triglyceride, and creatinine serum levels; and decreased glomerular filtration rates. Leukopenia, thrombocytopenia, anemia, rash, stomatitis, arthralgia, diarrhea, hypertension, and hypokalemia may also occur [11]. Owing to its lipophilic nature, SRL is widely distributed in lipid membranes, resulting in a large volume of distribution and a relatively long half-life [12,13].

The most commonly used first-generation SRL-eluting stent in clinical practice is Cypher stent (SES; Cordis, Warren, NJ, USA) [14]. This DES consists of a stainless-steel platform with a strut thickness of 140 mm and 140 mg/cm^2 sirolimus with a poly(ethylene-co-vinyl acetate) and poly(n-butyl methacrylate) polymer. A pharmacokinetic study showed that the maximum concentration of SRL in blood is observed between 3 and 4 h after stent implantation, with a peak concentration of 0.57 ng/mL in patients receiving one stent [15]. The average terminal-phase elimination half-life of SRL is 213 h.

It is known that SRL can induce fibrinoid vascular necrosis [16]. A reduction in the SRL dose can reduce this toxic effect in an animal model. A lower SRL dose (25, 40, and 100 μg), released from a nonpolymeric hydroxyapatite-coated stent built by MIV Therapeutics (Atlanta, GA, USA), results in less vascular healing, presenting significantly fewer fibrinoids without increasing neointima formation as compared with Cypher stent (111 μg of SRL) [16]. The REDOX trial showed that lower doses (40% or 70% compared with Cypher stent) of SRL effectively maintain luminal patency for up to at least 12 months, as determined at a follow-up [17].

The stent platform, the antiproliferative drug, and the drug carrier polymer are the three major components that determine stent safety and efficacy [18]. The Cypher stent has not been manufactured since 2011 because the first-generation stents displayed an increased risk for late thrombotic events [19]; this risk was considered to be associated with the permanent presence of the non-erodible polymers poly(ethylene-co-vinyl acetate) and poly(n-butyl methacrylate). Therefore, a second generation of DES was developed using more biocompatible polymer coatings. SRL was also used to develop a new generation of DESs with ultrathin (<70 μm) struts coated with biodegradable polymers, such as Orsiro and BioMime [20]. Another type of SRL-eluting stent is MiStent, which has ultrathin struts composed of crystalline SRL and a biodegradable polymer poly(lactic-co-glycolic acid). Clinical trials have shown the superiority of ultrathin strut biodegradable polymer-coated DESs compared with durable polymer-coated DESs in maintaining a good safety profile [21].

The Tetriflex (Sahajanand Medical Technology, Surat, India) is a new-generation, ultrathin (60 μm), biodegradable SRL-eluting stent with coating composed of a combination of hydrophilic and hydrophobic polymers for controlled and prolonged release of SRL. Clinical studies have demonstrated the safety and effectiveness of the Tetriflex stent [22].

One of the novel DESs is Svelte SRL-eluting stent containing fully bioresorbable polymer and pre-mounted on a single lumen, fixed-wire delivery system [23]. The pharmacokinetics of the Svelte stent mimics the SRL release of the Cypher stent. The Svelte amino-acid based polymer can be enzymatically degraded with complete resorption in 12 months. The OPTMIZE Trial, including 1630 subjects, will estimate the safety and clinical efficacy of the Svelte stent [23].

The DESs used in clinics are manufactured using different techniques for coating metallic vascular stents, including ultrasonic coating [24], dip-coating [25], spray-coating [2], air-brush [26], electrohydrodynamic jetting, plasma-treated coating, electrotreated coating [27], and electrospinning (ES) [28]. Coating techniques have an influence on drug release kinetics and clinical outcomes [27].

ES is based on the generation of polymer fibers from a blend of polymers and drugs in a strong electric field. The advantages of the ES method are the ability to obtain fibers of different diameters and ultrastructures laid in a 3D matrix, as well as its suitability for a broad range of materials, cost effectiveness, and mechanical support from the stented area that covers the inter-strut area (e.g., the InspireMD CGuard stents consisting of a fine mesh that inhibits neointima growth thanks to the prevention of cell migration from the damaged area) [29–31]. Drug release and dissolution of the bioactive substance can be controlled through the structure of polymer fibers, which is determined by the composition of the ES. In recent years, the ES technique and its modifications have been widely used in drug delivery and regenerative medicine. The FDA has approved heart valve prostheses and esophageal stent produced using the ES technique for clinical studies [32].

The materials for coating vascular stents (VSs) must be hemo- and biocompatible, and extensible by at least 2–2.5×. Synthetic materials produced from polycaprolactone (PCL) possess these properties and are widely used in medicine owing to their PCL biocompatibility, mechanical characteristics, and ease of manufacturing. The average degradation time for PCL is about 2–3 years [33]. The PCL matrices produced by ES are convenient for VSs owing to their mechanical properties [29] and good hemo- and biocompatibility [34,35]. The matrices produced from pure PCL possess poor wettability and low cell compatibility, but incorporation of human serum albumin (HSA) into the PCL matrices strongly interferes with its properties, thereby increasing the stiffness and biological properties of such matrices [29]. It is known that HSA reduces platelet adhesion and increases the thromboresistance and hemocompatibility of blood-exposed surfaces [36,37].

In a previous study of paclitaxel release from ES-produced matrices, we showed that drug release depends on the structure of the fibers and the composition of the surrounding medium [29].

In the current study, we investigated SRL release from ES-produced matrices using different solution compositions: 5% PCL/SRL/10% HSA, 5% PCL/SRL/10% HSA/3% dimethyl sulfoxide (DMSO), and 5% PCL/SRL. The physicochemical properties of the materials and the dependence of SRL release on the composition of the surrounding medium were also studied.

2. Materials and Methods

2.1. Production of ^{3}H-Labeled Sirolimus

^{3}H-SRL was synthesized by thermoactivated tritium exchange, as described earlier [38]. ^{3}H-SRL was purified from byproducts using reversed-phase chromatography (RP-HPLC) on a C18 column using a gradient of acetonitrile in water (25–100%). The purity of ^{3}H-SRL was evaluated by autoradiography following thin layer chromatography (TLC) on Kieselgel 60 F254 plates (Merck, Darmstadt, Germany, 25 Alufolien 20 cm × 20 cm) in a chloroform–methanol–water mixture (19:1:0.1, Rf = ~0.7). The radioactivity of the preparation was measured on a Tri-Carb 2800 TR β-counter (PerkinElmer, Waltham, MA, USA) in a "ULTIMA GOLD LTT" scintillator (Perkin Elmer, Waltham, MA, USA). The radioactivity of all samples was evaluated as reported earlier [29], that is, 0.1 mL of the sample was thoroughly mixed with 0.9 mL scintillator, and radioactivity was measured immediately.

2.2. Fabrication of the ES Matrices

The ES solutions were prepared using stock solutions of 9% PCL and 1% HSA (Sigma-Aldrich, St. Louis, MI, USA) in 1,1,1,3,3,3-hexafluoroisopropanol (HFIP, Sigma-Aldrich, USA). The HSA concentration in the matrices was 10% (w/w). SRL (150301, Fujian Kerui Pharmaceutical Co. Ltd., Fujian, China) was dissolved in HFIP or DMSO (Sigma-Aldrich, USA) and added to the ES solutions to a final concentration of 0.9 μg/cm^2 (0.7 μg per disk). Then, 3% (v/v) DMSO was added to the solution of polymers. ^{3}H-SRL was mixed with unlabeled SRL to reach radioactivity of 27,000 cpm/cm^2 (21,000 per 10 mm diameter disk). To obtain more-uniform fibers, the PCL/SRL ES solution contained 0.1 mm triethylamine (TEA). To obtain the matrices containing ^{3}H-SRL, we used a custom made ES device with an airproof chamber and an exhaust HEPA filter equipped with a Spellman SL 150 (30 kV, Spellman, Brockton, MA, USA) power supply. Matrices of 110–125 μm thickness were prepared using a drum collector (4 cm diameter and 5 cm length). Electrospinning conditions were the same for all matrices: voltage—26.5 kV; feed rate—1 mL/h; nozzle to collector distance—20 cm; collector rotation speed—300 rpm; 23–25 °C; 25–35% humidity. After fabrication, matrices were removed from the collector, dried in vacuum under 10 Pa for 12 h, and stored in sealed ziplock polyethylene containers at 4 °C.

2.3. Characterization of the Matrices

2.3.1. Mechanical Characterization

Strain–stress diagrams were obtained using a universal Zwick/Roell Z100 (Zwick Roell, Ulm, Germany) test bench as described in ISO 7198:1998 [39]. Four 1 × 5 cm rectangular sheets for each matrix were cut and placed between holders at a distance of 2–2.5 cm. Tensile strength testing was performed at a rate of 10 mm·min^{-1} at room temperature (21–23 °C).

2.3.2. Matrix Surface Characterization

The morphology of matrices were studied by scanning electron microscopy (SEM), as described earlier [35]. The size of the pores and the diameter of fibers were estimated from the SEM images according to ISO 7198:1998 [39]. To estimate the stability of the fiber structure, the matrices were incubated in PBS (Sigma-Aldrich, USA) or blood plasma (BP) at 37 °C for 27 days. After incubation, the matrices were rinsed with water, air-dried, and studied by SEM. To evaluate the structure after elongation, the matrices were fixed between clumps of the small device that expands the matrix and fixes it in an SEM camera. Matrices were fixed at conductive colloidal graphite (G303, Agar Scientific Ltd., Cambridge, UK) before gold sputter-coating.

2.3.3. Additional Characterization of the Matrices

The contact angle was measured on a Drop Shape Analyzer–DS A25 (Kruss GmbH, Hamburg, Germany) using water as a solvent (drop volume, 1 μL; shooting speed, 160 frames per second).

The porosity of the matrices was evaluated according to the following formula:

$$\text{Porosity (\%)} = [1 - \text{Da}/\text{Dp}] \times 100$$

where Da = matrix weight/matrix volume and Dp is the PCL density.

2.4. Sirolimus Release Kinetics

To evaluate SRL release, 10 mm diameter disks (~0.785 cm^2) were excised from the matrices by die cutting and placed in wells of a 48-well plate. Then, 400 μL of PBS or EDTA-stabilized BP was added to each well. The investigation was approved by the Local Ethical Committee of Center of Personalized Medicine, Institute of Chemical Biology and Fundamental Medicine of the Siberian Branch of the Russian Academy of Sciences (№1, 15.01.2016). The plate was sealed with a film (Microseal® 'B' PCR Plate Sealing Film, adhesive, Bio-Rad, Hercules, CA, USA) to prevent drying, followed by incubation on a Titramax 1000 shaker (Heidolph, Schwabach, Germany) at 37 °C; platform rotation speed of 200 rpm. Moreover, 0.01% sodium azide was used to prevent bacterial growth. Two series of SRL release kinetics were studied. The matrices were incubated in PBS or BP for 20 min, 60 min, 3 h, 9 h, 27 h, 3 days, 9 days, and 27 days without (Series 1) or with (Series 2) medium replacement. For Series 2, the supernatant was removed at each time point, fresh solution was added, and the matrix was incubated in a fresh solution until the next time point. After incubation, the matrix was washed with H_2O and air-dried at room temperature for SEM analysis. The radioactivity of the supernatants was measured in duplicate, as described above (Section 2.1).

The influence of matrix elongation on SRL release was evaluated as follows. The matrix was fixed in the clamps and then slowly stretched by a screw to double the measured distance. After stretching, the linear sizes of the matrix were measured and the amount of SRL per disk was calculated. The 10 mm diameter discs were then excised from the stretched matrix, and SRL release was estimated as described above. All experiments were performed in duplicate.

2.5. Statistical Analysis

The results were processed using Microsoft Excel 2010 and the Statistica 7.0 package (StatSoft Inc., Tulsa, OK, USA).

3. Results and Discussion

3.1. Preparation and Characterization of the SRL Matrices

^{3}H-SRL was obtained by thermoactivated tritium exchange. ^{3}H-SRL has the same Rf after RP-HPLC and TLC, suggesting a similar chemical structure of the compound. Purified ^{3}H-SRL preparation with a specific radioactivity of 0.2 mCi/mL (~0.055 Ci/mmol) was thus obtained. The compound was homogenous according to the TLC data, being detected as a single spot on the autoradiograph with the expected Rf corresponding to unlabeled SRL. ^{3}H-SRL was combined with unlabeled SRL to reach a dose of ~1.05 μg/cm^2 SRL and a radioactivity of ~27,000 cpm/cm^2.

ES is a very promising method based on the generation of polymer fibers from a blend of polymers and drugs in a strong electric field. The advantages of the ES method are the ability to obtain fibers with different diameters (from nm to μm) and ultrastructures, broad material suitability, and high cost effectiveness [30,31]. Several parameters have an influence on the structure and morphology of polymer fibers, including voltage, collector to needle distance, solution flow rate, polymer concentration, solvent type, dielectric constant, temperature, pressure, and moisture [31]. The solvent type, polymer concentration, and voltage strongly affect the resulting fiber diameter. Drug release and dissolution of the bioactive substance can be controlled by the structure of the polymer fibers. In this study, we used the same ES conditions to obtain all matrices.

Three different matrices were prepared using ES on a drum collector from solutions containing PCL, HSA, DMSO, SRL, and ^{3}H-SRL. Electrospinning conditions were the same for all matrices: voltage—26.5 kV, feed rate—1 mL/h; nozzle to collector distance—20 cm.

Polymers used for drug-eluting stent coatings need to have mechanical characteristics and biological properties similar to those of natural vessels. The PCL scaffolds exhibit good mechanical properties, flexibility, and nontoxicity [40]. The average degradation time for PCL is about 2–3 years [33]. Earlier, we studied the structure of fibers produced by ES from PCL–HSA solutions [35]; like other authors, we demonstrated that HSA increases the hemocompatibility of blood-exposed surfaces [35–37]. The mechanical properties of PCL/HSA-containing matrices improved compared with pristine PCL, and HSA was tightly bound, with the fiber surface exposed for a long period of time because it is only partially accessible for protease hydrolysis [35]. HSA contains carboxylic and amine groups that promote cell attachment and spreading via integrin-binding sites, and can thus provide endothelization [40]. As such, HSA is the most abundant protein in human blood responsible for fatty acid transport and is the major binding protein for neutral and acidic drugs [41,42].

Its non-immunogenicity, high biocompatibility, and good biodegradability (as well as its binding of drugs) make HSA a prospective compound for DES coatings.

The physical properties of the obtained matrices are shown in Table 1. The tensile strength of the prepared matrices depend on the composition of the ES solution, demonstrating a typical two-phase curve with an extended plastic deformation region, starting from 7–10% to 250–300% and varying from 10.4 to 15 MPa (Table 1). These properties provide elongation without breakage during stent expansion, as well as a low residual load after the stent installation, as previously discussed [29].

Table 1. Physical properties of the matrices. PCL, polycaprolactone; SRL, sirolimus; HSA, human serum albumin: DMSO, dimethyl sulfoxide.

Matrix Type	Thickness (μm)	Strength (MPa)	Mass Per Unit Area (mg/cm²)	Fiber Diameter (μm)	Pore Diameter (μm)	Porosity (%)	Contact Angle (°)
5% PCL/SRL/10% HSA	111	10.45 ± 1.48	2.51	0.27 ± 0.8	1.94 ± 0.33	63.3	117.63 ± 2.72
5% PCL/SRL/10% HSA/3%DMSO	125	12.41 ± 3.11	2.47	0.14 ± 0.05	1.12 ± 0.18	64.7	109.52 ± 2.22
5% PCL/SRL	114	15.04 ± 2.50	2.44	0.36 ± 0.09	2.11 ± 0.34	60.9	127.37 ± 3.41

The structure of the matrices was studied by SEM. All matrices were composed of randomly oriented fibers. The composition of the ES solution influenced the fiber diameter and its variability (Figure 1). The matrices obtained from the blend containing 5% PCL, SRL, and 0.1 mm TEA were more heterogeneous in fiber diameter. The average diameter of the fibers made from the PCL/SRL/HSA and PCL/SRL solution was 0.27 and 0.36 μm, respectively (Table 1, Figure 1). The fiber diameter significantly decreased when DMSO was added (0.14 μm for the PCL/SRL/HSA/DMSO matrix). The SEM at a higher resolution reveals the smooth surface of the PCL fibers. The decrease in the fiber diameter for matrices containing DMSO was previously shown for the PCL-based matrices containing paclitaxel [29].

According to SEM data, incubation in PBS for 27 days did not alter the structure of the PCL/SRL/HSA and PCL/SRL matrices (Figure 1), while fiber fusion and thickening were observed when the PCL/SRL/HSA/DMSO matrix was incubated in PBS or BP for 27 days. A negligible thickening effect was also observed for the PCL/SRL matrix, where the thickness of the PCL/SRL/HSA/DMSO fibers increased almost twofold. There was no significant change in the structure of the PCL/SRL/HSA matrix during incubation. No significant weight change after 27 days of incubation was observed for all matrix types (data not shown). This means that, during the assayed time, there was no substantial matrix degradation. The incorporation of hydrophilic biodegradable components like HSA into the PCL matrices can potentially increase their rate of biodegradation, but we did not observe such an effect for 27 days [40].

The porosity of the matrices was estimated from the matrix weight and polymer density in the range of 60–65% (Table 1). No significant differences were observed for the matrices prepared from different blends. The presence of DMSO in the electrospinning solution resulted in a decrease not only in fiber diameter, but also in pore size. The PCL/SRL/HSA/DMSO matrix containing 3% DMSO had a pore diameter approximately 1.7-fold less than those without DMSO (Table 1, Figure 1). These data correlate with those previously obtained for PCL-based matrices containing paclitaxel [29].

The wettability of the matrices affected the cell distribution in the matrix and cell proliferation [40]. The polar amine and carboxyl groups in HSA might have improved the surface wettability of the matrices. HSA has a high water-binding capacity and can increase the hydrophilicity of materials [40]. Measurement of the contact angle is a commonly used technique to determine wettability. The water contact angles of the matrices are indicated in Table 1. The HSA-containing matrices (PCL/SRL/HSA and PCL/SRL/HSA/DMSO) show a lower contact angle compared with that of PCL/SRL. According to the literature data, hydrophilic matrices are more suitable for implants owing to their lower ability to adsorb proteins and induce inflammation [43,44].

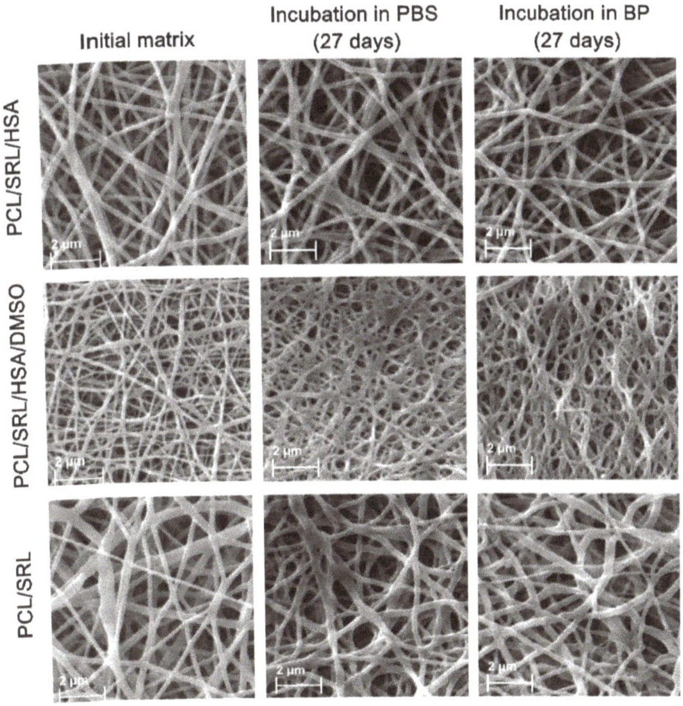

Figure 1. Scanning electron microscopy (SEM) images of the matrices incubated under different conditions. BP, blood plasma; PCL, polycaprolactone; SRL, sirolimus; HSA, human serum albumin: DMSO, dimethyl sulfoxide.

3.2. Study of SRL Release

The SRL release from PCL-based matrices was studied with (2) or without (1) a medium replacement to mimic SRL release in a biological system. The kinetic curves shown for PBS without any medium replacement were similar for all matrices (Supplementary materials, Table S1), although SRL release from the PCL/SRL/HSA/DMSO matrix was slightly slower, with the maximal release of 70% observed at 27 days (Figure 2A). The maximal release observed for the PCL/SRL/HSA and PCL/SRL matrices was 76% and 71%, respectively. When the matrices were incubated with replacement of PBS throughout, the rate of SRL release from the PCL/SRL/HSA/DMSO matrix was significantly slower, with a maximal level of 75% for 27 days (Figure 2B). Meanwhile, the PCL/SRL/HSA and PCL/SRL matrices released 75% SRL during the first three days, and reached maximal levels of 90% and 85%, respectively.

It was shown earlier that paclitaxel is distributed both throughout the volume and on the surface of the fibers in PCL-based matrices [29]. In this work, the SRL distribution seems to be very similar to that of paclitaxel, because we used the same method to obtain matrices. Desorption of SRL from the matrix surface limits the rate of establishing the equilibrium between SRL on the surface and SRL in the solution. This process is the result of the low solubility of SRL in water (2.6 mg/L) [45]. Removing the medium with dissolved SRL led to the additional desorption of SRL from the matrix surface into the solution, thus the amount of dissolved SRL increased. The presence of ions and SRL-binding molecules can also increase the solubility of SRL in PBS and BP.

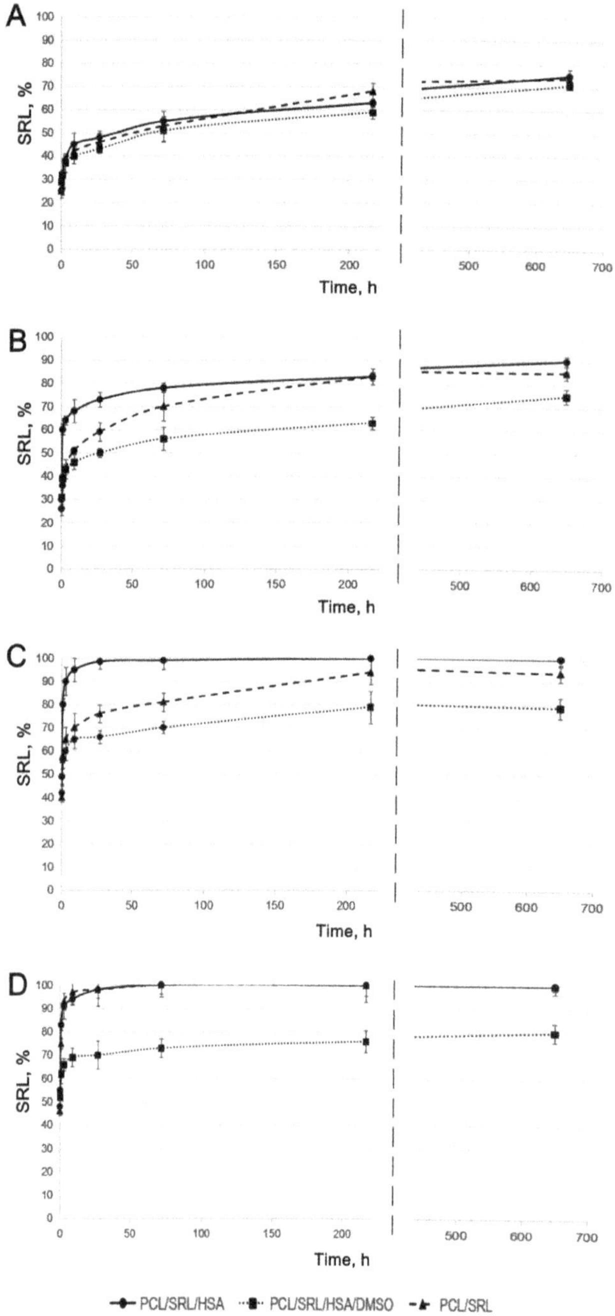

Figure 2. The kinetic curves of SRL release from the matrices. The matrices were incubated (**A**) in PBS without any medium replacement; (**B**) in PBS with medium replacement; (**C**) in BP without any medium replacement; and (**D**) in BP with medium replacement. The data are presented as the means of four independent measurements; the error of the mean did not exceed 7%.

When incubated in BP, SRL was completely released from the PCL/SRL/HSA matrix after 27 h (Figure 2C). The kinetic curve for the PCL/SRL matrix showed a slower SRL release, with saturation achieved over 6–9 days, and a maximal level of 96% was observed at 27 days.

HSA acts as the main transporter of SRL in blood and tissues. HSA has two ligand-specific binding sites for drugs, site I and site II (high-affinity binding and low-affinity binding, respectively) [42,46,47]. SRL binds to albumin at site I more often than at site II. At low BP concentrations, SRL binds to site I. At higher PB concentrations, both sites participate in SRL binding. The affinity association constant for SRL binding to HSA was determined to be 3.99×10^5 M^{-1} [42,46]. Apparently, the SRL resorption on the matrix surface is hindered by the binding of SRL with HSA in BP. It is also possible that other biomolecules in plasma interact with SRL.

Incubation with a BP replacement resulted in an increase in SRL release from the PCL/SRL matrix of up to 100%. According to the SEM data, there were no changes in the fiber structure of PCL/SRL matrices during the 27-day incubation in PBS or BP (Figure 1). This means that long-term SRL release is not associated with matrix structure reorganization or fiber degradation. According to the kinetic study, the PCL/SRL/HSA and PCL/SRL matrices are not capable of prolonged SRL delivery because of rapid SRL release (100% in three days) in BP. The changes in fiber structure and/or SRL distribution inside the fibers are necessary to obtain matrices suitable for prolonged SRL release. Addition of DMSO to the electrospinning blend enabled to obtain porous fibers and retain SRL in the fibers. DMSO is nontoxic, has a boiling point of 189 °C, and dissolves SRL. Moreover, as shown previously, the addition of 3% DMSO to the ES blend significantly reduced paclitaxel release in both PBS and BP [29]. As indicated above, the addition of 3% DMSO to the ES blend resulted in both decreased fiber diameter and pore size. In these matrices, the fiber diameters nearly doubled over the 27 days of incubation owing to the water absorption by HSA. The change in structure of the freshly obtained fibers and those of the fibers after incubation of the PCL/SRL/HSA/DMSO matrices, as well as for the initial and long-term incubation, induced redistribution of SRL in the fibers and obviously affected SRL release. The lowest rate of SRL release in BP (as well as that in PBS) was observed for the matrix containing DMSO. The matrix produced from a 5% PCL/SRL/10% HSA/3% DMSO blend released a maximum of 80% SRL over 27 days, both with and without medium replacement. The data obtained demonstrate the importance of the PCL/SRL/HSA/DMSO matrix as a prospective material for DES coating.

3.3. *Influence of Matrix Elongation on SRL Release*

Considering the twofold elongation of the matrices covering the vascular stents during their installation, the effect of the deformation of the matrices on drug release was studied. To study the influence of twofold elongation on the matrix structures, we used a device that expands the matrix and fixes it in an SEM camera (Figure 3A). This type of deformation differs from the deformation during stent implantation, whereupon uniform deformation of the tube walls occurs, but can simulate the influence of deformation on fiber location and structure. Different areas of the matrices were studied by SEM owing to unequal deformation of the matrices (the maximal deformation was in the center and the minimal deformation was near the clamps) (Figure 3B). In the central area (Figure 3B, 1 and 2; Figure 3C columns 1 and 2), SEM demonstrated that twofold elongation did not lead to a disruption of the fibers, although it was accompanied by the orientation of the fibers along the force vector and compaction of the fiber stack. The structure of the matrix near the clamps with minimal stretching (Figure 3B, 4; Figure 3C column 4) resembles an intact matrix (Figure 1). The intermediate area (Figure 3B, 3; Figure 3C column 3) differs in that fibers are only slightly oriented. After twofold elongation of the matrices and removal of the load, the linear dimensions were reduced by 10–17% (compared with twofold elongation), depending on the type of matrix. According to the SEM data, all obtained matrices are able to withstand elongation during stent installation.

As shown in Figure 4, SRL release from the parent and the twofold expanded matrices were very similar for all matrices under study (Supplementary materials, Table S1). The PCL/SRL/HSA and PCL/SRL matrices present very similar curves of SRL release. SRL was completely released from the

PCL/SRL/HSA and PCL/SRL/TEA matrices, with maximal release in the first 27 h, whereas SRL release from the PCL/SRL/HSA/DMSO matrix was significantly slower, with a maximal level of 80% for 27 days. Data on the SRL release from the matrices with BP replacement demonstrate that twofold elongation barely changes the release kinetics of the matrices (when normalized to the linear/weight/volume characteristics of the matrices). Nevertheless, it is necessary to account for the changes in the areas of the matrices after stent installation when calculating the dose of the drug introduced into the matrix, considering the approximately 1.8-fold increase in the linear size, and thus the decrease in the specific content of the drug in the covered square.

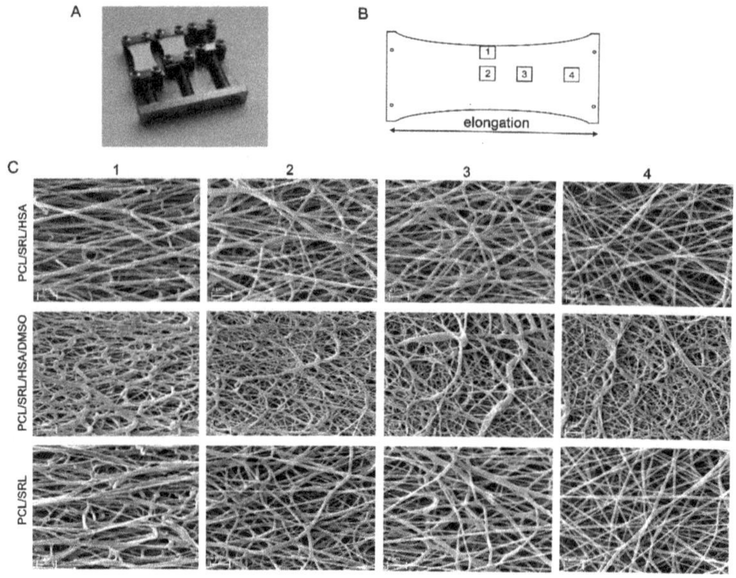

Figure 3. SEM images of the expanded matrices. (**A**) A device allowing one to expand the matrices and fix them in a sputter coater and SEM camera. (**B**) Location of the areas studied by SEM. (**C**) SEM images of different areas of the matrices.

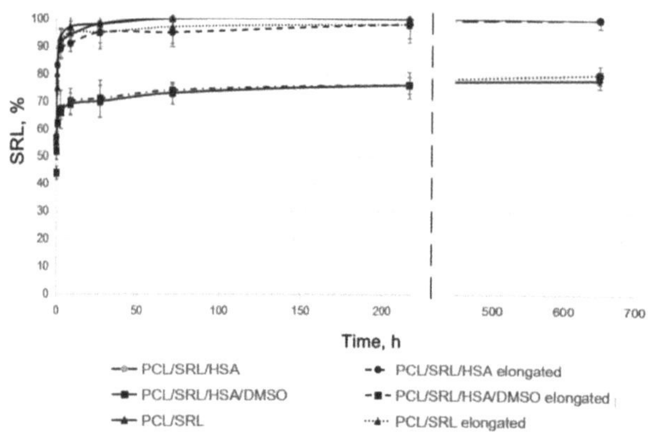

Figure 4. The kinetic curves of SRL release from the expanded matrices. Incubation of the matrices in BP with medium replacement. The data are presented as the means of four independent measurements; error of the mean did not exceed 7%.

4. Conclusions

The structure and physicochemical characteristics of the ES matrices obtained from the solutions of PCL with SRL in HFIP and their blends with HSA, DMSO, and TEA were studied. The matrix produced from 5% PCL/SRL/10% HSA/3% DMSO was shown to be the most suitable for bare-metal stent coating because it is sufficiently strong, exhibited long-term SRL release kinetics, and is thus expected to maintain a concentrations of SRL in the vascular wall that are suitable for antiproliferative activity over a long period. The use of BP increased SRL release. We showed that all matrices can release SRL without fiber degradation. The twofold elongation of the matrices barely changed the SRL release kinetics. Therefore, PCL-based matrices containing HSA, SRL, and DMSO are a promising material for vascular stent coating with prolonged delivery of SRL.

Supplementary Materials: The following are available online at http://www.mdpi.com/1996-1944/13/12/2692/s1, Table S1: Release curves comparison using Fit factors f1 and f2; Table S2: Release curves comparison using Fit factors f1 and f2. Comparison of kinetic curves of SRL release from the parent and the two-fold expanded matrices. Incubation of the matrices in BP with medium replacement.

Author Contributions: Z.K.N., investigation, matrix characterization and SRL release study, and writing—original draft; B.P.C., investigation, matrix production, and SEM study; V.S.C., investigation and contact angle measurements; I.V.R., investigation and SRL labeling; A.A.K., methodology and resources; P.P.L., supervision, project administration, methodology, funding acquisition, and writing—review & editing. All authors have read and agreed to the published version of the manuscript.

Funding: This research was funded by the Russian Scientific Foundation, grant number 18-15-00080, and partially supported by the Russian State funded budget project of ICBFM SB RAS number AAAA-A17-117020210026-2 (support of SEM and H^3-labeling).

Conflicts of Interest: The authors declare no conflict of interest. The funders had no role in the design of the study; in the collection, analyses, or interpretation of data; in the writing of the manuscript; or in the decision to publish the results.

References

1. Costa, M.A.; Simon, D.I. Molecular basis of restenosis and drug-eluting stents. *Circulation* **2005**, *111*, 2257–2273. [CrossRef] [PubMed]
2. Puranik, A.S.; Dawson, E.R.; Peppas, N.A. Recent advances in drug eluting stents. *Int. J. Pharm.* **2013**, *441*, 665–679. [CrossRef] [PubMed]
3. Scott, N.A. Restenosis following implantation of bare metal coronary stents: Pathophysiology and pathways involved in the vascular response to injury. *Adv. Drug Deliv. Rev.* **2006**, *58*, 358–376. [CrossRef] [PubMed]
4. Abizaid, A. Sirolimus-eluting coronary stents: A review. *Vasc. Health. Risk Manag.* **2007**, *3*, 191–201. [CrossRef] [PubMed]
5. Cheng-Lai, A.; Frishman, W.H. Sirolimus-eluting coronary stents: Novel devices for the management of coronary artery disease. *Am. J. Ther.* **2004**, *11*, 218–228. [CrossRef] [PubMed]
6. Piscione, F.; Piccolo, R.; Cassese, S.; Galasso, G.; Chiariello, M. Clinical impact of sirolimus-eluting stent in ST-segment elevation myocardial infarction: A meta-analysis of randomized clinical trials. *Catheter Cardiovasc. Interv.* **2009**, *74*, 323–332. [CrossRef]
7. Sehgal, S.N. Sirolimus: Its discovery, biological properties, and mechanism of action. *Transpl. Proc.* **2003**, *35*, 7S–14S. [CrossRef]
8. Haeri, A.; Osouli, M.; Bayat, F.; Alavi, S.; Dadashzadeh, S. Nanomedicine approaches for sirolimus delivery: A review of pharmaceutical properties and preclinical studies. *Artif. Cells Nanomed. Biotechnol.* **2018**, *46*, 1–14. [CrossRef]
9. Sehgal, S.N.; Camardo, J.S.; Scarola, J.A.; Maida, B.T. Rapamycin (sirolimus, rapamune). *Curr. Opin. Nephrol. Hypertens.* **1995**, *4*, 482–487. [CrossRef]
10. Hafizi, S.; Mordi, V.N.; Andersson, K.M.; Chester, A.H.; Yacoub, M.H. Differential effects of rapamycin, cyclosporine A, and FK506 on human coronary artery smooth muscle cell proliferation and signalling. *Vascul. Pharmacol.* **2005**, *41*, 167–176. [CrossRef]
11. Moes, D.J.; Guchelaar, H.J.; de Fijter, J.W. Sirolimus and everolimus in kidney transplantation. *Drug Discov. Today* **2015**, *20*, 1243–1249. [CrossRef] [PubMed]

12. MacDonald, A.; Scarola, J.; Burke, J.T.; Zimmerman, J.J. Clinical pharmacokinetics and therapeutic drug monitoring of sirolimus. *Clin. Ther.* **2000**, *22*, B101–B121. [CrossRef]

13. Mahalati, K.; Kahan, B.D. Clinical pharmacokinetics of sirolimus. *Clin. Pharmacokinet.* **2001**, *40*, 573–585. [CrossRef] [PubMed]

14. Tada, T.; Byrne, R.A.; Simunovic, I.; King, L.A.; Cassese, S.; Joner, M.; Fusaro, M.; Schneider, S.; Schulz, S.; Ibrahim, T.; et al. Risk of stent thrombosis among bare-metal stents, first-generation drug-eluting stents, and second-generation drug-eluting stents: Results from a registry of 18,334 patients. *JACC Cardiovasc. Interv.* **2013**, *6*, 1267–1274. [CrossRef] [PubMed]

15. Vetrovec, G.W.; Rizik, D.; Williard, C.; Snead, D.; Piotrovski, V.; Kopia, G. Sirolimus PK trial: A pharmacokinetic study of the sirolimus-eluting Bx velocity stent in patients with de novo coronary lesions. *Catheter Cardiovasc. Interv.* **2006**, *67*, 32–37. [CrossRef] [PubMed]

16. Van Der Giessen, W.J.; Sorop, O.; Serruys, P.W.; Peters- Krabbendam, I.; Van Beusekom, H.M.M. Lowering the dose of sirolimus, released from a nonpolymeric hydroxyapatite coated coronary stent, reduces signs of delayed healing. *J. Am. Coll. Cardiol. Intv.* **2009**, *2*, 284–290. [CrossRef]

17. Nakamura, M.; Abizaid, A.; Hirohata, A.; Honda, Y.; Sousa, J.E.; Fitzgerald, P.J. Efficacy of reduced-dose sirolimus-eluting stents in the human coronaryartery: Serial IVUS analysis of neointimal hyperplasia and luminal dimension. *Catheter Cardiovasc. Interv.* **2007**, *70*, 946–951. [CrossRef]

18. Onuma, Y.; Kogame, N.; Serruys, P.W. Unconfirmed very long-term clinical benefit of Xience over Cypher: Should we reset our vision? *JACC Cardiovasc Interv.* **2019**, *12*, 648–650. [CrossRef]

19. Daemen, J.; Wenaweser, P.; Tsuchida, K.; Abrecht, L.; Vaina, S.; Morger, C.; Kukreja, N.; Jüni, P.; Sianos, G.; Hellige, G.; et al. Early and late coronary stent thrombosis of sirolimuseluting and paclitaxel-eluting stents in routine clinical practice: Data from a large two-institutional cohort study. *Lancet* **2007**, *369*, 667–678. [CrossRef]

20. Bangalore, S.; Toklu, B.; Patel, N.; Feit, F.; Stone, G.W. Newer-generation ultrathin strut drug-eluting stents versus older second-generation thicker strut drug-eluting stents for coronary artery disease. *Circulation* **2018**, *138*, 2216–2226. [CrossRef]

21. Tijssen, R.Y.G.; Kraak, R.P.; Lu, H.; Mifek, J.G.; Carlyle, W.C.; Donohoe, D.J.; De Winter, R.J.; Koch, K.T.; Wykrzykowska, J.J. Evaluation of the MiStent sustained sirolimus eluting biodegradable polymer coated stent for the treatment of coronary artery disease: Does uniform sustained abluminal drug release result in earlier strut coverage and better safety profile? *Expert Rev. Med. Devices* **2017**, *14*, 325–334. [CrossRef] [PubMed]

22. Ajmera, P.; Pothineni, R.; Chawla, K.K.; Mantravadi, S.S.; Jariwala, P.V.; Vijan, V.; Vijan, V. Real-world use of ultrathin-strut biodegradable polymer-coated sirolimus-eluting stents in patients with coronary artery disease: 6-month clinical outcomes. *Vasc. Health Risk Manag.* **2019**, *18*, 439–447. [CrossRef] [PubMed]

23. Mauri, L.; Doros, G.; Rao, S.V.; Cohen, D.J.; Yakubov, S.; Lasala, J.; Wong, S.C.; Zidar, J.; Kereiakes, D.J. The OPTIMIZE randomized trial to assess safety and efficacy of the Svelte IDS and RX Sirolimus-eluting coronary stent Systems for the Treatment of atherosclerotic lesions: Trial design and rationale. *Am. Heart J.* **2019**, *216*, 82–90. [CrossRef] [PubMed]

24. Bae, I.H.; Lim, K.S.; Park, D.S.; Shim, J.W.; Lee, S.Y.; Jang, E.J.; Park, J.K.; Kim, J.H.; Jeong, M.H. Sirolimus coating on heparinized stents prevents restenosis and thrombosis. *J. Biomater. Appl.* **2017**, *31*, 1337–1345. [CrossRef] [PubMed]

25. Jelonek, K.; Jaworska, J.; Pastusiak, M.; Sobota, M.; Włodarczyk, J.; Karpeta-Jarzabek, P.; Kaczmarczyk, B.; Kasperczyk, J.; Dobrzyński, P. Effect of vascular scaffold composition on release of sirolimus. *Eur. J. Pharm. Biopharm.* **2018**, *132*, 41–49. [CrossRef] [PubMed]

26. Raval, A.; Parikh, J.; Engineer, C. Mechanism and in vitro release kinetic study of sirolimus from a biodegradable polymeric matrix coated cardiovascular stent. *Ind. Eng. Chem. Res.* **2011**, *50*, 9539–9549. [CrossRef]

27. Livingston, M.; Tan, A. Coating techniques and release kinetics of drug-eluting stents. *J Med. Device.* **2019**, *10*, 1. [CrossRef]

28. Zhang, Y.; Wang, J.; Xiao, J.; Fang, T.; Hu, N.; Li, M.; Deng, L.; Cheng, Y.; Zhu, Y.; Cui, W. An electrospun fiber-covered stent with programmable dual drug release for endothelialization acceleration and lumen stenosis prevention. *Acta Biomater.* **2019**, *94*, 295–305. [CrossRef]

29. Kuznetsov, K.A.; Stepanova, A.O.; Kvon, R.I.; Douglas, T.E.L.; Kuznetsov, N.A.; Chernonosova, V.S.; Zaporozhchenko, I.A.; Kharkova, M.V.; Romanova, I.V.; Karpenko, A.A.; et al. Electrospun produced 3D matrices for covering of vascular stents: Paclitaxel release depending on fiber structure and composition of the external environment. *Materials* **2018**, *11*, 2176. [CrossRef]
30. Rychter, M.; Baranowska-Korczyc, A.; Luleka, J. Progress and perspectives in bioactive agent delivery via electrospun vascular grafts. *RSC Adv.* **2017**, *7*, 32164. [CrossRef]
31. Mirbagheri, M.S.; Mohebbi-Kalhori, D.; Jirofti, N. Evaluation of mechanical properties and medical applications of polycaprolactone small diameter artificial blood vessels. *Int. J. Basic Sci. Med.* **2017**, *2*, 58–70. [CrossRef]
32. Rychter, M.; Milanowski, B.; Grzeskowiak, B.F.; Jarek, M.; Kempinski, M.; Coy, E.L.; Borysiak, S.; Baranowska-Korczyc, A.; Lulek, J. Cilostazol-loaded electrospun three-dimensional systems for potential cardiovascular application: Effect of fibers hydrophilization on drug release, and cytocompatibility. *J. Colloid Interface Sci.* **2019**, *15*, 310–327. [CrossRef] [PubMed]
33. Nair, L.S.; Laurencin, C.T. Biodegradable polymers as biomaterials. *Prog. Polym. Sci.* **2007**, *32*, 762–798. [CrossRef]
34. Michel, S.A.; Knetsch, M.L.; Koole, L.H. Adsorption of albumin on flax fibers increases endothelial cell adhesion and blood compatibility in vitro. *J. Biomater Sci. Polym. Ed.* **2014**, *25*, 698–712. [CrossRef] [PubMed]
35. Chernonosova, V.S.; Kvon, R.I.; Stepanova, A.O.; Larichev, Y.V.; Karpenko, A.A.; Chelobanov, B.P.; Kiseleva, E.V.; Laktionov, P.P. Human serum albumin in electrospun PCL fibers: Structure, release, and exposure on fibersurface. *Polym. Adv. Technol.* **2017**, *28*, 819–827. [CrossRef]
36. Shen, M.; Martinson, L.; Wagner, M.S.; Castner, D.G.; Ratner, B.D.; Horbett, T.A. PEO-like plasma polymerized tetraglyme surface interactions with leukocytes and proteins: In vitro and in vivo studies. *J. Biomater. Sci. Polym. Ed.* **2002**, *13*, 367–390. [CrossRef] [PubMed]
37. Denizli, F.K.; Guven, O. Competitive adsorption of blood proteins on gamma-irradiated-polycarbonate fifilms. *J. Biomater. Sci. Polym. Ed.* **2002**, *13*, 127–139. [CrossRef]
38. Sidorov, V.N.; Polak, Y.V.; Laktionov, P.P.; Roshcke, V.V.; Kist, A.G. Method of production of tritium labeled organic compounds and the device for its implementation. SU Patent 1823961 A3, 18 January 1991.
39. Cardiovascular Implants-Tubular Vacuum Prostheses. International Patent Application No. ISO/FDIS 7198:1998, July 1998.
40. Tiwari, A.P.; Joshi, M.K.; Park, C.H.; Kim, C.S. Nano-nets covered composite nanofibers with enhanced biocompatibility and mechanical properties for bone tissue engineering. *J. Nanosci. Nanotechnol.* **2018**, *18*, 529–537. [CrossRef]
41. Carter, D.C.; He, X.M.; Munson, S.H.; Twigg, P.D.; Gernert, K.M.; Broom, M.B. Three-dimensional structure of human serum albumin. *Science* **1989**, *244*, 1195–1198. [CrossRef]
42. Jr, T.P. Ligand binding by albumin. In *All about Albumin*; Academic Press: San Diego, CA, USA, 1995.
43. Chernonosova, V.S.; Gostev, A.A.; Gao, Y.; Chesalov, Y.A.; Shutov, A.V.; Pokushalov, E.A.; Karpenko, A.A.; Laktionov, P.P. Mechanical properties and biological behavior of 3D matrices produced by electrospinning from protein-enriched polyurethane. *BioMed Res. Int.* **2018**, *2018*, 1380606. [CrossRef]
44. Yasukawa, T.; Ogura, Y.; Sakurai, E.; Tabata, Y.; Kimura, H. Intraocular sustained drug delivery using implantable polymeric devices. *Adv. Drug Deliv. Rev.* **2005**, *57*, 2033–2046. [CrossRef] [PubMed]
45. Kim, M.-S.; Kim, J.-S.; Park, H.J.; Cho, W.K.; Cha, K.-H.; Hwan, S.-J. Enhanced bioavailability of sirolimus via preparation of solid dispersion nanoparticles using a supercritical antisolvent process. *Int. J. Nanomed.* **2011**, *6*, 2997–3009.
46. Rahman, M.; Prianka, F.; Shohel, M.; Mazid, M.A. Interaction of palmitic acid with metoprolol succinate at the binding sites of bovine serum albumin. *Adv. Pharm. Bull.* **2014**, *4*, 379–383. [PubMed]
47. Khodaei, A.; Bolandnazar, S.; Valizadeh, H.; Hasani, L.; Zakeri-Milani, P. Interactions between sirolimus and anti-inflammatory drugs: Competitive binding for human serum albumin. *Adv. Pharm. Bull.* **2016**, *6*, 227–233. [CrossRef] [PubMed]

materials

MDPI

Article

Multivariate Design of 3D Printed Immediate-Release Tablets with Liquid Crystal-Forming Drug—Itraconazole

Witold Jamróz [1,*], Jolanta Pyteraf [1], Mateusz Kurek [1,*], Justyna Knapik-Kowalczuk [2,3], Joanna Szafraniec-Szczęsny [1], Karolina Jurkiewicz [2,3], Bartosz Leszczyński [4], Andrzej Wróbel [4], Marian Paluch [2,3] and Renata Jachowicz [1]

1 Department of Pharmaceutical Technology and Biopharmaceutics, Jagiellonian University Medical College, Medyczna 9, 30-688 Krakow, Poland; jolanta.pyteraf@uj.edu.pl (J.P.); joanna.szafraniec@uj.edu.pl (J.S.-S.); renata.jachowicz@uj.edu.pl (R.J.)

2 Division of Biophysics and Molecular Physics, Institute of Physics, University of Silesia, Uniwersytecka 4, 40-007 Katowice, Poland; justyna.knapik-kowalczuk@us.edu.pl (J.K.-K.); karolina.jurkiewicz@us.edu.pl (K.J.); marian.paluch@us.edu.pl (M.P.)

3 Silesian Center for Education and Interdisciplinary Research, University of Silesia, 75 Pulku Piechoty 1a, 41-500 Chorzow, Poland

4 Marian Smoluchowski Institute of Physics, Jagiellonian University, Łojasiewicza 11, 30-348 Krakow, Poland; bartosz.leszczynski@uj.edu.pl (B.L.); andrzej.wrobel@uj.edu.pl (A.W.)

* Correspondence: witold.jamroz@uj.edu.pl (W.J.); mateusz.kurek@uj.edu.pl (M.K.); Tel.: +48-12-62-05-600 (W.J. & M.K.)

Received: 8 October 2020; Accepted: 2 November 2020; Published: 4 November 2020

Abstract: The simplicity of object shape and composition modification make additive manufacturing a great option for customized dosage form production. To achieve this goal, the correlation between structural and functional attributes of the printed objects needs to be analyzed. So far, it has not been deeply investigated in 3D printing-related papers. The aim of our study was to modify the functionalities of printed tablets containing liquid crystal-forming drug itraconazole by introducing polyvinylpyrrolidone-based polymers into the filament-forming matrices composed predominantly of poly(vinyl alcohol). The effect of the molecular reorganization of the drug and improved tablets' disintegration was analyzed in terms of itraconazole dissolution. Micro-computed tomography was applied to analyze how the design of a printed object (in this case, a degree of an infill) affects its reproducibility during printing. It was also used to analyze the structure of the printed dosage forms. The results indicated that the improved disintegration obtained due to the use of Kollidon®CL-M was more beneficial for the dissolution of itraconazole than the molecular rearrangement and liquid crystal phase transitions. The lower infill density favored faster dissolution of the drug from printed tablets. However, it negatively affected the reproducibility of the 3D printed object.

Keywords: 3D printing; fused deposition modeling; hot-melt extrusion; solid dosage forms; itraconazole

1. Introduction

Additive manufacturing has huge potential to revolutionize the methods of drug delivery system formation. It was proven for mass-scale drug production by Aprecia Pharmaceuticals, which registered the first 3D printed drug, Spritam®, in 2015. However, the use of additive manufacturing also enables the preparation of small batches of customized, on-demand-prepared formulations—for example, in the treatment of patients with rare diseases or for clinical trials. The great applicability of 3D printing (3DP) in the pharmaceutical field results from the simplicity of object shape modification, which allows the production of dosage forms of complex shape and internal structure, containing one or more active pharmaceutical ingredients (APIs) [1,2]. Moreover, the differences in shape and infill density of tablets, which cannot be achieved in compressed tablets, lead to alternation in the surface-to-volume ratio and allow us to produce printlets with desired drug dosages and dissolution profiles [3–5]. Although the issue of the correlation between the internal structure of printed tablets and their properties, particularly the dissolution characteristics, has been explored by several research teams, there is still deficiency in studies on the actual microstructure and quality of printed objects and the mechanisms driving the release of the drug from printed dosage forms [6–9].

In the case of nearly all 3DP methods, the object is built layer by layer based on the computer aided design (CAD) model. However, various printing technologies vary between each other regarding used materials and process conditions such as temperature. The 3D printing methods can operate with a powder, which is bound with a liquid binder or sintered with a laser, a photosensitive resin, a thermoplastic material, or a semi-solid formulation extruded through the printer nozzle. Several techniques, such as stereolithography [10–14], selective laser sintering [8,15] digital light processing [16,17], binder jetting, [18,19], and extrusion-based methods including direct powder extrusion [20,21], semi-solid extrusion [22–26], and fused deposition modeling (FDM) [27,28], have been investigated for application in the pharmaceutical industry. The 3DP methods which can be introduced in the high-scale manufacturing process should be characterized by the high-speed production of uniform objects [29,30]. In the case of most of the abovementioned printing methods, process conditions may cause amorphization of the active ingredient, which increases its solubility [31,32].

Various dosage forms, such as orodispersible films [33], mucoadhesive films [34], immediate and modified-release tablets [35,36], capsules [37,38], implants [39], or even formulations imitating sweets [40], have been recently developed using fused deposition modeling. In the printed dosage forms, drug release modification is obtained mostly by selecting either the filament-forming polymers characterized by suitable pH-dependent solubility [41] or the printlet shape and geometry, i.e., the presence of channels [42], empty cavities (floating tablets) [43,44], variations in the infill degree or shape as well as the use of shape-memory polymers to prepare retentive drug delivery systems (4D printing) [45,46]. Despite the fabrication of dosage forms by means of 3DP, this technique can be used for capsular shell fabrication [47] to control the API's dissolution process as well as mold preparation to create custom-made, patient-oriented drugs [48]. The 3D printed molds can be also used in a range of science and technology sectors including electrochemical electrical applications—for example, flexible sensor prototypes [49,50].

The application of FDM printing technology in the manufacturing of dosage forms requires the use of previously prepared drug-loaded filament. Filaments are produced mostly in the hot-melt extrusion process (HME), which is also the method applied to increase drug solubility. During this process, a mixture of drug and thermoplastic polymer is heated and blended, and the molten mass is pushed through a nozzle to form a filament [51]. Instead of drugs, other substances can be used in the HME process, e.g., insoluble hydroxyapatite for filament fabrication, which can be used in bone tissue engineering [52]. One of the most important advantages is that this process does not require the use of organic solvents, such as the preparation of amorphous solid dispersion (ASD) by spray drying. However, HME operates at high temperatures, which are required to melt the formulation components [53]. In some cases, it is necessary to add plasticizers to the formulation to lower the process temperature in order to protect the thermolabile active ingredient and improve filament printability [36,54,55]. The combination of HME and FDM can induce phase transitions, including amorphization, which results in increased drug solubility. Further drug dissolution modification can be also achieved by changing the shape and surface of the printed dosage form [26].

Itraconazole (ITR) is an oral antifungal agent used in the treatment of systemic and superficial fungal infections, commercially available in the form of 65 mg and 100 mg capsules, 200 mg tablets, and 10 mg/mL solutions. It is a highly lipophilic, weakly alkaline drug with very low water solubility of 1 ng/mL at pH 7 and 4 μg/mL at pH 1. ITR is classified as a Biopharmaceutics Classification System (BCS) class II substance [56], which means it has solubility-limited bioavailability. The drug exhibits three polymorphs varying in stability and solubility [57]. Moreover, ITR can form liquid crystals, which are particularly interesting from the perspective of pharmaceutical sciences. Liquid crystals can adopt various molecular arrangements (nematic and smectic in the case of ITR), which affect the free energy of the system and thus the dissolution performance. Due to the relatively high glass transition temperature (T_g = 59 °C), ITR can be also transformed into a stable amorphous state, usually in the form of amorphous solid dispersions with polymers or co-amorphous systems with small molecules [58].

Soluplus® [59–61], Eudragit® L [62], polyvinylpyrrolidone (PVP) [63], Kollidon® VA64 [64,65], polyvinyl alcohol (PVA) [65,66], as well as semi-synthetic cellulose derivatives such as hydroxypropyl cellulose [67] and hydroxypropyl methylcellulose acetate succinate [53,54,68–70], are examples of pharmaceutical polymers tested for preparing itraconazole amorphous solid dispersions (ASD) and also suitable as filament-forming polymers for FDM. Although many papers described the formation of amorphous solid dispersions with ITR, including the use of the hot-melt extrusion process [61], only two considered the formation of dosage forms using 3D printing. Kimura et al. reported that it is possible to use fused deposition modeling to prepare zero-order sustained-release floating tablets containing itraconazole [43]. They were able to control floating time by printing tablets with empty cavities inside and to modify the drug dissolution rate by changing the tablet surface and wall thickness. Goyanes et al. prepared tablets containing amorphous solid dispersions of itraconazole in different grades of hydroxypropylcellulose using direct powder extrusion 3D printing—a novel, single-step 3D printing process. In contrast to FDM, this 3D printer tool head is equipped with single screw extruder, which allows it to print directly using mixed powders or pellets, without preparing filaments [20].

In this paper, we describe for the first time the liquid crystal phase transitions of itraconazole in 3D printed tablets. The drug was combined with polymers, formed into filaments via hot-melt extrusion and then printed using fused deposition modeling technology. The filaments were based on poly(vinyl alcohol), a water-soluble semi-crystalline polymer known for its superior printability. The two PVP-based polymers were also added to the filament-forming mixture to introduce the additional functionalities into the printed matrices. Kollidon® VA64 was supposed to modify the physicochemical properties—the molecular arrangement in particular (analyzed using thermal analysis and X-ray diffractometry)—and Kollidon® CL-M was added to modify drug dissolution due to the improved tablet disintegration. We performed deep micro-computed tomography (μ-CT) analysis as the first attempt to analyze how the design of a printed object (degree of an infill) affects its reproducibility during printing. It was also used to analyze the structure of the printed dosage forms to support the dissolution data. To clearly understand the advantages of extrusion and printing processes, drug dissolution from printed formulations was compared with tablets having similar composition, obtained by the compression of either raw powders or milled filament.

2. Materials and Methods

2.1. Materials

Itraconazole (ITR, 1-(butan-2-yl)-4-{4-[4-(4-{[(2R,4S)-2-(2,4-dichlorophenyl)-2-[(1H-1,2,4-triazol-1-yl)methyl]-1,3-dioxolan-4-yl]methoxy}phenyl)piperazin-1-yl]phenyl}-4,5-dihydro-1H-1,2,4-triazol-5-one, 99.8%, Henan Tianfu Chemical Co., Ltd., Zhengzhou, China) served as a model drug. Poly(vinyl alcohol) (PVA, Parteck® MXP, Merck®- KGaA, Darmstadt, Germany), copovidone (K/VA, Kollidon® VA64, BASF®, Ludwigshafen, Germany), crospovidone (K/CL, Kollidon® CL-M, BASF®, Ludwigshafen, Germany) were utilized as the matrix-forming polymers to prepare both filaments and 3D printed tablets. Talc (Fagron®, Kraków, Poland) and magnesium stearate (Avantor® Performance Materials, Gliwice, Poland) were added to tablets prepared by compression in tablet press. Hydrochloric acid (Merck® KGaA, Darmstadt, Germany) and potassium chloride (Avantor® Performance Materials, Gliwice, Poland) were used as dissolution media ingredients. Water used in all experiments was produced by Elix 15UV Essential reversed osmosis system (Merck® KGaA, Darmstadt, Germany).

2.2. *Preparation of Drug-Loaded Filaments*

Filaments were extruded using a 40D, 12-mm co-rotating twin-screw extruder (RES-2P/12A Explorer, Zamak Mercator®, Skawina, Poland) equipped with a gravimetric feeder MCPOWDER® (Movacolor®, Sneek, The Netherlands) and an air-cooled conveying belt (Zamak Mercator®, Skawina, Poland). The mixtures of itraconazole and matrix-forming polymers, of the composition presented in Table 1, and the total mass equal to 200 g were extruded through a 1.75 mm die at 160 °C. The feeding rate was set to approximately 70 g/h, which resulted in the linear filament extrusion speed of 25 m/h. The barrel temperature varied from 40 to 190 °C. The optimized temperature profile and screw configuration are presented in Figure 1.

Table 1. Composition of the filaments.

Formulation	Itraconazole	Poly(vinyl alcohol)	Copovidone	Crospovidone
PVA		80%	-	-
PVA_K/VA	20%	56%	24%	-
PVA_K/CL		76%	-	4%

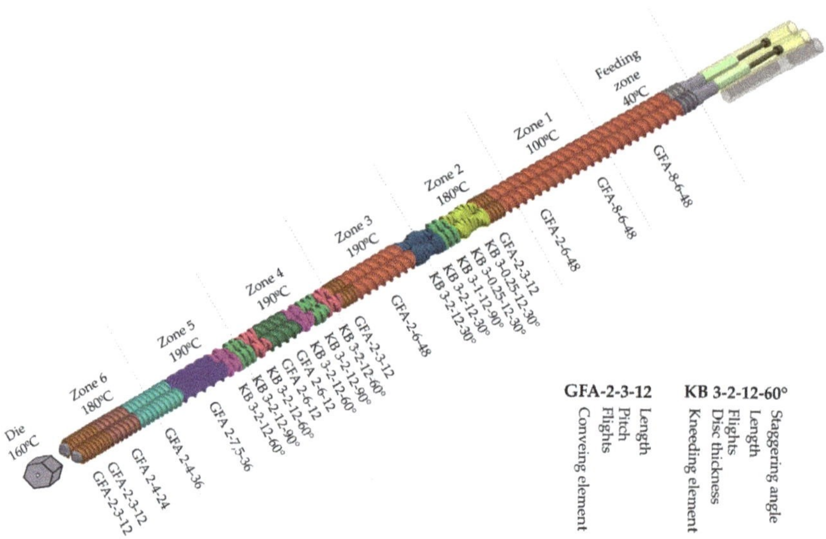

Figure 1. Screw configuration and temperature profile.

2.3. *Evaluation of Filament Properties*

The diameter uniformity of the obtained filament was evaluated using a Mitutoyo® micrometer screw (Kawasaki, Japan). Six randomly selected points were measured. Mechanical properties were assessed in stretching test performed with an EZ-SX tensile tester (Shimadzu®, Kioto, Japan). The measurements were performed six times for each type of filament. Randomly selected pieces of filament, 100 mm in length, were placed in the tensile tester's jaws and stretched up to breakage. Hardness and elasticity of the filaments were determined based on the measurements of tensile strength and Young's modulus.

2.4. *Determination of Itraconazole Content in the Obtained Filament*

Six randomly selected and accurately weighed pieces of filament were placed in conical flasks filled with 25 mL of a mixture of methanol and 0.1 M HCl of pH 1.2 (1:1 v/v) and shaken for 24 h using a Memmert® water bath (WNB 22, Schwabach, Germany). The drug concentration was assayed at λ = 255 nm using a Shimadzu® UV-1800 spectrophotometer (Kioto, Japan). The specificity of the analytical method was verified. There was no sign of interference between the drug and excipients at the analytical wavelength.

2.5. Preparation of 3D Printed Tablets

The Blender® 2.79b software was used to design the models of the oblong tablets (Blender Foundation, Amsterdam, The Netherlands). The basic model was 20 mm long and 10 mm wide. The model height varied from 2.4 to 3.65 mm, which was related to the number of 3D printed layers. Voxelizer® slicing software (version 1.4.18, ZMorph®, Wroclaw, Poland) was applied to define the height and the width of the single layer path. The 3D model was imported in stl format and divided into layers before printing. The thickness of the first layer was equal to 0.2 mm to improve the adhesion of the print to the printer bed, whereas the height of the subsequent layers was 0.15 mm. The path width was equal to the diameter of the printing nozzle, i.e., 0.4 mm. One outline and rectilinear infill (density of 20%, 35%, and 60%) were designed for the printing process. Each tablet was composed of 50 mg of ITR and 200 mg of polymer carriers (Table 1). The tablets were printed by an FDM ZMorph® 2.0 S personal fabricator (Wroclaw, Poland) equipped with a 1.75 mm commercially available printhead. Printing temperature was 205 °C. The tablets were printed with a 10–15 mm/s printing speed. The temperature of building platform was 40 °C.

2.6. Preparation of Tablets by Filament Compression (HME Tablets)

For comparison purposes, filament milled in a Tube Mill 100 control (IKA®, Staufen, Germany) and raw compounds were compressed in a Korsch® EK0 single-punch tablet press (Berlin, Germany). The composition of the tablets was similar to 3D printed tablets; each tablet was composed of 50 mg of ITR and 200 mg of polymer mixture. Additionally, the blends contained 12.5 mg of a talc and magnesium stearate mixture (9:1 w/w), which played the role of glidant and lubricant, respectively.

2.7. Preparation of Directly Compressed Tablets (DC Tablets)

Powder blends composed of 3DP tablet ingredients with the addition of the talc and magnesium stearate mixture (9:1 w/w) were compressed using Korsch® EK0 single-punch tablet press (Berlin, Germany) for comparison purposes, to investigate the impact of technological processes on the ITR dissolution profile.

2.8. Micro-Computed Tomography

Micro-computed tomography (μ-CT) analysis was performed using a SkyScan® 1172 microtomograph (Bruker®, Billerica, MA, USA). It was applied to examine the structure of the 3DP tablets with 20%, 35%, and 60% of infill and to verify the repeatability of printing process (the data collected for three tablets with 35% of infill were compared). The image pixel size was 6.9 μm for measurements of all samples. A cone beam reconstruction software program (Nrecon SkyScan®, Bruker®, Billerica, MA, USA) based on the Feldkamp algorithm was used for the reconstruction of the projections. A CT-Analyser® (SkyScan®, Bruker®, Billerica, MA, USA) was used for binarization purposes. The procedure was based on density distribution histograms collected for the whole sample volume. A CT-Analyser® was also used for the characterization of the morphological features of the tablets, their volume, and surface. CTVox® software (Bruker®, Billerica, MA, USA) was applied to present the 3D results.

2.9. Differential Scanning Calorimetry (DSC)

Thermodynamic properties of neat ITR, PVA, K/VA, K/CL, and their mixtures in the form of filaments and 3DP tablets were examined using a DSC 1 STAR^e System (Mettler-Toledo®, Greifensee,

Switzerland) equipped with an HSS8 ceramic sensor with 120 thermocouples and liquid nitrogen cooling station. Zinc and indium standards were used for the temperature and enthalpy calibration. The samples were measured in an aluminum, pinned crucible (40 mL). The samples were heated with a rate of 10 K/min. The experiments were performed in nitrogen atmosphere with a gas flow of 60 mL/min.

2.10. X-Ray Powder Diffraction (XRD)

A Rigaku Denki® D/MAX Rapid II-R (Tokyo, Japan) equipped with a rotating Ag anode and an image plate detector in the Debye–Scherrer geometry was used for the X-ray diffraction measurements. Graphite (002) crystal was used to monochromatize the incident radiation ($\lambda_{K\alpha}$ = 0.5608 Å). The width of the X-ray beam at the sample was 0.3 mm. The samples were pulverized before the experiment and measured at room temperature, in glass capillaries with a diameter of 1.5 mm and wall thickness of 0.01 mm. The background intensity from empty capillary was subtracted. The obtained two-dimensional diffraction patterns were converted into one-dimensional functions of intensity versus the scattering vector.

2.11. Dissolution Studies

The dissolution of ITR from tablets was determined in 1000 mL of 0.1 M HCl with the addition of KCl, in the pharmacopeial paddle apparatus (Vision® G2 Elite 8, Hanson Research®, Chatsworth, CA, USA) equipped with a VisionG2 AutoPlus autosampler. Stainless steel, spring-like sinkers were used to prevent tablet floating. The samples were filtered and analyzed on-line at 255 nm at predetermined periods using a UV-1800 spectrophotometer (Shimadzu®, Kioto, Japan) equipped with flow-through cuvettes. Three repetitions for each sample were carried out. The results represent the averaged results and the standard deviations (mean ± SD).

2.12. Solubility Study

An excess of physical mixture (PM), extrudate (HME), and printed systems (3DP) were dispersed in 20 mL of 0.1 MHCl and shaken at ambient temperature using a KS 130 basic orbital shaker (IKA®, Staufen im Breisgau, Germany). After 48h, the samples were filtered through a 0.45 μm Chromafil® Xtra CA-45/25 membrane filter and analyzed spectrophotometrically at λ = 255 nm (UV-1800 Shimadzu®, Kioto, Japan). The reported data represent the averages from three series of measurements with standard deviations (SD).

3. Results

3.1. Evaluation of the Filaments

All prepared filaments were made using a PVA as a filament-forming polymer, a semi-crystalline polymer of molecular weight equal to 32 kDa with 87–89% hydrolysis grade, having a glass transition temperature, melting point, and degradation temperature of 40–45 °C, 170 °C, and ≥250 °C, respectively [66]. The obtained itraconazole-loaded filaments were opaque and creamy in color. The diameter of the filaments was kept at a constant level; however, in the case of the PVA_K/CL filament, the diameter variations were higher than 0.05 mm, which is considered as a maximum acceptable deviation from the declared diameter [71]. The itraconazole content and its uniformity were satisfactory. All the API-loaded filaments were tested for their tensile strength and elasticity, which were found to be critical quality attributes in term of printability. The results are presented in Table 2. It was found that the addition of copovidone and crosslinked PVP resulted in a decrease in the tensile strength and Young's modulus of the filaments. All the prepared filaments were able to be printed with a ZMorph® 2.0 S 3D printer immediately after extrusion and after storage in zipper storage bags.

Table 2. Hot-melt extruded filament characteristics.

Filament Composition	Diameter ± SD (mm)	Itraconazole Content ± SD (%)	Tensile Strength ± SD (MPa)	Young's Modulus ± SD (MPa)
PVA	1.70 ± 0.02	19.67 ± 0.43	49.0 ± 10.3	2641.1 ± 144.4
PVA_K/CL	1.68 ± 0.07	19.60 ± 0.34	52.6 ± 19.8	2771.1 ± 347.2
PVA_K/VA	1.69 ± 0.05	19.20 ± 0.33	28.2 ± 7.1	2042.1 ± 256.3

In Figure 2, the differences in the mechanical characteristics are presented. The Young's modulus corresponds to the slope of the curve in the elastic behavior region.

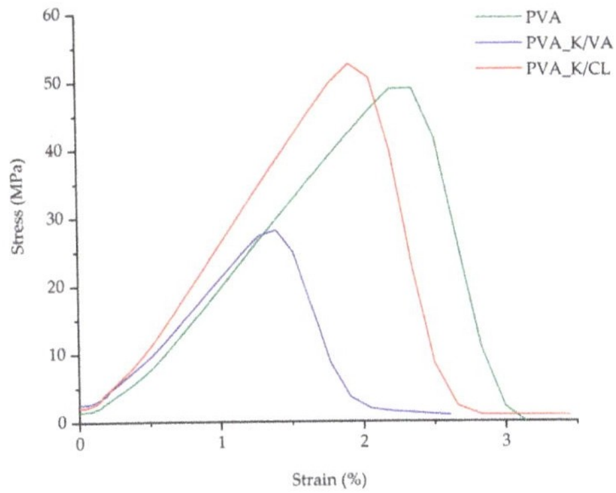

Figure 2. Comparison of the mechanical strength and resilience of the filaments.

3.2. *Thermal and Structural Properties of the Filaments and 3DP Tablets*

To investigate how the employed polymers modify the thermal properties of neat ITR, the systems, prepared in the form of both filaments and 3DP tablets, were measured (after pulverization) by means of DSC. The samples were examined in the temperature range from 273 to 453 K at a heating rate of 10 K/min. In Figure 3, the obtained DSC traces are compared to the thermogram of the neat, quench-cooled ITR. Because the used PVA polymer has a lower glass transition temperature than ITR (Table 3) (T_g of neat PVA and ITR are equal to 313 and 332 K, respectively), the plasticization effect was observed. Interestingly, the DSC thermograms of the same compositions with different forms, filament or 3DP, differ from each other. As can be seen in Figure 3, the thermograms of the 3DP tablets are characterized by: (i) a shift in glass transition temperature towards lower values when compared with filament and (ii) the appearance of an additional, very broad endothermal event in the vicinity of 320 to 420 K. The observed differences suggest that the 3DP tablets also contain water in addition to API and polymers. Water exerts a plasticization effect on the samples and evaporates at temperatures from in the range of 320 to 420 K.

When the neat ITR is heated above its glass transition temperature, on the DSC thermogram, one can distinguish two endothermal processes associated with the liquid crystal (LC) phase transitions. The thermal event located at 348 K reflects transition from smectic (Sm) to nematic (N) LC alignment, while at 364 K, ITR loses the nematic order and becomes an isotropic (I) liquid. The performed experiments reveal that the employed polymers shift to lower temperatures for both Sm-N and N-I phase transition. The determined, based on calorimetric studies, values of T_g, $T_{Sm\text{-}N}$, and $T_{N\text{-}I}$ for all investigated systems are compared in Table 3. It is worth noting that in one of the examined systems (PVA_K/VA), regardless of the applied technological process, the lack of the nematic phase was observed (i.e., the N-I endothermal event was not registered by means of DSC).

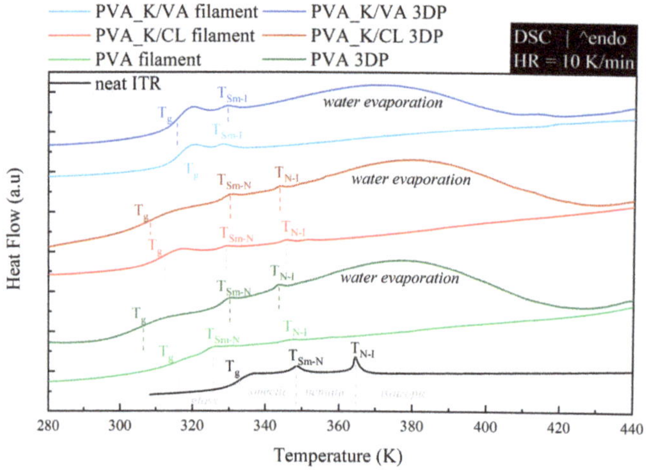

Figure 3. DSC thermograms of neat ITR and its mixtures with PVA, PVA_K/VA, and PVA_K/CL prepared in two forms: filament and 3DP tablet.

Table 3. Comparison of values of T_g, $T_{Sm\text{-}N}$, and $T_{N\text{-}I}$ of neat ITR and its mixtures with PVA, PVA_K/VA, and PVA_K/CL which were prepared in two forms: filament and 3DP tablet.

Sample	T_g (K)	$T_{Sm\text{-}N}$ (K)	$T_{N\text{-}I}$ (K)
Neat ITR	332	348	364
PVA filament	315	326	347
PVA 3DP tablet	306	330	344
PVA_K/CL filament	312	329	346
PVA_K/CL 3DP tablet	308	330	344
PVA_K/VA filament	317	328 ($T_{Sm\text{-}I}$)	
PVA_K/VA 3DP tablet	315	330 ($T_{Sm\text{-}I}$)	

In order to investigate whether the employed polymers indeed modify the ITR's LC alignment, both the neat ITR as well as the pulverized 3DP tablets were measured by wide-angle X-ray diffraction (XRD) technique. The comparison of the scattering patterns collected at room temperature for neat ITR and pulverized tablets containing either PVA, PVA_K/VA, or PVA_K/CL is presented in Figure 4. The presented XRD patterns demonstrate that the polymers affect the LC order in ITR. As can be seen, samples containing PVA or PVA_K/CL reveal less intense peaks at around 0.22, 0.45, and 0.68 $Å^{-1}$, which are indicators of smectic layering [72]. In the case of the system containing K/VA, the reduction in the intensity of the peaks at 0.22 and 0.68 $Å^{-1}$ is combined with the disappearance of the peak at 0.45 $Å^{-1}$. These results indicate that the layered structure in ITR is medicated by the employed additives.

3.3. Micro-Computed Tomography Studies of Tablets

The dimensions and masses of 3DP tablets corresponded to predefined values. The average tablet mass ranged from 239.73 to 253.05 mg. Tablet length varied from 19.85 to 20.15 mm, whereas height ranged from 1.78 to 3.65 mm. The real layer height was from 0.142 to 0.158 mm and was calculated by dividing the tablet height by the number of layers, given the fact that the first layer was 0.2 mm (Table 4). Digital photos of 3D printed tablets can be found in the Supplementary Materials associated with this article (Figures S1–S3, Supplementary Materials).

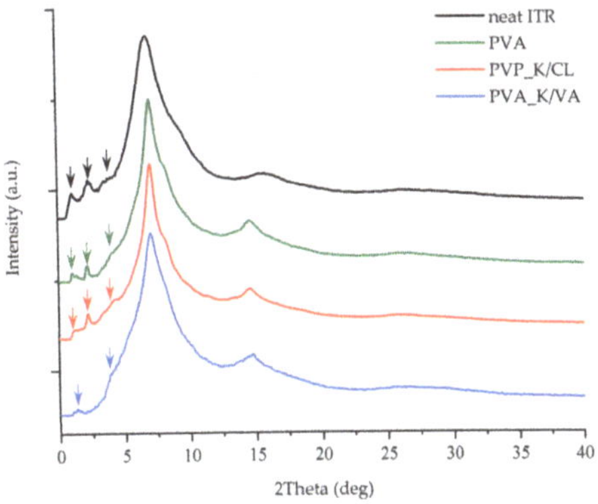

Figure 4. XRD diffraction patterns of neat ITR and its mixtures with PVA, PVA_K/VA, and PVA_K/CL in an initial form of 3DP tablet.

Table 4. Parameters of 3D printed tablets.

Polymers	Infill (%)	Mass (mg)	Width (mm)	Length (mm)	Height (mm)	Number of Layers	Real Layer Height (mm)
PVA	35	252.82 ± 4.16	10.18 ± 0.03	20.15 ± 0.03	2.34 ± 0.03	16	0.142
PVA_K/VA	35	253.05 ± 3.67	10.08 ± 0.01	20.09 ± 0.01	2.89 ± 0.05	20	0.142
PVA_K/CL	35	250.12 ± 4.52	9.98 ± 0.02	19.85 ± 0.12	2.67 ± 0.03	17	0.154
PVA_K/CL	20	244.12 ± 5.77	9.96 ± 0.03	19.86 ± 0.09	3.65 ± 0.03	24	0.150
PVA_K/CL	60	239.73 ± 3.01	9.99 ± 0.05	20.05 ± 0.03	1.78 ± 0.02	11	0.158

Based on the 3D tablet images obtained from Voxelizer slicing software (Figure 5) and predefined settings of the path size, the theoretical volume of 3DP PVA_K/CL tablets was calculated. The values varied from 184.4 mm^3 for T_20 tablets to 195.6 mm^3 for T_60 and 195.9 mm^3 for T_35 tablets.

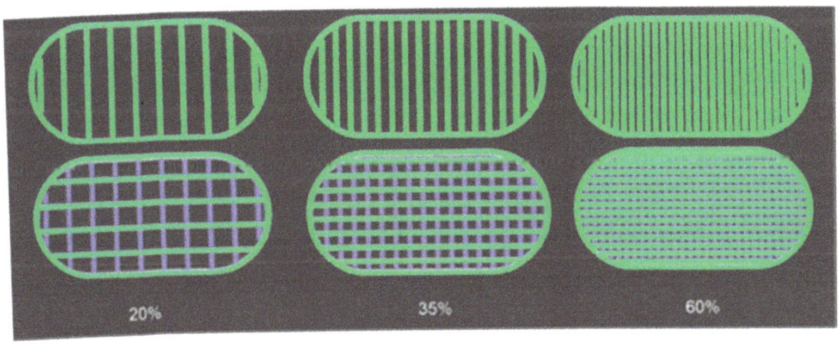

Figure 5. Images of PVA_K/CL tablet layers obtained from Voxelizer software.

The morphology of the PVA_K/CL printed tablets was verified by the μCT scans. Tablets with 20% of infill had the highest object volume (236 mm^3) and the highest open pore volume (485 mm^3). Medium pore size (structure separation) was 1.11 mm, whereas the average structure thickness was 0.25 mm. Tablets with 60% of infill were characterized by the lowest values of object volume (202 mm^3) and pore volume (134 mm^3) as well as structure separation (0.19 mm) and pore size (0.25 mm). Aforementioned parameters for tablets with 35% of infill can be placed between T_20 and T_60 values (Table 5, Figure 6).

Table 5. Comparison of μCT scan data of 3DP PVA_K/CL tablets with 20% (T_20), 35% (T_35), and 60% (T_60) infill ratio.

Description	Unit	T_20	T_35	T_60
Object volume	mm^3	236	220	202
Percent object volume	%	33	41	60
Structure thickness	mm	0.25	0.20	0.19
Structure separation	mm	1.11	0.62	0.25
Volume of open pore space	mm^3	485	312	134
Open porosity	%	67.2	58.5	39.9

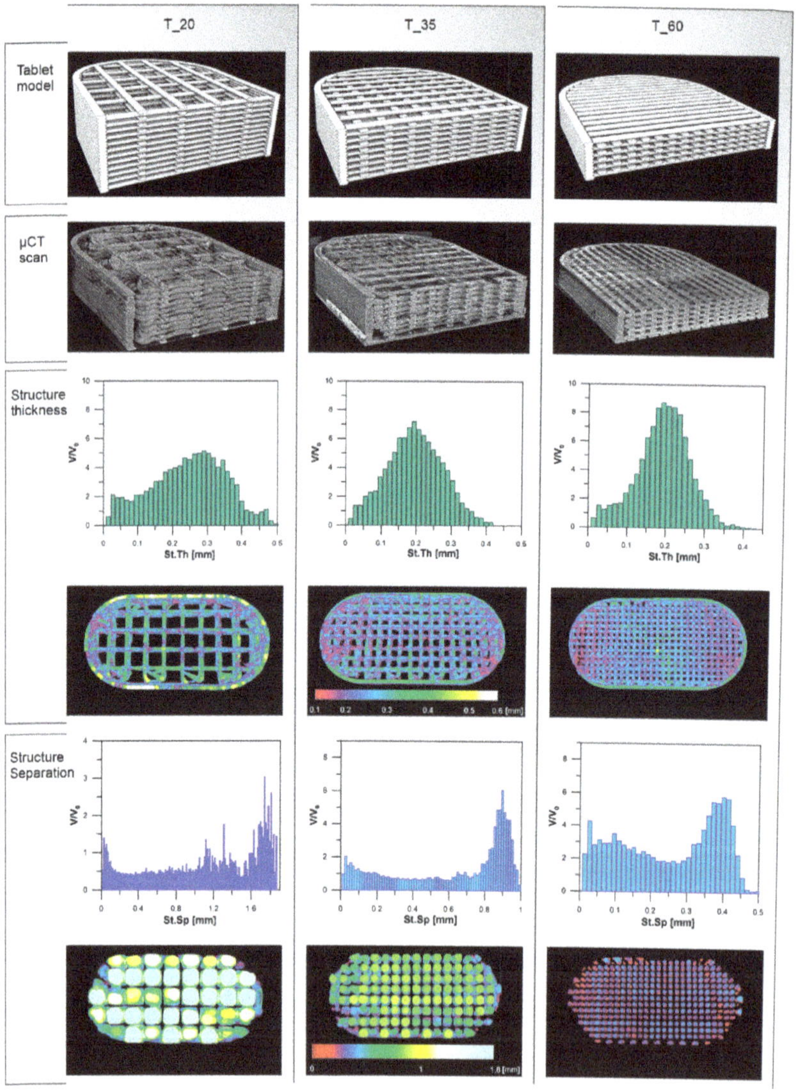

Figure 6. 3D tablet models, μ-CT scan images of 3DP tablets, and structure thickness and structure separation of 3DP tablets.

Parameters of tablets with 35% of infill are similar and no important differences between the three analyzed tablets can be distinguished (Table 6, Figure 7).

Table 6. Comparison of μCT scan data of 3DP PVA_K/CL tablets with 35% of infill ratio.

Description	Unit	T_35_1	T_35_2	T_35_3
Object volume	mm^3	220	224	213
Percent object volume	%	41	42	39
Structure thickness	mm	0.20	0.19	0.17
Structure separation	mm	0.62	0.62	0.62
Volume of open pore space	mm^3	312	307	327
Open porosity	%	58.5	57.7	60.4

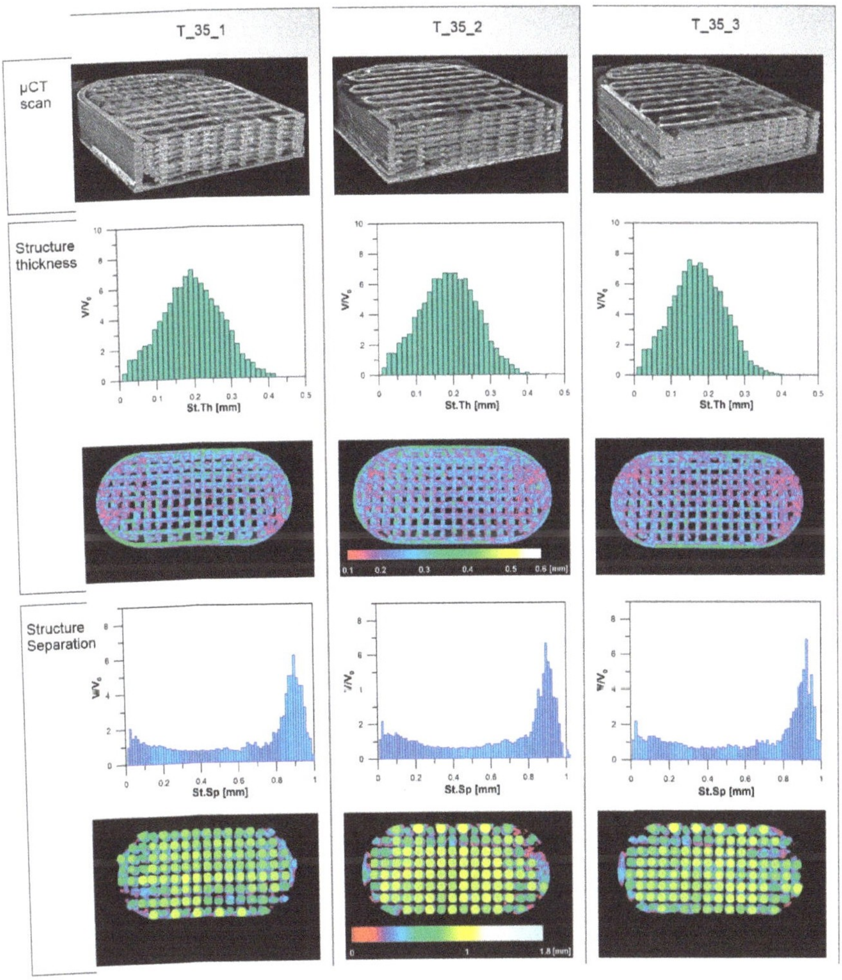

Figure 7. Comparison of 3DP tablets with 35% of infill.

3.4. Dissolution Studies

Itraconazole dissolution from 3D printed tablets with 35% infill was compared with the dissolution profiles obtained for the tablets made from milled extrudate (HME tablets) and directly compressed tablets (DC tablets) to evaluate the impact of the excipients and hot-melt extrusion on the dissolution of the API. Determined itraconazole solubility limits were equal to 5.8, 22.2, and 29.3 µg/mL for physical mixture, extrudate, and 3D printed matrix, respectively. The solubility limits were calculated as the percentage of ITR dose in tablets (11.6%, 44.4%, and 58.6% for physical mixture, extrudate, and 3D printed tablet, respectively) and are marked in Figure 8 to make the interpretation of the dissolution easier. It was found that the performed technological processes, namely hot-melt extrusion and 3D printing, affected the dissolution profile of itraconazole. The highest amount of the drug was dissolved from 3D printed tablets. The amount of ITR released from milled extrudate was significantly lower, while the smallest amount was released from directly compressed tablets (Figure 8). After 2 h of the dissolution test, 75.8%, 51.3%, and 11.0% of the itraconazole was released from the PVA-based 3D printed, hot-melt extruded, and directly compressed tablets, respectively. This relationship was confirmed for all the prepared formulations. It must be highlighted that in the case of all 3D printed formulations, i.e., PVA, PVA_K/VA, and PVA_K/CL, the amount of dissolved itraconazole was far above the solubility limit and the supersaturation lasted as long as the dissolution test was performed.

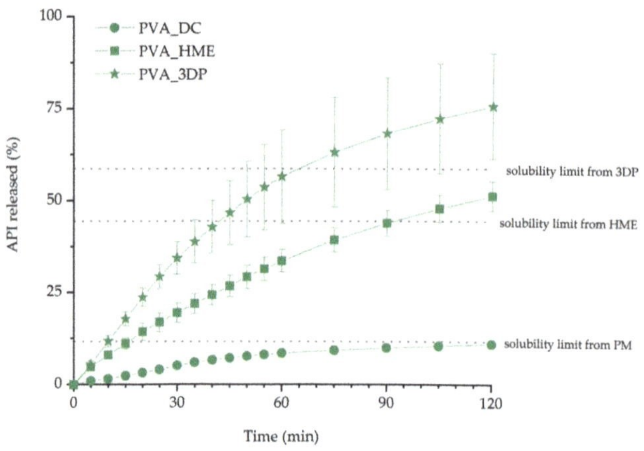

Figure 8. The influence of the technological process on the release profiles of itraconazole from PVA-based tablets (infill density equal to 35%).

The impact of copovidone and crospovidone addition to the PVA formulation on the release profile was also evaluated (Figure 9). The best dissolution profile was noticed for PVA_K/CL 3D printed tablets. After 45 min, 91.5% of the API was dissolved from PVA_K/CL 3D printed tablets, while only 64.3% and 46.7% of the drug was released from 3D printed tablets with Kollidon® VA64 and PVA-based tablets, respectively.

The impact of the infill density on the dissolution characteristics was evaluated for the PVA_K/CL formulation (Figure 10) as it was selected as the most promising formulation from all the prepared 3D printed tablets. Three rectilinear infills with different densities, namely 20%, 35%, and 60%, were evaluated. The results confirmed that the lower infill density favored faster dissolution of the API. After 45 min of the dissolution test, 96.9%, 89.7%, and 80.9% of the itraconazole was released from 3D printed tablets with 20%, 35%, and 60% infill, respectively.

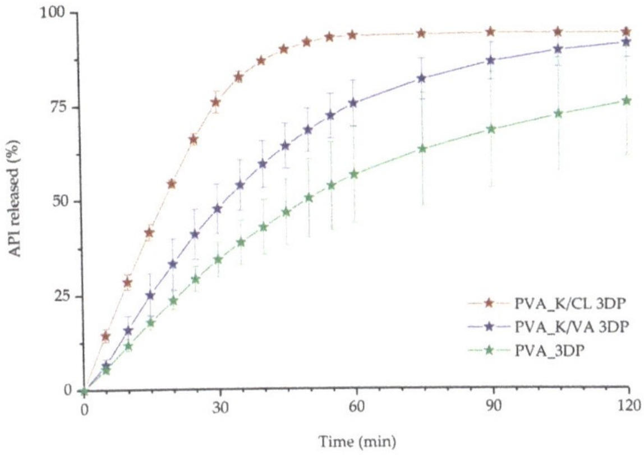

Figure 9. The influence of the excipients on dissolution profiles of itraconazole from 3DP tablets (infill density equal to 35%).

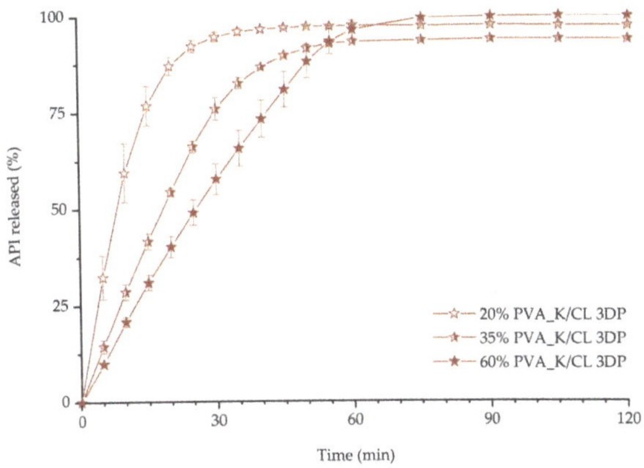

Figure 10. The influence of infill percentage on dissolution profiles of itraconazole from 3DP tablets.

4. Discussion

The filament extrusion went smoothly, and it can be carried out as a continuous manufacturing process. As a result of the optimization of the barrel temperature profile, generated torque, which may be considered as one of the major limitations during the extrusion, was as low as 2.82 ± 0.09 Nm during the filament extrusion process. All the prepared filaments were of satisfying quality and were printable using a ZMorph® 2.0 S 3D printer. PVA-based filaments were characterized by the most uniform diameter which may result from the simplest composition of the filament. Copovidone (K/VA) was added to the filament formulation to improve the solubility of the drug in the polymer matrix as it was shown by Włodarski et al. [65], while crosslinked PVP (K/CL) was added to improve disintegration and API dissolution from the extrudates and 3D printed tablets. The elasticity of the filaments was evaluated based on the Young's modulus values. The values obtained for itraconazole-loaded filaments were in the range 2042.1–2641.1 MPa and they were comparable to the results obtained by Feuerbach et al. for Resomer filaments [73]. The filament elasticity was not significantly affected by the addition of either copovidone or crospovidone to the formulation, while the values of the Young's modulus varied in the narrow range. However, it was found that the filament with the addition of copovidone was characterized by slightly higher elasticity than the one composed of only PVA or PVA-K/VA filaments. The obtained Young's modulus values for all prepared filaments suggest that they are suitable for fused deposition modeling 3D printing. The tensile strength was in the range from 28.2 to 52.6 MPa; the lowest value was obtained for the filament with the addition of Kollidon®VA64. Its introduction to the polymer matrix caused a more than 1.7-fold decrease in tensile strength in comparison with the itraconazole-loaded PVA filament. This may result from the Kollidon®VA64 extrudate's brittleness, which was confirmed by Fuenmayor et al. [74]; however, it was still durable enough to be printed.

A set of 10 × 20 mm^2 oblong tablets with different infill densities was printed with good repeatability. The tablets were uniform in shape and mass. The dimensions of the 3DP tablets were similar to predefined 3D objects. The adjustment of tablet height and, in consequence, the number of layers was related to the filament properties to obtain tablets with comparable mass (Table 4). The differences in tablet mass did not exceed 12.5 mg (±5%) from the theoretical value of 250 mg.

The theoretical tablets' volume was compared to the real object volume of 3DP PVA_K/CL tablets with infill of 20%, 35%, and 60% (Table 5), determined during the µCT scan. In the case of 20% of infill tablets (T_20), the real tablet volume was almost 1.3 times higher than calculated. This is related to the morphology of tablets with low infill density. The substantial distance between infill cross-points, in which two adjacent layers adhere, resulted in overhangs without support. It led to path disorder and an increase in vertical layer dimension. Therefore, subsequent cohesion in some spaces between cross-points was observed (Figure 6). In the case of T_35, the difference in tablet volume was smaller (1.12 times higher) whereas the volumes of T_60 tablets were similar (1.03 times higher). This improvement was related to the higher density of tablets' infill with increasing number of cross-points.

The phenomenon of path expansion between the cross-points can also be explained by the deviations of the structure thickness parameter in comparison with the theoretical value of 0.15 mm. This effect was observed for all degrees of infill; however, it was less pronounced in the systems with the higher infill density (Figure 6). The biggest difference was noticed in the case of T_20 tablets, for which the mean structure thickness was 100 µm higher than the theoretical layer height. For T_35 and T_60 tablets, the structure thickness was 40–50 µm higher. The differences in the structure thickness distribution are presented in Figure 6. The widest span of structure thickness was noticed for T_20 tablets and the structures with 0.25–0.35 mm thickness had the greatest volume within 3DP objects. On the contrary, the T_60 tablets exhibited the narrowest span, with structures of thickness varying between 0.15 and 0.25 mm highly represented within the object (Figure 6). Structure thickness distribution among a set of T_35 tested tablets was similar and showed good repeatability of printed dosage forms with 35% of infill (Table 6, Figure 7). Moreover, identical mean structure separation was observed within all T_35 tablets (Table 6) The porosity within T_35 tablets was similar, and histograms of structure separation distribution revealed that pores with size 0.8–1.0 mm are highly represented (Figure 7). Decreasing the tablet infill from 35% to 20% resulted in porosity changes. Pores with larger sizes, between 1.2 and 1.75 mm, are visible on the histograms and the total porosity increased from 58.5% to 67.2%. In the case of T_60 tablets, pore size did not exceed 0.5 mm and total porosity was almost 1.7 times smaller (39.9%) than T_20 (Figure 6). It should be emphasized that the volume of the open pore space within the 3DP T_20 tablet (485 mm^3) is twice as high as the volume of the solid part of the tablet (236 mm^3), whereas the volume of the open pore space of T_60 (134 mm^3) is 1.5 times smaller than the solid part of the tablet (202 mm^3). The tablet open space will promote the penetration of dissolution media through the tablet's internal structure and will have an impact on its disintegration and dissolution behavior. The influence of the internal structure of 3D printed objects on their properties was highlighted and widely discussed by Nazir et al. in the comprehensive review of the various 3D printed lattice and cellular structures, their advantages and limitations [75].

The results of the dissolution studies indicate that the 3D printing process improved itraconazole release when compared with tablets made by compression with either milled extrudate or a simple powder blend. This should be attributed to the developed internal structure and resulting extended surface area as well as the molecular rearrangement in the structure of API within the polymer matrix. Itraconazole release was faster from tablets containing added copovidone than PVA alone because the hot-melt extrusion and following 3D printing led to the formation of more disordered systems, which was confirmed by the lower intensity of the characteristic peaks in the XRD diffractograms and the lack of the nematic phase confirmed by DSC in the PVA_K/VA 3DP tablets. The release of itraconazole from filaments and 3D printed tablets containing only PVA was lower than from the corresponding systems containing the additive of PVP-based polymers since its structure was more ordered, as indicated by the presence of smectic and nematic domains. It is worth mentioning that the improved drug dissolution results from the applied technological processes, not just the addition of the polymers. The results of the dissolution from directly compressed tablets revealed that the presence of the polymers themselves did not enhance the dissolution of itraconazole as the amount of dissolved API did not exceed 12% of the initial dose.

The results indicated that the addition of the disintegrant, i.e., crospovidone, to the 3D printed tablets is beneficial in terms of ITR dissolution. The addition of the disintegrant to the formulation led to a higher increase in API dissolution than adding a copovidone to achieve molecularly disordered material. With the presented results, we have demonstrated that the PVA_K/CL formulation is the most promising in terms of immediate-release tablet preparation, as it is characterized by the best dissolution profile. Subsequent optimization was performed to evaluate the possibility of further improvement of itraconazole release. The optimization included changes in the infill density, as it was confirmed by many research groups that infill density significantly affects the dissolution rate of the API [76]. The tablets with infill density of 20%, 35%, and 60% were successfully 3D printed and tested. As predicted, lower infill density resulted in faster dissolution. However, the micro-computed tomography imaging revealed that during the printing of the tablets with 20% infill, there was an issue with maintaining the internal structure geometry, which also manifested in higher deviations in the amount of dissolved itraconazole in the first 20 min of the dissolution test (Figure 10). The tablets with 60% infill were characterized by the slowest itraconazole release. This is directly connected with the difficulty of water penetration into the tablet due to the smaller pores and channels in the internal structure. Therefore, we chose 35% infill as the best formulation to evaluate the tablet shape, dose, and internal structure reproducibility in the 3D printing process. In all cases of 3D printed tablets, long-lasting supersaturation of the itraconazole was achieved. It is well-known that the persisting state of supersaturation may lead to bioavailability improvements, which is especially beneficial in terms of poorly soluble drugs such as itraconazole [77].

5. Conclusions

Our study has shown the detailed methodology for the development of immediate-release 3D printed tablets with liquid crystal-forming itraconazole. The development stage included both the optimization of the formulation composition and the correlation between the geometry of the printed object, namely the degree of infill, with shape reproducibility and drug dissolution.

The use of well-printable PVA polymer alongside the functionalized excipients, i.e., polyvinylpyrrolidone derivatives, during the hot-melt extrusion process covered not only the optimization of the mechanical properties of the filament and its printability but also the function of the polymer matrix in terms of intended drug release profiles. The results of the dissolution study and physicochemical analysis indicated that improved disintegration obtained due to the use of Kollidon®CL-M was more beneficial than the molecular rearrangement and liquid crystal phase transitions. The lower infill density favored faster dissolution of the drug from printed tablets.

Micro-computed tomography was utilized to confirm that the design of printed objects was properly reconstructed. The comprehensive analysis revealed that the infill density, which is often considered as a way to control or improve drug dissolution, should be utilized with a deep understanding of its effect on the 3D printed objects' reproducibility. In the case of low infill densities, reproducibility issues, i.e., path disorder, increased layer dimension, and the path cohesion between cross-points, may occur. On the contrary, dense infill limits the surface area available for dissolution media and slows down the dissolution of the API.

In the case of the presented results, the most appropriate properties, i.e., good reproducibility during the object printing combined with superior drug dissolution, were achieved for the filament composed of 20% of itraconazole, 76% of PVA, and 4% of crospovidone acting as a disintegrant.

Supplementary Materials: The following are available online at http://www.mdpi.com/1996-1944/13/21/4961/s1, Figure S1: 3D printed itraconazole-loaded tablet with 20% infill density; Figure S2: 3D printed itraconazole-loaded tablet with 35% infill density; Figure S3: 3D printed itraconazole-loaded tablet with 60% infill density.

Author Contributions: Conceptualization, W.J., M.P. and R.J.; validation, J.S.-S.; investigation, W.J., J.P., M.K., J.K.-K., K.J., B.L. and A.W.; writing—original draft preparation, W.J.; writing—review and editing, J.P., M.K., J.S.-S., J.K.-K., M.P. and R.J.; visualization, W.J. and M.K.; supervision, R.J.; project administration, R.J. All authors have read and agreed to the published version of the manuscript.

Funding: This research was funded by the National Science Centre (Poland), grant number: OPUS 16 No. 2018/31/B/ST8/01327. K.J. is grateful for the financial support from the Foundation for Polish Science within the START program.

Conflicts of Interest: The authors declare no conflict of interest.

References

1. Pandey, M.; Choudhury, H.; Fern, J.L.C.; Kee, A.T.K.; Kou, J.; Jing, J.L.J.; Her, H.C.; Yong, H.S.; Ming, H.C.; Bhattamisra, S.K.; et al. 3D printing for oral drug delivery: A new tool to customize drug delivery. *Drug Deliv. Transl. Res.* **2020**, *10*, 986–1001. [CrossRef] [PubMed]
2. Pereira, B.C.; Isreb, A.; Forbes, R.T.; Dores, F.; Habashy, R.; Petit, J.-B.; Alhnan, M.A.; Oga, E.F. 'Temporary Plasticiser': A novel solution to fabricate 3D printed patient-centred cardiovascular 'Polypill' architectures. *Eur. J. Pharm. Biopharm.* **2019**, *135*, 94–103. [CrossRef]
3. Solanki, N.G.; Tahsin, M.; Shah, A.V.; Serajuddin, A.T. Formulation of 3D Printed Tablet for Rapid Drug Release by Fused Deposition Modeling: Screening Polymers for Drug Release, Drug-Polymer Miscibility and Printability. *J. Pharm. Sci.* **2018**, *107*, 390–401. [CrossRef]
4. Jamróz, W.; Szafraniec, J.; Kurek, M.; Jachowicz, R. 3D Printing in Pharmaceutical and Medical Applications—Recent Achievements and Challenges. *Pharm. Res.* **2018**, *35*, 1–22. [CrossRef]
5. Tagami, T.; Nagata, N.; Hayashi, N.; Ogawa, E.; Fukushige, K.; Sakai, N.; Ozeki, T. Defined drug release from 3D-printed composite tablets consisting of drug-loaded polyvinylalcohol and a water-soluble or water-insoluble polymer filler. *Int. J. Pharm.* **2018**, *543*, 361–367. [CrossRef]
6. Korte, C.; Quodbach, J. 3D-Printed Network Structures as Controlled-Release Drug Delivery Systems: Dose Adjustment, API Release Analysis and Prediction. *AAPS PharmSciTech* **2018**, *19*, 3333–3342. [CrossRef] [PubMed]
7. Jamróz, W.; Kurek, M.; Szafraniec-Szczęsny, J.; Czech, A.; Gawlak, K.; Knapik-Kowalczuk, J.; Leszczyński, B.; Wróbel, A.; Paluch, M.; Jachowicz, R. Speed it up, slow it down ... An issue of bicalutamide release from 3D printed tablets. *Eur. J. Pharm. Sci.* **2020**, *143*, 105169. [CrossRef] [PubMed]
8. Allahham, N.; Fina, F.; Marcuta, C.; Kraschew, L.; Mohr, W.; Gaisford, S.; Basit, A.W.; Goyanes, A. Selective Laser Sintering 3D Printing of Orally Disintegrating Printlets Containing Ondansetron. *Pharmaceutics* **2020**, *12*, 110. [CrossRef] [PubMed]
9. Ong, J.J.; Awad, A.; Martorana, A.; Gaisford, S.; Stoyanov, E.; Basit, A.W.; Goyanes, A. 3D printed opioid medicines with alcohol-resistant and abuse-deterrent properties. *Int. J. Pharm.* **2020**, *579*, 119169. [CrossRef] [PubMed]
10. Martinez, P.R.; Goyanes, A.; Basit, A.W.; Gaisford, S. Influence of Geometry on the Drug Release Profiles of Stereolithographic (SLA) 3D-Printed Tablets. *AAPS PharmSciTech* **2018**, *19*, 3355–3361. [CrossRef] [PubMed]
11. Pere, C.P.P.; Economidou, S.N.; Lall, G.; Ziraud, C.; Boateng, J.S.; Alexander, B.D.; Lamprou, D.A.; Douroumis, D. 3D printed microneedles for insulin skin delivery. *Int. J. Pharm.* **2018**, *544*, 425–432. [CrossRef]
12. Healy, A.V.; Fuenmayor, E.; Doran, P.; Geever, L.M.; Higginbotham, C.L.; Lyons, J.G. Additive Manufacturing of Personalized Pharmaceutical Dosage Forms via Stereolithography. *Pharmaceutics* **2019**, *11*, 645. [CrossRef]
13. Karakurt, I.; Aydoğdu, A.; Çıkrıkcı, S.; Orozco, J.; Lin, L. Stereolithography (SLA) 3D printing of ascorbic acid loaded hydrogels: A controlled release study. *Int. J. Pharm.* **2020**, *584*, 119428. [CrossRef]

14. Uddin, J.; Scoutaris, N.; Economidou, S.N.; Giraud, C.; Chowdhry, B.Z.; Donnelly, R.F.; Douroumis, D. 3D printed microneedles for anticancer therapy of skin tumours. *Mater. Sci. Eng. C* **2020**, *107*, 110248. [CrossRef]
15. Fina, F.; Goyanes, A.; Madla, C.M.; Awad, A.; Trenfield, S.J.; Kuek, J.M.; Patel, P.; Gaisford, S.; Basit, A.W. 3D printing of drug-loaded gyroid lattices using selective laser sintering. *Int. J. Pharm.* **2018**, *547*, 44–52. [CrossRef]
16. Kadry, H.; Wadnap, S.; Xu, C.; Ahsan, F. Digital light processing (DLP) 3D-printing technology and photoreactive polymers in fabrication of modified-release tablets. *Eur. J. Pharm. Sci.* **2019**, *135*, 60–67. [CrossRef] [PubMed]
17. Yang, Y.; Zhou, Y.; Lin, X.; Yang, Q.; Yang, G. Printability of External and Internal Structures Based on Digital Light Processing 3D Printing Technique. *Pharmaceutics* **2020**, *12*, 207. [CrossRef] [PubMed]
18. Infanger, S.; Haemmerli, A.; Iliev, S.; Baier, A.; Stoyanov, E.; Quodbach, J. Powder bed 3D-printing of highly loaded drug delivery devices with hydroxypropyl cellulose as solid binder. *Int. J. Pharm.* **2019**, *555*, 198–206. [CrossRef]
19. Sen, K.; Manchanda, A.; Mehta, T.; Ma, A.W.; Chaudhuri, B. Formulation design for inkjet-based 3D printed tablets. *Int. J. Pharm.* **2020**, *584*, 119430. [CrossRef]
20. Goyanes, A.; Allahham, N.; Trenfield, S.J.; Stoyanov, E.; Gaisford, S.; Basit, A.W. Direct powder extrusion 3D printing: Fabrication of drug products using a novel single-step process. *Int. J. Pharm.* **2019**, *567*, 118471. [CrossRef]
21. Fanous, M.; Gold, S.; Muller, S.; Hirsch, S.; Ogorka, J.; Imanidis, G. Simplification of fused deposition modeling 3D-printing paradigm: Feasibility of 1-step direct powder printing for immediate release dosage form production. *Int. J. Pharm.* **2020**, *578*, 119124. [CrossRef]
22. Öblom, H.; Sjöholm, E.; Rautamo, M.; Sandler, N. Towards Printed Pediatric Medicines in Hospital Pharmacies: Comparison of 2D and 3D-Printed Orodispersible Warfarin Films with Conventional Oral Powders in Unit Dose Sachets. *Pharmaceutics* **2019**, *11*, 334. [CrossRef]
23. Cui, M.; Pan, H.; Fang, D.; Qiao, S.; Wang, S.; Pan, W. Fabrication of high drug loading levetiracetam tablets using semi-solid extrusion 3D printing. *J. Drug Deliv. Sci. Technol.* **2020**, *57*, 101683. [CrossRef]
24. Karavasili, C.; Gkaragkounis, A.; Moschakis, T.; Ritzoulis, C.; Fatouros, D.G. Pediatric-friendly chocolate-based dosage forms for the oral administration of both hydrophilic and lipophilic drugs fabricated with extrusion-based 3D printing. *Eur. J. Pharm. Sci.* **2020**, *147*, 105291. [CrossRef]
25. El Aita, I.; Breitkreutz, J.; Quodbach, J. Investigation of semi-solid formulations for 3D printing of drugs after prolonged storage to mimic real-life applications. *Eur. J. Pharm. Sci.* **2020**, *146*, 105266. [CrossRef]
26. Elbl, J.; Gajdziok, J.; Kolarczyk, J. 3D printing of multilayered orodispersible films with in-process drying. *Int. J. Pharm.* **2020**, *575*, 118883. [CrossRef]
27. Jamróz, W.; Kurek, M.; Czech, A.; Szafraniec, J.; Gawlak, K.; Jachowicz, R. 3D printing of tablets containing amorphous aripiprazole by filaments co-extrusion. *Eur. J. Pharm. Biopharm.* **2018**, *131*, 44–47. [CrossRef]
28. Gioumouxouzis, C.I.; Tzimtzimis, E.; Katsamenis, O.L.; Dourou, A.; Markopoulou, C.; Bouropoulos, N.; Tzetzis, D.; Fatouros, D.G. Fabrication of an osmotic 3D printed solid dosage form for controlled release of active pharmaceutical ingredients. *Eur. J. Pharm. Sci.* **2020**, *143*, 105176. [CrossRef]
29. Nazir, A.; Jeng, J.-Y. A high-speed additive manufacturing approach for achieving high printing speed and accuracy. *Proc. Inst. Mech. Eng. Part C J. Mech. Eng. Sci.* **2019**, *234*, 2741–2749. [CrossRef]
30. Shaw, L.L.; Islam, M.; Li, J.; Li, L.; Ayub, S.M.I. High-Speed Additive Manufacturing Through High-Aspect-Ratio Nozzles. *JOM* **2018**, *70*, 284–291. [CrossRef]
31. Wang, J.; Goyanes, A.; Gaisford, S.; Basit, A.W. Stereolithographic (SLA) 3D printing of oral modified-release dosage forms. *Int. J. Pharm.* **2016**, *503*, 207–212. [CrossRef]
32. Fina, F.; Goyanes, A.; Gaisford, S.; Basit, A.W. Selective laser sintering (SLS) 3D printing of medicines. *Int. J. Pharm.* **2017**, *529*, 285–293. [CrossRef]
33. Jamróz, W.; Kurek, M.; Łyszczarz, E.; Szafraniec, J.; Knapik-Kowalczuk, J.; Syrek, K.; Paluch, M.; Jachowicz, R. 3D printed orodispersible films with Aripiprazole. *Int. J. Pharm.* **2017**, *533*, 413–420. [CrossRef]
34. Speer, I.; Preis, M.; Breitkreutz, J. Novel Dissolution Method for Oral Film Preparations with Modified Release Properties. *AAPS PharmSciTech* **2018**, *20*, 7. [CrossRef]

35. Gioumouxouzis, C.I.; Baklavaridis, A.; Katsamenis, O.L.; Markopoulou, C.K.; Bouropoulos, N.; Tzetzis, D.; Fatouros, D.G. A 3D printed bilayer oral solid dosage form combining metformin for prolonged and glimepiride for immediate drug delivery. *Eur. J. Pharm. Sci.* **2018**, *120*, 40–52. [CrossRef]
36. Öblom, H.; Zhang, J.; Pimparade, M.; Speer, I.; Preis, M.; Repka, M.; Sandler, N. 3D-Printed Isoniazid Tablets for the Treatment and Prevention of Tuberculosis—Personalized Dosing and Drug Release. *AAPS PharmSciTech* **2019**, *20*, 1–13. [CrossRef]
37. Smith, D.; Kapoor, Y.; Hermans, A.; Nofsinger, R.; Kesisoglou, F.; Gustafson, T.P.; Procopio, A. 3D printed capsules for quantitative regional absorption studies in the GI tract. *Int. J. Pharm.* **2018**, *550*, 418–428. [CrossRef]
38. Melocchi, A.; Uboldi, M.; Parietti, F.; Cerea, M.; Foppoli, A.; Palugan, L.; Gazzaniga, A.; Maroni, A.; Zema, L. Lego-Inspired Capsular Devices for the Development of Personalized Dietary Supplements: Proof of Concept With Multimodal Release of Caffeine. *J. Pharm. Sci.* **2020**, *109*, 1990–1999. [CrossRef]
39. Fu, J.; Yu, X.; Jin, Y. 3D printing of vaginal rings with personalized shapes for controlled release of progesterone. *Int. J. Pharm.* **2018**, *539*, 75–82. [CrossRef]
40. Scoutaris, N.; Ross, S.A.; Douroumis, D. 3D Printed "Starmix" Drug Loaded Dosage Forms for Paediatric Applications. *Pharm. Res.* **2018**, *35*, 34. [CrossRef] [PubMed]
41. Kempin, W.; Domsta, V.; Brecht, I.; Semmling, B.; Tillmann, S.; Weitschies, W.; Seidlitz, A. Development of a dual extrusion printing technique for an acid- and thermo-labile drug. *Eur. J. Pharm. Sci.* **2018**, *123*, 191–198. [CrossRef]
42. Sadia, M.; Arafat, B.; Ahmed, W.; Forbes, R.T.; Alhnan, M.A. Channelled tablets: An innovative approach to accelerating drug release from 3D printed tablets. *J. Control. Release* **2018**, *269*, 355–363. [CrossRef]
43. Kimura, S.-I.; Ishikawa, T.; Iwao, Y.; Itai, S.; Kondo, H. Fabrication of Zero-Order Sustained-Release Floating Tablets via Fused Depositing Modeling 3D Printer. *Chem. Pharm. Bull.* **2019**, *67*, 992–999. [CrossRef]
44. Giri, B.R.; Song, E.S.; Kwon, J.; Lee, J.-H.; Park, J.-B.; Kim, D.S. Fabrication of Intragastric Floating, Controlled Release 3D Printed Theophylline Tablets Using Hot-Melt Extrusion and Fused Deposition Modeling. *Pharmaceutics* **2020**, *12*, 77. [CrossRef]
45. Melocchi, A.; Uboldi, M.; Inverardi, N.; Briatico-Vangosa, F.; Baldi, F.; Pandini, S.; Scalet, G.; Auricchio, F.; Cerea, M.; Foppoli, A.; et al. Expandable drug delivery system for gastric retention based on shape memory polymers: Development via 4D printing and extrusion. *Int. J. Pharm.* **2019**, *571*, 118700. [CrossRef]
46. Melocchi, A.; Inverardi, N.; Uboldi, M.; Baldi, F.; Maroni, A.; Pandini, S.; Briatico-Vangosa, F.; Zema, L.; Gazzaniga, A. Retentive device for intravesical drug delivery based on water-induced shape memory response of poly(vinyl alcohol): Design concept and 4D printing feasibility. *Int. J. Pharm.* **2019**, *559*, 299–311. [CrossRef]
47. Melocchi, A.; Parietti, F.; Maccagnan, S.; Ortenzi, M.A.; Antenucci, S.; Briatico-Vangosa, F.; Maroni, A.; Gazzaniga, A.; Zema, L. Industrial Development of a 3D-Printed Nutraceutical Delivery Platform in the Form of a Multicompartment HPC Capsule. *AAPS PharmSciTech* **2018**, *19*, 3343–3354. [CrossRef]
48. Jiang, H.; Yu, X.; Fang, R.; Xiao, Z.; Jin, Y. 3D printed mold-based capsaicin candy for the treatment of oral ulcer. *Int. J. Pharm.* **2019**, *568*, 118517. [CrossRef]
49. He, S.; Feng, S.; Nag, A.; Afsarimanesh, N.; Han, T.; Mukhopadhyay, S.C. Recent Progress in 3D Printed Mold-Based Sensors. *Sensors* **2020**, *20*, 703. [CrossRef]
50. Nag, A.; Feng, S.; Mukhopadhyay, S.; Kosel, J.; Inglis, D. 3D printed mould-based graphite/PDMS sensor for low-force applications. *Sens. Actuators A Phys.* **2018**, *280*, 525–534. [CrossRef]
51. Sarode, A.L.; Sandhu, H.; Shah, N.; Malick, W.; Zia, H. Hot melt extrusion (HME) for amorphous solid dispersions: Predictive tools for processing and impact of drug-polymer interactions on supersaturation. *Eur. J. Pharm. Sci.* **2013**, *48*, 371–384. [CrossRef] [PubMed]
52. Corcione, C.E.; Gervaso, F.; Scalera, F.; Montagna, F.; Maiullaro, T.; Sannino, A.; Maffezzoli, A. 3D printing of hydroxyapatite polymer-based composites for bone tissue engineering. *J. Polym. Eng.* **2017**, *37*, 741–746. [CrossRef]
53. Solanki, N.G.; Lam, K.; Tahsin, M.; Gumaste, S.G.; Shah, A.V.; Serajuddin, A.T. Effects of Surfactants on Itraconazole-HPMCAS Solid Dispersion Prepared by Hot-Melt Extrusion I: Miscibility and Drug Release. *J. Pharm. Sci.* **2019**, *108*, 1453–1465. [CrossRef]

54. Solanki, N.G.; Gumaste, S.G.; Shah, A.V.; Serajuddin, A.T. Effects of Surfactants on Itraconazole -Hydroxypropyl Methylcellulose Acetate Succinate Solid Dispersion Prepared by Hot Melt Extrusion. II: Rheological Analysis and Extrudability Testing. *J. Pharm. Sci.* **2019**, *108*, 3063–3073. [CrossRef]
55. Jennotte, O.; Koch, N.; Lechanteur, A.; Evrard, B. Three-dimensional printing technology as a promising tool in bioavailability enhancement of poorly water-soluble molecules: A review. *Int. J. Pharm.* **2020**, *580*, 119200. [CrossRef]
56. Albadarin, A.B.; Potter, C.B.; Davis, M.T.; Iqbal, J.; Korde, S.; Pagire, S.; Paradkar, A.; Walker, G.M. Development of stability-enhanced ternary solid dispersions via combinations of HPMCP and Soluplus® processed by hot melt extrusion. *Int. J. Pharm.* **2017**, *532*, 603–611. [CrossRef] [PubMed]
57. Zhang, S.; Lee, W.Y.T.; Chow, A.H.L. Crystallization of Itraconazole Polymorphs from Melt. *Cryst. Growth Des.* **2016**, *16*, 3791–3801. [CrossRef]
58. Heczko, D.; Kamińska, E.; Jurkiewicz, K.; Tarnacka, M.; Merkel, K.; Kamiński, K.; Paluch, M. The impact of various azole antifungals on the liquid crystalline ordering in itraconazole. *J. Mol. Liq.* **2020**, *307*, 112959. [CrossRef]
59. Zhong, Y.; Jing, G.; Tian, B.; Huang, H.; Zhang, Y.; Gou, J.; Tang, X.; He, H.; Wang, Y. Supersaturation induced by Itraconazole/Soluplus® micelles provided high GI absorption in vivo. *Asian J. Pharm. Sci.* **2016**, *11*, 255–264. [CrossRef]
60. Singh, A.; Bharati, A.; Frederiks, P.; Verkinderen, O.; Goderis, B.; Cardinaels, R.; Moldenaers, P.; Van Humbeeck, J.; Mooter, G.V.D. Effect of Compression on the Molecular Arrangement of Itraconazole-Soluplus Solid Dispersions: Induction of Liquid Crystals or Exacerbation of Phase Separation? *Mol. Pharm.* **2016**, *13*, 1879–1893. [CrossRef]
61. Solanki, N.; Gupta, S.S.; Serajuddin, A.T. Rheological analysis of itraconazole-polymer mixtures to determine optimal melt extrusion temperature for development of amorphous solid dispersion. *Eur. J. Pharm. Sci.* **2018**, *111*, 482–491. [CrossRef]
62. Miller, D.A.; DiNunzio, J.C.; Yang, W.; McGINITY, J.W.; Williams, R.O.; Williams, R.O. Targeted Intestinal Delivery of Supersaturated Itraconazole for Improved Oral Absorption. *Pharm. Res.* **2008**, *25*, 1450–1459. [CrossRef]
63. Meng, F.; Meckel, J.; Zhang, F. Investigation of itraconazole ternary amorphous solid dispersions based on povidone and Carbopol. *Eur. J. Pharm. Sci.* **2017**, *106*, 413–421. [CrossRef]
64. Parikh, T.; Serajuddin, A.T.M. Development of Fast-Dissolving Amorphous Solid Dispersion of Itraconazole by Melt Extrusion of its Mixture with Weak Organic Carboxylic Acid and Polymer. *Pharm. Res.* **2018**, *35*, 127. [CrossRef]
65. Wlodarski, K.; Zhang, F.; Liu, T.; Sawicki, W.; Kipping, T. Synergistic Effect of Polyvinyl Alcohol and Copovidone in Itraconazole Amorphous Solid Dispersions. *Pharm. Res.* **2018**, *35*, 16. [CrossRef] [PubMed]
66. Zheng, M.; Bauer, F.; Birk, G.; Lubda, D. Polyvinyl Alcohol in Hot Melt Extrusion to Improve the Solubility of Drugs. 2013. Available online: https://www.sigmaaldrich.com/content/dam/sigma-aldrich/0/content/pdf/PS-PVA-HME-Improve-Solubility-03-2017_EN_MS.pdf (accessed on 27 October 2020).
67. Malaquias, L.F.; Schulte, H.L.; Chaker, J.A.; Karan, K.; Durig, T.; Marreto, R.N.; Gratieri, T.; Gelfuso, G.M.; Cunha-Filho, M. Hot Melt Extrudates Formulated Using Design Space: One Simple Process for Both Palatability and Dissolution Rate Improvement. *J. Pharm. Sci.* **2018**, *107*, 286–296. [CrossRef]
68. Lang, B.; McGINITY, J.W.; Williams, R.O.; Williams, R.O. Dissolution Enhancement of Itraconazole by Hot-Melt Extrusion Alone and the Combination of Hot-Melt Extrusion and Rapid Freezing—Effect of Formulation and Processing Variables. *Mol. Pharm.* **2013**, *11*, 186–196. [CrossRef]
69. Feng, D.; Peng, T.; Huang, Z.; Singh, V.; Shi, Y.; Wen, T.; Lu, M.; Quan, G.; Pan, X.; Wu, C. Polymer-Surfactant System Based Amorphous Solid Dispersion: Precipitation Inhibition and Bioavailability Enhancement of Itraconazole. *Pharmaceutics* **2018**, *10*, 53. [CrossRef]
70. Solanki, N.G.; Kathawala, M.; Serajuddin, A.T. Effects of Surfactants on Itraconazole-Hydroxypropyl Methylcellulose Acetate Succinate Solid Dispersion Prepared by Hot Melt Extrusion III: Tableting of Extrudates and Drug Release From Tablets. *J. Pharm. Sci.* **2019**, *108*, 3859–3869. [CrossRef]
71. Ponsar, H.; Wiedey, R.; Quodbach, J. Hot-Melt Extrusion Process Fluctuations and Their Impact on Critical Quality Attributes of Filaments and 3D-Printed Dosage Forms. *Pharmaceutics* **2020**, *12*, 511. [CrossRef]
72. Knapik, J.; Jurkiewicz, K.; Kocot, A.; Paluch, M. Rheo-dielectric studies of the kinetics of shear-induced nematic alignment changes in itraconazole. *J. Mol. Liq.* **2020**, *302*, 112494. [CrossRef]

73. Feuerbach, T.; Callau-Mendoza, S.; Thommes, M. Development of filaments for fused deposition modeling 3D printing with medical grade poly(lactic-co-glycolic acid) copolymers. *Pharm. Dev. Technol.* **2018**, *24*, 487–493. [CrossRef] [PubMed]
74. Fuenmayor, E.; Forde, M.; Healy, A.V.; Gately, N.; Lyons, J.G.; McConville, C.; Major, I. Material Considerations for Fused-Filament Fabrication of Solid Dosage Forms. *Pharmaceutics* **2018**, *10*, 44. [CrossRef] [PubMed]
75. Nazir, A.; Abate, K.M.; Kumar, A.; Jeng, J.-Y. A state-of-the-art review on types, design, optimization, and additive manufacturing of cellular structures. *Int. J. Adv. Manuf. Technol.* **2019**, *104*, 3489–3510. [CrossRef]
76. Kyobula, M.; Adedeji, A.; Alexander, M.R.; Saleh, E.; Wildman, R.D.; Ashcroft, I.; Gellert, P.R.; Roberts, C.J. 3D inkjet printing of tablets exploiting bespoke complex geometries for controlled and tuneable drug release. *J. Control. Release* **2017**, *261*, 207–215. [CrossRef] [PubMed]
77. Park, H.; Ha, E.-S.; Kim, M.-S. Current Status of Supersaturable Self-Emulsifying Drug Delivery Systems. *Pharmaceutics* **2020**, *12*, 365. [CrossRef]

materials

MDPI

Article

How Does the CO_2 in Supercritical State Affect the Properties of Drug-Polymer Systems, Dissolution Performance and Characteristics of Tablets Containing Bicalutamide?

Agata Antosik-Rogóż [1,*], Joanna Szafraniec-Szczęsny [1,2], Krzysztof Chmiel [3], Justyna Knapik-Kowalczuk [3], Mateusz Kurek [1], Karolina Gawlak [2], Vittorio P. Danesi [4], Marian Paluch [3] and Renata Jachowicz [1]

1 Department of Pharmaceutical Technology and Biopharmaceutics, Faculty of Pharmacy, Jagiellonian University Medical College, Medyczna 9, 30-688 Krakow, Poland; joanna.szafraniec@uj.edu.pl (J.S.-S.); mateusz.kurek@uj.edu.pl (M.K.); renata.jachowicz@uj.edu.pl (R.J.)

2 Department of Physical Chemistry and Electrochemistry, Faculty of Chemistry, Jagiellonian University, Gronostajowa 2, 30-387 Krakow, Poland; gawlak@chemia.uj.edu.pl

3 Institute of Physics, Faculty of Science and Technology, University of Silesia, SMCEBI, 75 Pułku Piechoty 1a, 41-500 Chorzow, Poland; krzysztof.chmiel@smcebi.edu.pl (K.C.); justyna.knapik-kowalczuk@smcebi.edu.pl (J.K.-K.); marian.paluch@smcebi.edu.pl (M.P.)

4 Department of Drugs Sciences, University of Pavia, viale Taramelli, 12-27100 Pavia, Italy; vittoriopaolo.danesi01@universitadipavia.it

* Correspondence: agata.antosik@uj.edu.pl; Tel.: +48-12-6205-606

Received: 8 June 2020; Accepted: 22 June 2020; Published: 25 June 2020

Abstract: The increasing demand for novel drug formulations has caused the introduction of the supercritical fluid technology, CO_2 in particular, into pharmaceutical technology as a method enabling the reduction of particle size and the formation of inclusion complexes and solid dispersions. In this paper, we describe the application of $scCO_2$ in the preparation of binary systems containing poorly soluble antiandrogenic drug bicalutamide and polymeric excipients, either Macrogol 6000 or Poloxamer®407. The changes in the particle size and morphology were followed using scanning electron microscopy and laser diffraction measurements. Differential scanning calorimetry was applied to assess thermal properties, while X-ray powder diffractometry was used to determine the changes in the crystal structure of the systems. The dissolution of bicalutamide was also considered. Binary solid dispersions were further compressed, and the attributes of tablets were assessed. Tablets were analyzed directly after manufacturing and storage in climate chambers. The obtained results indicate that the use of supercritical CO_2 led to the morphological changes of particles and the improvement of drug dissolution. The flowability of blends containing processed binary systems was poor; however, they were successfully compressed into tablets exhibiting enhanced drug release.

Keywords: Bicalutamide; Poloxamer® 407; Macrogol 6000; supercritical carbon dioxide; solid dispersions; dissolution rate; amorphization

1. Introduction

The growing number of active pharmaceutical ingredients (APIs) are characterized by poor water solubility. Currently, more than 70% of APIs under development are considered as poorly water soluble and almost 40% present in the pharmaceutical industry possesses aqueous solubility less than 100 µg/mL. Most of them belong to the second class of the Biopharmaceutical Classification System (BCS). These compounds are characterized by low solubility and high intestinal membrane permeability [1].

Given that the oral bioavailability of active pharmaceutical ingredients depends on their solubility and/or dissolution rate, many methods of solubility improvement have been developed [2]. They include chemical modifications, such as prodrugs [3,4] or salt formation [5], physical processes as micronization or polymorph transitions, as well as the preparation of drug-carrier systems, including solid dispersions (SD) [6,7] and complexes with cyclodextrins [8–11]. Solid dispersions are defined as drug-carrier systems in which a hydrophobic active pharmaceutical ingredient is either molecularly dispersed in a hydrophilic matrix or exists as micro-fine crystals [12]. The matrix can be made from low molecular weight compounds, or more frequently, hydrophilic polymers. They provide the dissolution rate improvement, particle size reduction, reduction of agglomeration, wettability improvement that are achieved by solid dispersion [13,14]. They possess different functional groups, molecular weight, melting temperatures (T_m), and glass transition temperatures (T_g). They can be crystalline (poly(vinyl alcohol)), semicrystalline (poly L-lactic acid), or amorphous (polyvinylpyrrolidone). Moreover, each polymer is characterized by different abilities to maintain molecular dispersion of active substances [15]. Drug-polymer solid dispersions are formed using evaporation methods, such as spray drying, and melting methods, including extrusion. These techniques often require high temperatures, which can influence the stability of the active substance, or high amounts of solvents, which may increase product toxicity and affect the environment [16]. These undesirable effects can be avoided by using supercritical fluid technology for solid dispersion preparation. The most commonly used gas is carbon dioxide, which is a non-toxic, non-flammable and widely available substance. In the supercritical state, i.e., above the critical temperature (31.4 °C) and pressure (74 bar), it is viscous and permeable like gaseous substances, while the density is comparable with liquids. It can act as a solvent and/or as a plasticizer for many substances. Also, it is easy to remove from the product after processing, so it is possible to obtain a solvent-free product without a dedicated solvent removal process [16–19].

Currently, tablets are the most popular dosage form. This is due to the ease of administration, neutral taste, appropriate stability and low cost of manufacturing [20–23]. Given that compression is an integral part of tableting, the effect of applied the compression force on the physical stability of the drug needs to be considered, especially in the case of the compounds undergoing mechanically induced activation [24,25]. One of such is bicalutamide, a non-steroidal antiandrogen, which belongs to class II of the Biopharmaceutics Classification System, exhibiting low aqueous solubility (below 5 µg/mL) and high lipophilicity (logP = 2.92) [26]. Th drug undergoes polymorph transition (from physically stable form I to metastable form II) or amorphization upon mechanical treatment as a consequence of a need for the relaxation of stress field applied during milling [27,28]. Drug amorphization was previously described upon milling and spray drying with polyviylpyrrolidone (PVP) [29,30]. Given that glassy bicalutamide recrystallizes easily upon grinding or scratching, the effect of increased pressure on physical stability needs to be investigated deeply. Although many papers describe the effect of temperature and humidity on the stability of amorphous drugs, the effect of pressure applied to solid dispersions during tableting was rarely described [31]. Also, only a few papers describe how the tableting and storage of solid dispersions affect the dissolution performance of bicalutamide.

The polymers chosen for this study were polyethylene glycol (Macrogol 6000, PEG6000,) and Poloxamer® 407 (PLX407). These excipients are commonly used for solid dispersion preparation due to their solubilizing effects, surface adsorption, and wetting enhancement. Macrogols are widely used as carriers due to their low melting point, fast solidification, and capability of forming solid drug solutions [32,33]. However, the presence of a crystalline form may result in unstable formulations and lower dissolution rates [32]. For some drug-active substances, the improvement in solubility enhancement by Poloxamer is more efficient than by Macrogols [34]. Poloxamers, also called pluronics, are amphiphilic block copolymers consisting of hydrophilic ethylene oxide blocks (EO) attached to a central polypropylene oxide unit (PO) of hydrophobic character. They are widely used as carriers, mostly due to a low melting point. Moreover, the differences in the solubility of the constituent blocks of poloxamer macromolecule lead to thermo-responsive self-assembly in an aqueous environment, which is the additional advantage for solubility enhancement [35].

In our studies, we evaluated the use of the supercritical carbon dioxide ($scCO_2$) method as a way to obtain solid dispersions with bicalutamide, which were further (after mixing with excipients) compressed into tablets. The effect of tableting and storage in both normal and accelerated conditions was evaluated together with drug dissolution performance.

2. Materials and Methods

2.1. Materials

Bicalutamide (BCL, 99.3% purity, Hangzhou Hyper Chemicals Limited, Hangzhou, China) was used as an active substance: molecular formula $C_{18}H_{14}F_4N_2O_4S$, molecular weight 430.4 g/mol, solubility in water 3.7 μg/mL, logP = 2.92, pKa = 12.6, melting point 196 °C. Macrogol 6000 (PEG6000, Clariant, Muttenz, Switzerland): molecular formula $H(OCH_2CH_2)_nOH$, average n = 100, nonionic substance, melting point 65 °C, and Poloxamer® 407 (PLX407, BASF, Ludwigshafen am Rhein, Germany), molecular formula $HO(C_2H_4O)_a(C_3H_6O)_b(C_2H_4O)_aH$ a = 101, b = 56; average molecular weight 9840–14,600 g/mol, melting point 55 °C, freely soluble polymers were used as carriers. Carbon dioxide (CO_2; 99.99% purity, Pszczyna, Linde Gaz, Poland) was used as a supercritical fluid. Sodium lauryl sulfate (SLS; 98.8% purity; BASF) was used to prepare dissolution medium. Distilled water was used to prepare all the solutions. Cellulose microcrystalline (JRS Pharma, Rosenberg, Germany), sodium starch glycolate (DFE Pharma, Goch, Germany) and magnesium stearate (Merck, Darmstadt, Germany) were used as excipients to formulate tablets. All excipients used to formulate tablets were of pharmaceutical grade.

2.2. Methods

2.2.1. Solid Dispersion Preparation Using Supercritical Carbon Dioxide ($scCO_2$)

Bicalutamide and carriers, i.e., Macrogol 6000 or Poloxamer® 407, were mixed in a 1:1 weight ratio and loaded into a high-pressure reactor BR-300 (Berghof Products + Instruments GmbH, Eningen unter Achalm, Germany) equipped with a magnetic stirrer MR Hei-Standard (Heidolph Instruments, Schwabach, Germany), a heating mantle and a CHY 700T thermometer (CHY Firemate Co., Tainan City, Taiwan). To obtain supercritical conditions, carbon dioxide was pumped and compressed into a high-pressure reactor by a syringe pump SFT-10 (Supercritical Fluid Technologies Inc., Newark, NJ, USA). The process was conducted at 50 °C and 150 bar with Macrogol 6000 as a carrier, and at 60 °C and 160 bar with Poloxamer®407. In both cases, the magnetic stirrer was set at 500 rpm. The process parameters were monitored during the whole operation. The total mass of powder introduced into the high-pressure reactor was equal to 2 g. After the process, solid dispersions were pulverized through 1.8 mm sieve.

2.2.2. Scanning Electron Microscopy (SEM)

The morphological characterization of the samples was conducted using a Phenom Pro desktop electron microscope (Phenom World, Thermo Fisher Scientific, Waltham, MA, USA). Powdered samples were placed on the conductive adhesive tape glued to the specimen mount and measured in a holder for non-conductive samples. The acceleration voltage was equal to 10 kV. The stream of argon was used to remove the powder loosely bounded to the tape before the measurement.

2.2.3. Laser Diffraction Measurements

A Malvern Mastersizer 3000 (Malvern, UK) equipped with a HydroEV unit was used to determine particle size distribution. The samples were analyzed by the wet method using cyclohexane (reflective index, RI = 1.426) as a dispersant. Fraunhofer diffraction theory was applied to find the relationship between the particle size and the light intensity distribution pattern. The reported data represents the medians (D50 values) calculated as averages from ten series of measurements of each sample and the distribution span.

2.2.4. Powder X-Ray Diffraction (PXRD)

The diffraction patterns of the samples, i.e., pure substances, solid dispersions, and tablets, were registered at ambient temperature using an X-ray Rigaku Mini Flex II diffractometer (Tokyo, Japan) with a 5°/min step. Samples were scanned from 3° to 45°. Monochromatic Cu Kα radiation (λ = 1.5418 Å) was used.

2.2.5. Differential Scanning Calorimetry (DSC)

The thermal properties of the solid dispersions, raw bicalutamide, and carriers were examined using a differential scanning calorimetry (DSC) 1 STARe System (Mettler-Toledo, Greifensee, Switzerland). The measuring device was equipped with an HSS8 ceramic sensor with 120 thermocouples and a liquid nitrogen cooling station. The instrument was calibrated for temperature and enthalpy using zinc and indium standards. The glass transition temperature, crystallization, and melting were determined as the midpoint of the glass transition step, the onset of the exothermic peak, and the peak of the endothermic event, respectively. The samples were measured in an aluminum crucible (40 μL). The sample mass used for DSC experiments varied between 7 to 10 mg. All measurements were carried out with and without annealing (T = 323 K; t = 10 min). The experiments were performed from 273 to 478 K, with a heating rate equal to 10 K/min.

2.2.6. Preparation of Tablet Blends

The composition of each three formulations containing raw BCL or solid dispersions are presented in Section 3.3. Bicalutamide or solid dispersions were manually mixed with excipients: cellulose microcrystalline as a filler for 3 min, sodium starch glycolate as a disintegrant for 3 min and magnesium stearate as a lubricant for 1 min, in the porcelain dish.

2.2.7. Tableting Process and Tablet Characterization

The compression of tablets blends was performed using a single punch tableting machine (Korsch EK0, Berlin, Germany) equipped with the round flat punches which are 8 mm in diameter; each tablet type contained 50 mg of the active substance. Punches were equipped with tensiometers connected to a computer by Esam Traveler 1 Data Acquisition System Master Unit, type 0508-S-ESA (ESA Messtechnik GmbH, München, Germany). To obtain tablets that possess satisfactory properties, the pressure force applied on a blend containing raw bicalutamide and solid dispersions was equal to 7 kN (140 MPa) and 2 kN (40 MPa), respectively.

The mass uniformity of tablets was measured based on the method described in European Pharmacopoeia (Ph. Eur.) 9.0. The hardness, thickness and diameter of 20 randomly selected tablets were measured using hardness tester apparatus (VanKel VK200, Agilent Technologies, Santa Clara, CA, USA) and expressed as a mean value with standard deviation. The friability test was performed using a Pharma Test apparatus (Pharma Test Apparatebau AG, Hainburg, Germany). The disintegration time of the tablets was measured in distilled water at 37 °C using an Electrolab ED-2 SAPO (Electrolab, Mumbai, India) disintegration apparatus at 30 strokes/min.

2.2.8. Dissolution Studies

The dissolution studies were carried out using the method recommended by the Food and Drug Administration (FDA) for BCL tablets, i.e., United States Pharmacopeia (USP) paddle apparatus (II type), and the parameters, i.e., 50 rpm, 1000 mL of 1% SLS solution, 37 ± 0.5 °C. The analysis was performed in a Vision Elite 8 apparatus (Hanson Research, Chatsworth, CA, USA) equipped with a Vision G2 AutoPlus Autosampler, and an AutoFill sample collector (Hanson Research, Chatsworth, CA, USA). Samples, i.e., raw BCL, physical mixture, and solid dispersions obtained by supercritical carbon dioxide methods, containing the equivalent of 50 mg of the active substance, as well as prepared tablets, were poured into the beakers. A volume of 5 mL of the samples was withdrawn from each dissolution vessel at predefined intervals, filtered, diluted, and assayed spectrophotometrically at λ = 272 nm (UV-1800 Shimadzu, Shimadzu Corporation, Kyoto, Japan). After the collection of each sample, the dissolution medium was replaced with the same volume of pure one. The tests were carried out in triplicate, and the results represent averages with their standard deviations.

2.2.9. Contact Angle Measurements

A DSA255 drop shape analyzer (Krüss, Hamburg, Germany) was used to measure the contact angles using a sessile drop technique. A droplet of distilled water of volume equal to 2 µL was deposited on the surface of prepared tablets.

2.2.10. Stability Studies

Tablets were packed into PVC/Al blisters and placed in climate chambers (Memmert GmbH + Co. KG, Schwabach, Germany) and stored in either 25 °C/60% relative humidity (long-term conditions) or 40 °C/75% relative humidity (accelerated conditions). After six-month storage, tablets were analyzed for the following parameters: mass, hardness, thickness, contact angle, and dissolution rate of bicalutamide. The solid dispersion was analyzed after its preparation. The tablets' properties were evaluated just after the preparation and after the stability studies.

3. Results and Discussion

3.1. *Solid-State Characteristics of Solid Dispersions*

Based on the SEM results (Figure 1), raw bicalutamide exhibited elongated crystals of smooth surfaces. They were hexagonal and 30–160 µm long (Figure 1A). The median of the particle size distribution determined using a laser diffraction measurements laid was 81.7 µm (Table 1). The analysis of microphotographs of carriers indicated that Macrogol 6000 exhibited irregular particles of smooth surface and length not exceeding 100 µm (Figure 1B). In the case of the particles of Poloxamer® 407, they were smooth and spherical, of a diameter varying between 50 µm and 250 µm (Figure 1C). The medians of the particle size were 113 µm and 180 µm for PEG6000 and PLX407, respectively (Table 1). However, the distribution span for PEG6000 was much higher than for the other compounds, which indicated a wide distribution. The analysis of the particle size distribution revealed the presence of dust particles in each sample visible as a distribution tail within the region below 10 µm (Figure 2). In the case of bicalutamide and PLX407 the distributions were narrow and symmetrical (despite said tails), while in the case of Macrogol, the particle size was distributed uneven, with a maximum shift towards big particles. The distribution curve is skewed to the left, which can explain the differences between the size calculated from the SEM picture and laser diffraction measurements. The particles visualized using an electron microscope represent the fraction described by the extended distribution shoulder visible on the left side of the plot (pictures showing greater particles at the same magnification were not representative enough and are not presented here), while the laser diffraction assessed all the particles present in the sample.

After the treatment of BCL-PEG6000 binary mixtures by a supercritical carbon dioxide at 130 bar and 50 °C in the high-pressure reactor, solid dispersions were pulverized and white, free-flowing powders were obtained. In the case of solid dispersion with Poloxamer® 407, prepared at higher process parameters, i.e., pressure 140 bar, temperature 60 °C. The obtained sample also was white; however, its structure was foamy. After pulverized, white, free flowing powder was obtained.

Solid dispersions morphology was analyzed using SEM. The obtained results showed that the sample containing Macrogol 6000 exhibited irregular particles of a smooth surface with well-visible incorporated particles of different contrast. They were identified as the crystals of the active substance (Figure 1D). The particles of the solid dispersion with Poloxamer® 407 were rough, with visible inclusions of BCL crystals (Figure 1E).

Table 1. Particles size distribution of raw bicalutamide and carriers, results obtained by the laser diffraction method.

Substance	D_{50} (μm)	Span
Bicalutamide	81.7	1.70
Macrogol 6000	113.0	3.18
Poloxamer® 407	180.0	2.02
BCL-PEG6000	1120	2.01
BCL-PLX407	356	2.90

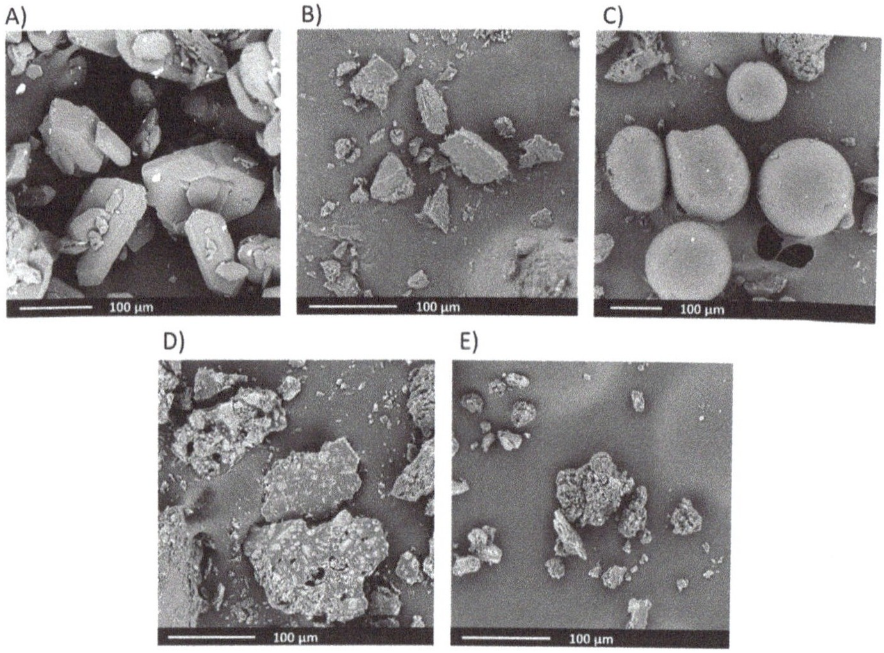

Figure 1. SEM image of raw bicalutamide (**A**), PEG6000 (**B**), PLX407 (**C**), BCL-PEG6000, (**D**) and BCL-PLX407 (**E**) solid dispersions.

The laser diffraction measurements revealed that solid dispersions exhibited a wide particle size distribution, in both cases skewed to the left, with two visible modes (Figure 2). The median of BCL-PEG6000 solid dispersion particles lay above 1 mm, which could be a result of the intense laser light diffraction on big particles and the presence of a well-resolved mode of the distribution curve. In the case of solid dispersion with Poloxamer® 407, the median of the particle size distribution was equal to 356 μm (Figure 2).

The thermal properties were obtained using differential scanning calorimetry (DSC). The presence of an endothermic peak with an onset at 196 °C corresponds to the melting temperature of form I polymorph of bicalutamide (Figure 3). After the treatment with supercritical CO_2, decreases in bicalutamide crystallinity in both obtained solid dispersions were noticed. This was confirmed by a decrease in the melting temperature of the active substance. For the solid dispersion prepared with Macrogol 6000, the melting of bicalutamide was recorded as an onset at 152.6 °C. The presence of the second endothermic peak with the onset at 60 °C corresponds to the melting point of the carrier. In the case of solid dispersions with Poloxamer® 407, the glass transition temperature was recorded at −6.2 °C. Bicalutamide crystallization occurred at 145.4 °C, while it melted at 167.4 °C. This indicated that the treatment with supercritical CO_2 led to drug amorphization. However, the system was characterized by very low physical stability in ambient conditions, expressed as a negative value of T_g. It resulted from low values of PLX407 T_g, −65 °C, combined with T_g of amorphous form of bicalutamide equal to 55 °C [36,37].

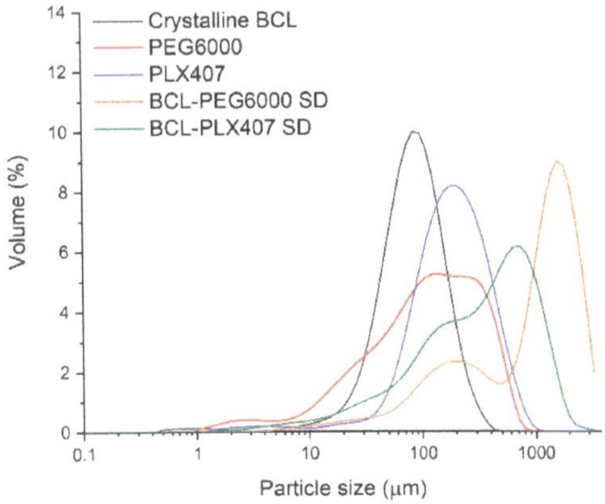

Figure 2. The particle size distribution of raw bicalutamide, polymers, and solid dispersions.

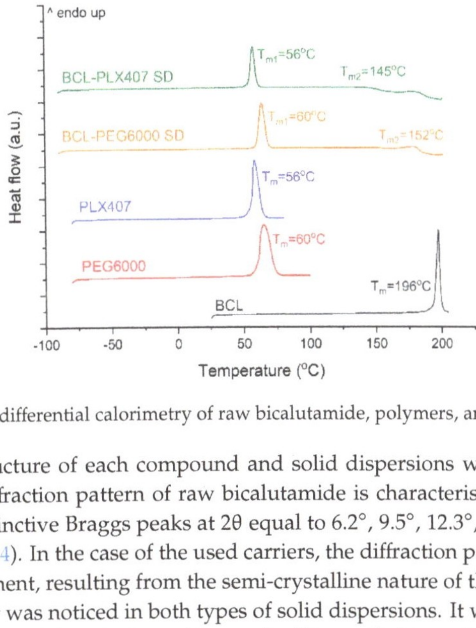

Figure 3. Scanning differential calorimetry of raw bicalutamide, polymers, and solid dispersions.

The molecular structure of each compound and solid dispersions was assessed using X-ray diffractometry. The diffraction pattern of raw bicalutamide is characteristic for crystalline form I polymorph, rich in distinctive Braggs peaks at 2θ equal to 6.2°, 9.5°, 12.3°, 14.3°–14.6°, 17.0°, 19.0°, 20.0°, and 24.0° (Figure 4). In the case of the used carriers, the diffraction patterns showed only two signals for each component, resulting from the semi-crystalline nature of the polymers. A decrease in the drug crystallinity was noticed in both types of solid dispersions. It was demonstrated by the decrease in the intensity and broadening of the peaks (Figure 4). Interestingly, the positions of the peaks shifted in comparison with the initial form of the drug. This indicated the recrystallization into the second polymorph of BCL, with the Braggs peaks positioned at 12.8°, 17.4°, 19.5°, 23.7–24.3°, and 25.4° [27].

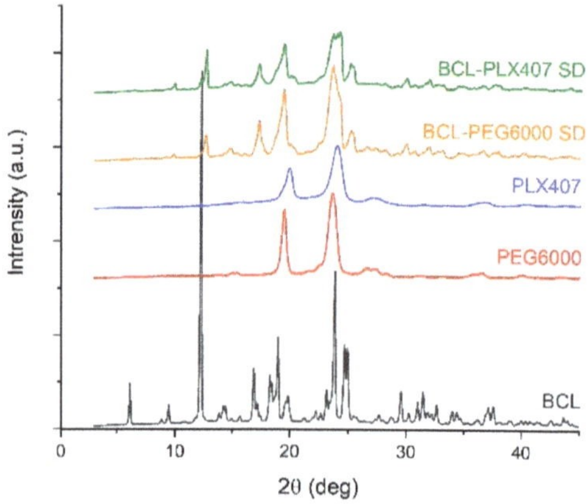

Figure 4. X-ray diffraction of raw bicalutamide and carriers, i.e., Macrogol 6000 and Poloxamer® 407, and solid dispersions.

3.2. *Bicalutamide Dissolution*

The preparation of solid dispersions using the supercritical carbon dioxide method with Macrogol 6000 and Poloxamer®407 led to a decrease in bicalutamide crystallinity, which resulted in dissolution improvement. The dissolution profiles of bicalutamide were very similar in both types of solid dispersions (Figure 5). After 1 h, 74.80 ± 1.66% of the drug dissolved from BCL-PEG6000 solid dispersion, and 77.43 ± 6.01% from the BCL-PLX407 one. The amount of bicalutamide dissolved from solid dispersions was approximately nine times greater than that determined for crystalline bicalutamide (form I polymorph), as 8.85 ± 1.02% of pure drug dissolved. No significant differences in the drug dissolution characteristics were noticed after the physical mixing of the active substance with any of the two polymers (approximately 12% of drug dissolved from each mixture). This confirms that the hydrophilic character of the carriers was not a crucial factor driving drug dissolution, and the dissolution improvement was related to the changes that occurred after the treatment with supercritical CO_2. During the process, the drug substance was dissolved in the polymer in supercritical CO_2, and the magnetic stirrer ensured uniform mixing of the ingredients. In the process of carbon dioxide removal, the sample solidified as a result of the rapid decrease in pressure. It provided homogenous distribution of the drug substance in the carrier.

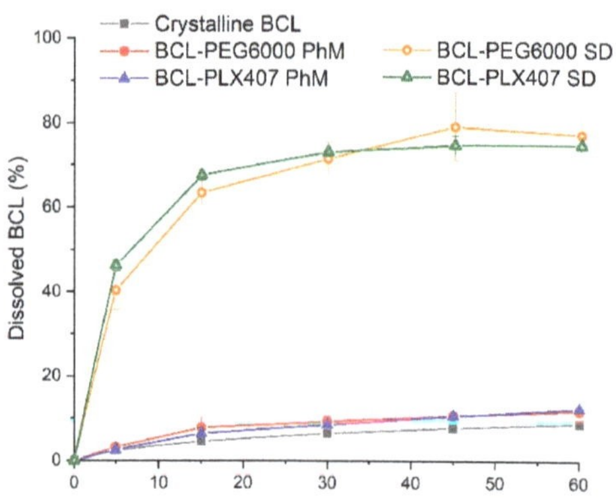

Figure 5. Dissolution of bicalutamide from solid dispersions with Macrogol 6000 and Poloxamer®407 in comparison with physical mixtures and crystalline bicalutamide (form I polymorph).

3.3. Characteristics of Tablets Containing BCL-PEG6000, BCL-PLX407 Solid Dispersions and Raw BCL

The three types of tablets were prepared using the direct compression method. The BCL-PEG6000 and BCL-PLX407 solid dispersions and raw bicalutamide were mixed with cellulose microcrystalline, sodium starch glycolate, and magnesium stearate to formulate tablets (Table 2). The obtained results revealed that hardness, thickness and friability were satisfactory.

Table 2. The composition of tablets containing BCL-PEG6000 and BCL-PLX407 solid dispersions and raw bicalutamide.

Substance	Tablets					
	Raw BCL		Solid Dispersion			
			BCL-PEG6000		BCL-PLX407	
	Content of Compounds in One Tablet					
	mg	%	mg	%	mg	%
BCL	50.00	41.67	-	-	-	-
BCL-Polymer (SD)	-	-	100.00	58.82	100.00	58.82
Cellulose Microcrystalline	64.00	53.33	64.00	37.65	64.00	37.65
Sodium starch glycolate	4.80	4.00	4.80	2.82	4.80	2.82
Magnesium stearate	1.20	1.00	1.20	0.71	1.20	0.71

The analysis of hardness showed that the tablets containing solid dispersions were characterized by higher hardness than those with raw bicalutamide (Table 3). This resulted from the binding properties of the used carriers, which also affected the disintegration time. Tablets containing solid dispersions disintegrated for longer than tablets with raw bicalutamide. Moreover, the tablets with BCL-PLX407 solid dispersion exhibited a six-fold longer disintegration time than those with BCL-PEG6000 solid dispersion, 24 min 22 s and 4 min 23 s, respectively. The difference was caused by the properties of poloxamer, which behaves like a binding substance [38]. During the compression process, particles tend to aggregate and increase the number of direct contact points between them, which extends tablet disintegration. Despite the surface activity and solubilizing activity of the polymer, a great content of poloxamer in the sample can retard tablet disintegration due to the gelling properties of the polymer [38–40].

Table 3. The properties of tablets containing solid dispersions (SD) and crystalline bicalutamide.

System \ Parameter	Mass (mg)	Thickness (mm)	Hardness (kp)	Friability (%)	Disintegration Time (min:s)	Contact Angle (°)
SD BCL-PEG6000	169.3 ± 7.0	3.60 ± 0.04	2.87 ± 0.63	0.0	4:23	50.8
SD BCL-PLX407	168.1 ± 5.8	3.43 ± 0.04	2.28 ± 0.64	0.4	24:22	55.8
Raw BCL	121.5 ± 3.6	2.08 ± 0.04	1.97 ± 0.25	0.6	00:31	85.4

The use of hydrophilic polymers in solid dispersion resulted in an increase in the wettability of the surface in comparison with the tablets containing raw bicalutamide. The tablets containing solid dispersions were more hydrophilic. They exhibited lower values of contact angles than the tablets with crystalline BCL:50.8° and 55.8° for the tablets containing PEG6000 and PLX407, respectively.

The X-ray diffraction analysis indicated that the active substance was amorphous. The presence of a broad halo instead of the Braggs peaks characteristic for crystalline drug substances, visible in Figure 4, indicated that bicalutamide was amorphous in all tested tablets (Figure 6). We assumed that this resulted from the propensity of bicalutamide to undergo activation upon mechanical treatment [10,12,27]. This means that the material gains some amount of energy which needs to be released. The relaxation of the stress field, applied during compression, occurs in several ways, including amorphization, as in the case of investigated tablets.

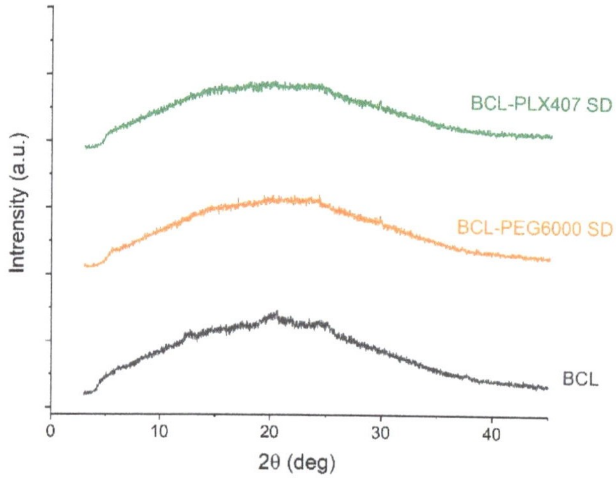

Figure 6. Diffraction patterns of tablets containing raw bicalutamide and solid dispersions.

As shown in Figure 7, bicalutamide release from the tablets indicated that the use of solid dispersions in tablets significantly enhanced drug release. This resulted from the improved wettability and amorphization of bicalutamide upon compression. After 1 h, 80.16 ± 8.56% and 56.61 ± 4.84% of the drug released from the tablets with PEG6000 and PLX407 solid dispersions, respectively. From tablets containing raw bicalutamide, 17.99 ± 1.70% dissolved, which was two times greater in comparison with the drug alone. Although the dissolution of bicalutamide from both solid dispersions was comparable at each time point, the differences in profiles were noticeable between solid dispersions (Figure 5) and corresponding tablets. After 30 min, 28% more bicalutamide was dissolved from tablets containing BCL-PEG6000 solid dispersion in comparison with the BCL-PLX407 tablets.

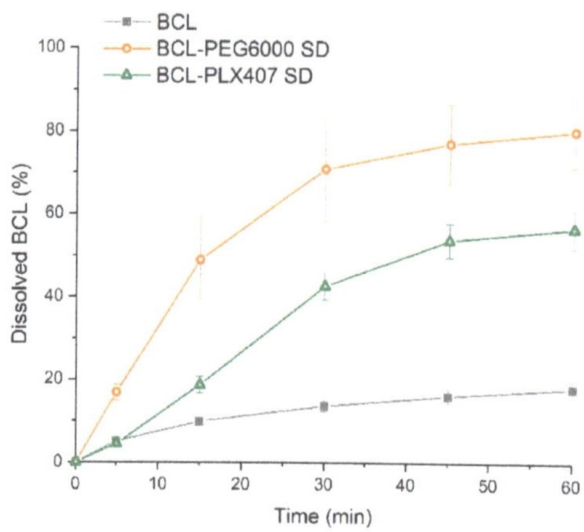

Figure 7. Dissolution profile of tablets containing solid dispersions and raw bicalutamide.

The prepared tablets were evaluated for stability upon storage. The results of the studies after six months revealed that the storage in both the long-term and accelerated conditions affected the properties of the tablets (Table 4) in comparison with tablets analyzed immediately after preparation (please refer to Table 3).

Table 4. Properties of tablets after six months of storage.

System \ Parameter	Mass (mg)	Hardness (kp)	Thickness (mm)	Disintegration Time (min:s)	Contact Angle (°)
		25 °C, 60% RH			
SD BCL-PEG	168.3 ± 3.5	2.47 ± 0.15	3.66 ± 0.01	04:57	48.9
SD BCL-PLX	174.5 ± 5.1	2.63 ± 0.51	3.53 ± 0.05	27:10	58.9
		40 °C, 75% RH			
SD BCL-PEG	177.0 ± 4.4	1.80 ± 0.36	3.73 ± 0.08	04:59	59.8
SD BCL-PLX	175.9 ± 5.5	1.25 ± 0.07	3.76 ± 0.06	14:39	53.8

In both types of tablets, drug release decreased after storage, regardless of the storage conditions (Figure 8A,B). However, less noticeable changes occurred for tablets stored at 25 °C and 60% RH. The amount of dissolved bicalutamide was 70.42% and 50.82% in the case of tablets containing solid dispersion with PEG6000 and PLX407, respectively. On the contrary, 32.01% of bicalutamide was released from BCL-PEG6000 tablets stored in accelerated conditions, which was a 2.5-fold decrease, in comparison with the freshly prepared tablets.

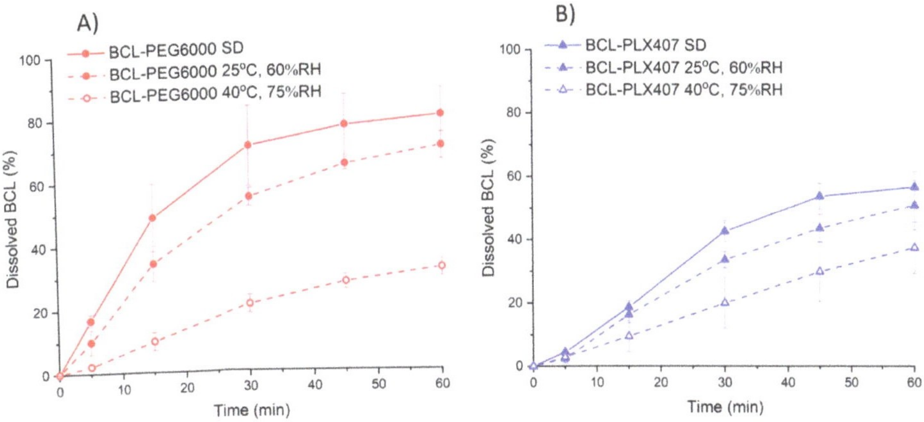

Figure 8. Release profiles of bicalutamide from tablets containing solid dispersion with either PEG6000 (**A**) or PLX4087 (**B**) after stability studies, in comparison with tablets analyzed after preparation.

Moreover, after the six-month stability studies, no phase transition of the drug substance occurred (Figure 9). The diffraction patterns did not change in comparison with those measured for tablets directly after compression (Figure 6). The bicalutamide remained amorphous, as confirmed by the halo visible in the diffractograms (Figure 9). This indicated that water absorbed from the air in the climate chambers did not cause drug recrystallization.

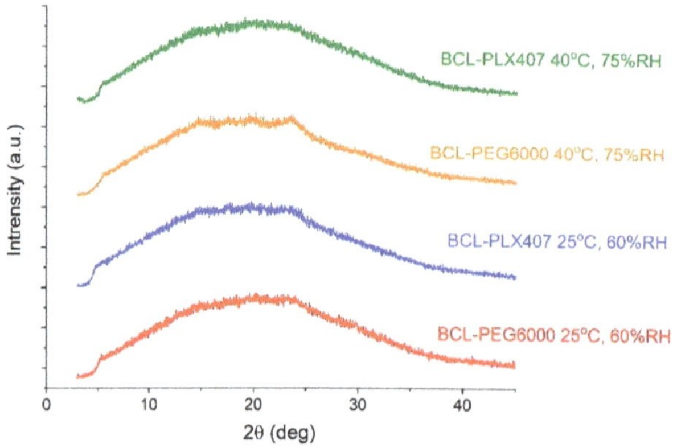

Figure 9. Diffraction patterns of bicalutamide in tablets with solid dispersions after stability studies.

4. Conclusions

This study has examined the effect of supercritical carbon dioxide treatment on the properties of binary systems containing bicalutamide and Macrogol 6000 or Poloxamer® 407. The dissolution and attributes of tablets were evaluated. The $scCO_2$ influenced not only the size and shape of the particles of the processed substances, but also the dissolution characteristics of bicalutamide from solid dispersions. Supercritical conditions induced a decrease in the crystallinity of the active substance. The obtained solid dispersions were successfully formulated in tablets. The tableting process induced the amorphization of the drug substance. This phenomenon confirmed the sensitivity of bicalutamide on mechanical activation. After 1h, the amount of bicalutamide dissolved from tablets with PEG6000 remained unchanged. In the case of tablets containing Poloxamer®407, the amount of released substance decreased by 20%. However, the compression affected the dissolution profiles of the drug substance in comparison with solid dispersion. The stability studies indicated that storage conditions affected the tablets' properties. The amount of the released drug decreased regardless of the storage conditions. However, in the stability studies after six months, the diffraction patterns did not change, which indicated that the physical state of the active substance was unchanged.

Author Contributions: A.A.-R.—writing original draft, investigation; J.S.-S.—writing, review and editing, investigation; K.C.—investigation; J.K.-K.—investigation; M.K.—investigation; K.G.—investigation; V.P.D.—investigation; M.P.—conceptualization, project administration; R.J.—conceptualization, review, supervision. All authors have read and agreed to the published version of the manuscript.

Funding: The authors acknowledge the Polish National Science Centre for the financial support (grant Symfonia 3 no 2015/16/W/NZ7/00404).

Conflicts of Interest: The authors declare no conflict of interest.

References

1. Thiry, J.; Kok, M.; Collard, L.; Frere, A.; Krier, F.; Fillet, M.; Evrard, B. Bioavailability enhancement of itraconazole-based solid dispersions produced by hot melt extrusion in the framework of the Three Rs rule. *Eur. J. Pharm. Sci.* **2017**, *99*, 1–8. [CrossRef]
2. Fujii, M.; Okada, H.; Shibata, Y.; Teramachi, H.; Kondoh, M.; Watanabe, Y. Preparation, characterization, and tableting of a solid dispersion of indomethacin with crospovidone. *Int. J. Pharm.* **2005**, *293*, 145–153. [CrossRef]
3. Abet, V.; Filance, F.; Recio, J.; Alvarez-Builla, J.; Burgos, C. Prodrug approach: An overview of recent cases. *Eur. J. Med. Chem.* **2017**, *127*, 810–827. [CrossRef]
4. Huttunen, K.M.; Raunio, H.; Rautio, J. Prodrugs—From Serendipity to Rational Design. *Pharm. Rev.* **2011**, *63*, 750–771. [CrossRef]

5. Serajuddin Abu, T.M. Salt formation to improve drug solubility. *Adv. Drug Del. Rev.* **2007**, *59*, 603–616. [CrossRef] [PubMed]
6. Benes, M.; Pekarek, T.; Beranek, J.; Havlicek, J.; Krejcik, L.; Simek, M.; Tkadlecova, M.; Dolezal, P. Methods for the preparation of amorphous solid dispersions—A comparative study. *J. Drug Del. Sci. Tech.* **2017**, *38*, 125–134. [CrossRef]
7. Newman, A.; Knipp, G.; Zografi, G. Assessing the Performance of Amorphous Solid Dispersions. *J. Pharm. Sci.* **2012**, *101*, 1355–1377. [CrossRef]
8. He, Y.; Ho, C. Amorphous Solid Dispersions: Utilization and Challenges in Drug Discovery and Development. *J. Pharm. Sci.* **2015**, *104*, 3237–3258. [CrossRef] [PubMed]
9. Huang, Y.; Dai, W.-G. Fundamental aspects of solid dispersions technology for poorly soluble drugs. *Acta Pharm. Sinica B* **2014**, *4*, 18–28. [CrossRef]
10. Jermain, S.V.; Brough, C.; Williams, R.O. Amorphous solid dispersions and nanocrystal technologies for poorly water-soluble drug delivery. *Int. J. Pharm.* **2018**, *535*, 379–392. [CrossRef]
11. Vo, C.; Park, C.; Lee, B.-J. Current trends and future perspectives of solid dispersions containing poorly water soluble drugs. *Eur. J. Pharm. Biopharm.* **2013**, *85*, 799–813. [CrossRef] [PubMed]
12. Craig, D. The mechanisms of drug release from solid dispersions in water-soluble polymers. *Int. J. Pharm.* **2002**, *231*, 131–144. [CrossRef]
13. Eloy, J.O.; Marchetti, J.M. Solid dispersions containing ursolic acid in Poloxamer 407 and PEG 6000: A comparative study of Fusion and solvent methods. *Powder Tech.* **2014**, *253*, 98–116. [CrossRef]
14. Altamimi, M.A.; Neau, S.H. Investigation of the in vitro performance difference of drug-Soluplus® and drug-PEG 6000 dispersions when using spray drying or lyophilization. *Saudi Pharm. J.* **2017**, *25*, 419–439. [CrossRef] [PubMed]
15. Weuts, I.; Van Dycke, F.; Voorspoels, J.; De Cort, S.; Stokbroekx, S.; Leemans, R.; Brewster, M.E.; Xu, D.; Segmuller, B.; Turner, Y.T.; et al. Physicochemical Properties of the Amorphous Drug, Cast Films, and Spray Dried Powders to Predict Formulation Probability of Success for Solid Dispersions Etravirine. *J. Pharm. Sci.* **2011**, *100*, 260–274. [CrossRef] [PubMed]
16. Milovanovic, S.; Djuris, J.; Dapcevic, A.; Madarevic, D.; Ibric, S.; Zizovic, I. Soluplus®, Eudragit®, HPMC-AS foams and solid dispersions for enhancement of Carvedilol dissolution rate prepared by supercriticial CO_2 process. *Polym. Test.* **2019**, *76*, 54–64. [CrossRef]
17. Nuchuchua, O.; Nejadnik, M.R.; Goulooze, S.C.; Ljescovic, N.J.; Every, H.A.; Jiskoot, W. Characterization of drug delivery particles produced by supercritical carbon dioxide technologies. *J. Supercrit Fluids* **2017**, *128*, 244–262. [CrossRef]
18. Abuzar, S.M.; Hyun, S.-M.; Kim, J.-H.; Park, H.J.; Kim, M.-S.; Park, J.-S.; Hwang, S.-J. Enhancing the solubility and bioavailability of poorly water-soluble drugs using supercritical antisolvent (SAS) process. *Int. J. Pharm.* **2018**, *532*, 1–13. [CrossRef]
19. Badens, E.; Masmoudi, Y.; Mouahid, A.; Crampon, C. Current situation and perspectives in drug formulation by using supercritical fluid technology. *J. Supercrit Fluids* **2018**, *134*, 274–283. [CrossRef]
20. Vasconcelos, T.; Sarmento, B.; Costa, P. Solid dispersion as strategy to improve oral bioavailability of poor water soluble drugs. *Drug Disc. Today* **2007**, *12*, 1068–1074. [CrossRef]
21. Wei, Q.; Keck, C.; Muller, R. Preparation and tableting of long-term stable amorphous rutin using porous silica. *Eur. J. Pharm. Biopharm.* **2017**, *113*, 97–107. [CrossRef] [PubMed]
22. Krasnyuk, I.I.; Koval'skii, I.V.; Nikulina, O.I.; Belyatskaya, A.V.; Krasnyuk, I.I.; Kharitonov, Y.Y.; Grikh, V.V.; Obidchenko, Y.A.; Vorob'ev, A.N. Preparation and investigation of tabletted medicinal formulations of a solid dispersion of rutin. *Pharm. Chem. J.* **2015**, *49*, 481–485. [CrossRef]
23. Abebe, A.; Akseli, I.; Sprockel, O.; Kottala, N.; Cuitino, A.M. Review of bilayer tablet technology. *Int. J. Pharm.* **2014**, *461*, 549–558. [CrossRef] [PubMed]
24. Wlodarski, K.; Tajber, L.; Sawicki, W. Physicochemical properties of direct compression tablets with spray dried and ball milled solid dispersions of tadalafil in PVP-VA. *Eur. J. Pharm. Biopharm.* **2016**, *109*, 14–23. [CrossRef] [PubMed]
25. Patel, S.; Kou, X.; Hou, H.; Huang, Y.; Strong, S.C.; Zhang, G.G.Z.; Sun, C.C. Mechanical Properties and Tableting Behavior of Amorphous Solid Dispersions. *J. Pharm. Sci.* **2017**, *106*, 217–223. [CrossRef]

26. Antosik, A.; Witkowski, S.; Woyna-Orlewicz, K.; Talik, P.; Szafraniec, J.; Wawrzuta, B.; Jachowicz, R. Application of supercritical carbon dioxide to enhance dissolution rate of bicalutamide. *Acta Pol. Pharm.* **2017**, *74*, 1231–1238.
27. Német, Z.; Sztatisz, J.; Demeter, Á. Polymorph transitions of bicalutamide: A remarkable example of mechanical activation. *J. Pharm. Sci.* **2008**, *97*, 3222–3232. [CrossRef]
28. Vega, D.R.; Polla, G.; Martinez, A.; Mendioroz, E.; Reinoso, M. Conformational polymorphism in bicalutamide. *Int. J. Pharm.* **2007**, *328*, 112–118. [CrossRef]
29. Szafraniec, J.; Antosik, A.; Knapik-Kowalczuk, J.; Kurek, M.; Syrek, K.; Chmiel, K.; Paluch, M.; Jachowicz, R. Planetary ball milling and supercritical fluid technology as a way to enhance dissolution of bicalutamide. *Int. J. Pharm.* **2017**, *533*, 470–479. [CrossRef]
30. Szafraniec, J.; Antosik, A.; Knapik-Kowalczuk, J.; Gawlak, K.; Kurek, M.; Szlęk, J.; Jamróz, W.; Paluch, M.; Jachowicz, R. Molecular Disorder of Bicalutamide-Amorphous Solid Dispersions Obtained by Solvent Methods. *Pharmaceutics* **2018**, *10*, 194. [CrossRef]
31. Szafraniec-Szczęsny, J.; Antosik-Rogóż, A.; Knapik-Kowalczuk, J.; Kurek, M.; Szefer, E.; Gawlak, K.; Chmiel, K.; Peralta, S.; Niwiński, K.; Pielichowski, K.; et al. Compression-Induced Phase Transition of Bicalutamide. *Pharmaceutics* **2020**, *12*, 438. [CrossRef] [PubMed]
32. Bley, H.; Fussnegger, B.; Bodmeier, R. Characterization and stability of solid dispersions based on PEG/polimer blends. *Int. J. Pharm.* **2010**, *390*, 165–173. [CrossRef] [PubMed]
33. Betageri, G.V.; Makarla, K.R. Enhancement of dissolution of glyburide by solid dispersion and lyophilization techcniques. *Int. J. Pharm.* **1995**, *126*, 155–160. [CrossRef]
34. Chutimaworapan, S.; Ritthidej, G.C.; Yonemochi, E.; Oguchi, T.; Yamamoto, K. Effect of Water-Soluble Carriers on Dissolution Characteristics of Nifedipine Solid Dispersions. *Drug Dev. Ind. Pharm.* **2000**, *26*, 1141–1150. [CrossRef]
35. Sancheti, P.P.; Vyas, V.M.; Shah, M.; Karekar, P. Development and characterization of bicalutamide poloxamer F68 solid dispersion system. *Pharmazie* **2008**, *8*, 571–575.
36. Szafraniec, J.; Antosik, A.; Knapik-Kowalczuk, J.; Chmiel, K.; Kurek, M.; Gawlak, K.; Odrobińska, J.; Paluch, M.; Jachowicz, R. The Self-Assembly Phenomenon of Poloxamers and Its Effect on the Dissolution of a Poorly Soluble Drug from Solid Dispersions Obtained by Solvent Methods. *Parmaceutics* **2019**, *11*, 130. [CrossRef]
37. Szczurek, J.; Rams-Baron, M.; Knapik-Kowalczuk, J.; Antosik, A.; Szafraniec, J.; Jamróz, W.; Dulski, M.; Jachowicz, R.; Paluch, M. Molecular Dynamics, Recrystallization Behavior, and Water Solubility of the Amorphous Anticancer Agent Bicalutamide and Its Polyvinylpyrrolidone Mixtures. *Mol. Pharm.* **2017**, *14*, 1071–1081. [CrossRef]
38. Kaul, G.; Huang, J.; Chatlapalli, R.; Ghosh, K.; Nagi, A. Quality-by-Design Case Study: Investigation of the Role of Poloxamer in Immediate-Release Tablets by Experimental Design and Multivariate Data Analysis. *AAPS Pharm. Sci. Tech.* **2011**, *12*, 1064–1076. [CrossRef]
39. D'souza, A.A.; Shegokar, R. Polyethylene glycol (PEG): A versatile polymer for pharmaceutical applications. *Expert Opin. Drug Del.* **2016**, *13*, 1257–1275. [CrossRef]
40. Szafraniec, J.; Antosik, A.; Knapik-Kowalczuk, J.; Chmiel, K.; Kurek, M.; Gawlak, K.; Paluch, M.; Jachowicz, R. Enhanced Dissolution of Solid Dispersions Containing Bicalutamide Subjected to Mechanical Stress. *Int. J. Pharm.* **2018**, *542*, 18–26. [CrossRef] [PubMed]

materials

MDPI

Article

Prototype Gastro-Resistant Soft Gelatin Films and Capsules—Imaging and Performance In Vitro

Bartosz Maciejewski [1], Vishnu Arumughan [2], Anette Larsson [2] and Małgorzata Sznitowska [1,*]

1 Department of Pharmaceutical Technology, Medical University of Gdansk, 80-416 Gdansk, Poland; bartosz.maciejewski@gumed.edu.pl

2 Department of Chemistry and Chemical Engineering, Chalmers University of Technology, 412 96 Gothenburg, Sweden; vishnu.arumughan@chalmers.se (V.A.); anette.larsson@chalmers.se (A.L.)

* Correspondence: msznito@gumed.edu.pl or malgorzata.sznitowska@gumed.edu.pl; Tel.: +48-58-349-1080

Received: 18 March 2020; Accepted: 7 April 2020; Published: 9 April 2020

Abstract: The following study is a continuation of the previous work on preparation of gastro-resistant films by incorporation of cellulose acetate phthalate (CAP) into the soft gelatin film. An extended investigation on the previously described binary Gelatin-CAP and ternary Gelatin-CAP-carrageenan polymer films was performed. The results suggest that the critical feature behind formation of the acid-resistant films is a spinodal decomposition in the film-forming mixture. In the obtained films, upon submersion in an acidic medium, gelatin swells and dissolves, exposing a CAP-based acid-insoluble skeleton, partially coated by a residue of other ingredients. The dissolution-hindering effect appears to be stronger when iota-carrageenan is added to the film-forming mixture. The drug release study performed in enhancer cells confirmed that diclofenac sodium is not released in the acidic medium, however, at pH 6.8 the drug release occurs. The capsules prepared with a simple lab-scale process appear to be resistant to disintegration of the shell structure in acid, although imperfections of the sealing have been noticed.

Keywords: gelatin; gastro-resistant; films; capsules; structure; drug release

1. Introduction

Gastro-resistant formulations are an example of the most common type of modified drug release systems. Gastro-resistant forms of drug administration allow to:

(1) minimize adverse effects such as nausea and bleeding associated with irritation of gastric mucosa that may be caused by some active substances;
(2) deliver drug intended for local action in intestines;
(3) protect the drug substance from degradation in an acidic environment of the stomach [1].

Gastro-resistant soft gelatin capsules can prove their usefulness in oral administration of drugs of irritating or acid-labile nature, often displaying at the same time enhanced bioavailability in a liquid form, which can be considered an advantage to coated tablets [2]. The most obvious examples of the substances that need to be formulated in gastro-resistant dosage forms are non-steroidal anti-inflammatory drugs (NSAIDs), which are irritating to gastric mucosa.

The products in the form of gastro-resistant capsules usually are designed as conventional hard capsule shells filled with the enteric-coated pellets or minitablets. Manufacturing of gastro-resistant soft capsules, however, is a challenge. Due to the liquid fill, modification of the drug release rate from soft capsules can be achieved only by modification of the capsule shell to make it resistant to acidic pH. This issue can be approached by-coating of standard capsules with acid-resistant polymers such as methacrylic acid—methyl acrylate copolymers (e.g., Eudragit L or S®) [3]. A less popular alternative is incorporation of gastro-resistant polymers in the shell material used to form the capsules [4]. Both approaches are technologically perplexing at some points, although modification of the shell material can be considered more beneficial from both economic and technological point of view. However, it is not yet utilized in commercial products. It is substantial to take into consideration that any changes in the composition of the film-forming mixture can result in significant alteration of the overall physicochemical properties of the prepared films, that can lead to the loss of their potential to be formed into capsules in a conventional manufacturing process.

A very important issue associated with the development of a new capsule shell composition is to identify the physiochemical phenomena that can be utilized in designing and manufacturing of modified release gelatin-based films. In our previous work, selection of the most effective modification of the shell material composition was performed, and their microstructure and barrier properties were described [5,6]. However, there are still a few unexplained issues in the description of the phenomena that lead to formation of the films, as well as the changes that the films undergo when exposed to various conditions. Therefore, in the present work, a more detailed investigation of the events associated with the formation of the gastro-resistant film was performed and further, the structural changes upon submersion of such films in acidic dissolution fluid is performed. For the purpose of better characterization of the films and film formation processes, several modern techniques may be employed. In the present research, a scanning electron microscopy (SEM), confocal laser scanning microscopy (CLSM), confocal Raman microscopy and quartz crystal microbalance with dissipation monitoring (QCM-D) were used. Additionally, the barrier properties of the films against oxygen were evaluated.

In comparison to tablets or hard capsules, the transfer of a new technology for soft capsules from the lab to the production site is much more complicated, and a scale-up procedure may be complicated and time-consuming. One of the main issues when soft capsules are developed is a poor access to a lab-scale equipment that could allow to assess the utility of the modified films for capsule formation. The most problematic is the fact that, at a commercial scale, specific rheological and mechanical properties of the film-forming material are required [7–9]. The fact that the shell-forming material has to be tested on a large scale, significantly increases the cost of technology development. In our present work, the lab-scale production process of the soft capsules is presented, utilizing a simple mold for suppositories, what allowed to evaluate the shell compatibility with the filling material.

2. Materials and Methods

2.1. Materials

Components of films and capsules: gelatin type B, bovine hide, 220 bloom (Sigma Aldrich, Saint Louis, MO, USA), glycerol 99.5% w/w (Chempur, Piekary Slaskie, Poland), Aquacoat®CPD (FMC Biopolymer, Philadelphia, PA, USA), aqueous dispersion of cellulose acetate phthalate (CAP), iota-carrageenan (Sigma Aldrich, Saint Louis, MO, USA), medium-chain triglycerides (MCT)—Miglyol 812 N (Caelo, Hilden, Germany), polyethylene glycol 400 (PEG 400)—Kollisolv PEG E400 (Sigma Aldrich, Saint Louis, MO, USA), cetearyl alcohol—TEGO Alkanol 1618 (Evonik, Essen, Germany). Diclofenac sodium was a gift from Polpharma Pharmaceutical Works (Starogard Gdanski, Poland).

Analysis: QCM-D: branched polyethyleneimine Mw 25,000 (Sigma Aldrich, Saint Louis, MO, USA), formaldehyde 37% (Sigma Aldrich, Saint Louis, MO, USA), ethanol 95% (Sigma Aldrich, Saint Louis, MO, USA). Disintegration and dissolution media (0.1 M HCl and phosphate buffer pH 6.8) were prepared according to the European Pharmacopeia 10th edition.

2.2. Capsule Formation

Schematic presentation of the capsule formation process is shown in Figure 1.

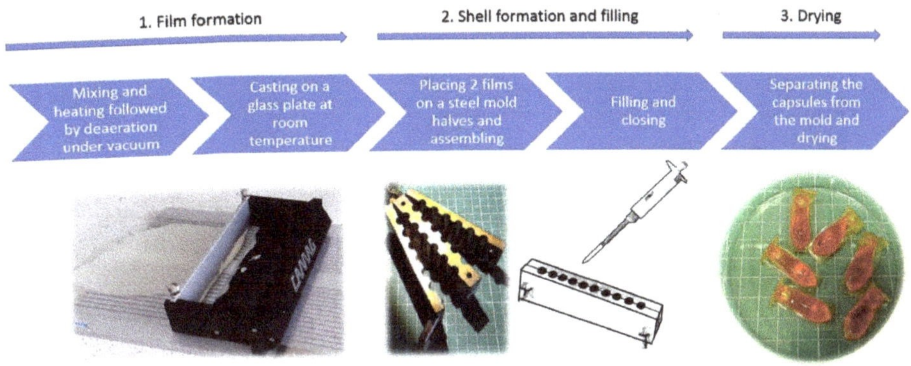

Figure 1. Capsule formation scheme.

The preparation method of the film-forming mixtures and films was described in detail in the previously published work [5,6]. Shortly, the mixture of components (Table 1) was stirred at 80 °C for 2 h, which was followed by deaeration under vacuum. Afterwards the mixture was casted on a glass plate using a plate coating device (Camag TLC Plate Coater, Camag, Muttenz, Switzerland) with a height of fluid layer of 1500 µm. After drying the thickness of the film was around 600 µm.

Table 1. Compositions of the films.

Composition Symbol	Mixture Components (g/100 g)				
	Gelatin	Aquacoat CPD	Iota-Carrageenan	Water	Glycerol
GEL [1]	41.2	–	–	40.0	18.8
GA [2]	30.9	34.4	–	15.9	18.8
GAC [2]	30.0	34.4	0.9	15.9	18.8

[1] non-modified film (reference); [2] modified binary (GA); and ternary (GAC) polymer films.

The capsules were prepared using GAC composition (Table 1), by placing 2 pieces of the film (immediately after casting) in a steel form for suppositories. After closing the form, the resulting reservoirs were filled with: (a) MCT oil, (b) PEG 400 or (c) cetearyl alcohol. For a better visual identification of a disintegration test endpoint, the filling material was colored with small amount of a hydrophilic or lipophilic dye. Afterwards, the filling orifice was manually closed with a strip of a film, and to ensure good sealing the capsules were placed for 5 min at 60 °C. Finally, the capsules were stored and dried at ambient temperature and of 15–25% RH for at least 24 h. The measured moisture content in the capsules was around 2.5% (Radwag WPS210S Moisture Analyzer, Radwag, Radom, Poland).

2.3. Microscopic Imaging

The imaging of samples was performed with use of a scanning electron microscopy (SEM), confocal laser scanning microscopy (CLSM), confocal Raman microscopy and optical microscopy.

The observation of film samples was performed before and after submersion in 0.1 M HCl at 37 °C, under constant stirring for 2 h (similar to the procedure of swelling test described in our previous work [5]). The films after submersion in acid were frozen in a liquid nitrogen and freeze-dried for 24 h. The investigation was performed with Jeol 7900F SEM (Jeol, Tokyo, Japan), Nikon Ti-E/A1 + CLSM (Nikon, Tokyo, Japan) and WITec Alpha 300 Access Raman microscope equipped with 785 nm laser (WITec, Ulm, Germany).

The imaging of the lab-manufactured capsules was performed using Phenom Pure SEM (Phenom World, Eindhoven, the Netherlands), and Nikon Eclipse 50i optical microscope (Nikon, Tokyo, Japan).

2.4. Gas Permeability

The films GEL, GA and GAC (Table 1) were subjected to oxygen permeability tests, performed with an coulometric detector technique according to method ASTM F 1927-14. The equipment used was OX-TRAN 2-20 (Mocon, Minneapolis, MN, USA). The investigated surface was 50 cm^2.

2.5. Quartz Crystal Microbalance with Dissipation Monitoring (QCM-D)

QCM-D was employed to investigate the affinity of CAP latex particles present in Aquacoat CPD to gelatin. The preparation step comprised coating of the gold-plated quartz crystal sensor with branched polyethyleneimine (PEI), then spin-coating the sensor with 1% gelatin solution (5 s at 2500 rpm and low acceleration, followed by 60 s at 8000 rpm and high acceleration). Afterwards the sensor was dried at 60 °C for 20 min. The gelatin on the sensor was subjected to crosslinking by submersion in 1.5% formaldehyde solution, in order to prevent it from dissolving in aqueous conditions. Finally, the sensor was dried at 60 °C for 60 min.

The sensors were mounted in a Qsense equipment (Qsense, Västra Frölunda, Sweden). A deionized water (at 25 °C) was flushed over the sensors until a stable baseline was obtained. Then the diluted (0.1%) Aquacoat CPD at 25 °C was pumped through the cells, and the changes in fundamental frequency overtones of the crystal were registered. After stabilization of the system, the cells were once again pumped with deionized water to remove all the substances that were not bound to the film.

Additionally, to assess the surface structure and stability of the gelatin films obtained in situ on the sensors, the Atomic Force Microscopy (AFM) was performed with NTEGRA Prima setup (NT-MDT Spectrum Instruments, Moscow, Russia), with a silicon probe (spring constant of 40 N m^{-1} and resonant frequency of 300 kHz) (Tap 300AI-G, Budget Sensors, Sofia, Bulgaria). The images were analyzed using a Gwyddion software (Version 2.55, Free Software Foundation, Boston, MA, USA).

2.6. Disintegration Time

Disintegration time test of the capsules filled with PEG-400, MCT oil or cetearyl alcohol, was performed. The test was performed using: (a) a tablet disintegration tester ED-2SAPO (Electrolab, Mumbai, India); (b) a paddle dissolution apparatus DT800 (Erweka, Langen, Germany), with a capsule placed in a steel sinker (the stirring rate was 50 rpm). The capsules were tested for 120 min in 0.1 M HCl, followed by pH 6.8 phosphate buffer until disintegration.

2.7. Drug Release Test

The study was performed using a vertical diffusion cell (Enhancer cell, Erweka, Langen, Germany) and a paddle dissolution apparatus DT800 (Erweka, Langen, Germany) equipped with a built-in autosampler. The stirring rates of 50, 100 and 150 rpm were used. The enhancer cell with the mounted modified gelatin film is shown in Figure 2.

The film selected for the test was gelatin + Aquacoat + carrageenan (GAC), the same as for disintegration tests. The diffusion cell was filled with 2.5 mL of a 1% diclofenac solution in PEG 400 (the amount of diclofenac sodium was 25 mg). Then the investigated film (cut to a circle of 3 cm in diameter) was carefully placed on the top of the solution and secured with a sealing ring and a screw cap; the active surface was 4.15 cm^2. The test was performed in 900 mL of 0.1 M HCl for 120 min followed by 900 mL phosphate buffer pH 6.8 for 60 min.

Sampling of the acceptor fluid was performed every 15 min in the acid phase, and every 5 min in the buffer phase. Quantification of diclofenac was performed spectrophotometrically at 276 nm wavelength. The study was performed in triplicates.

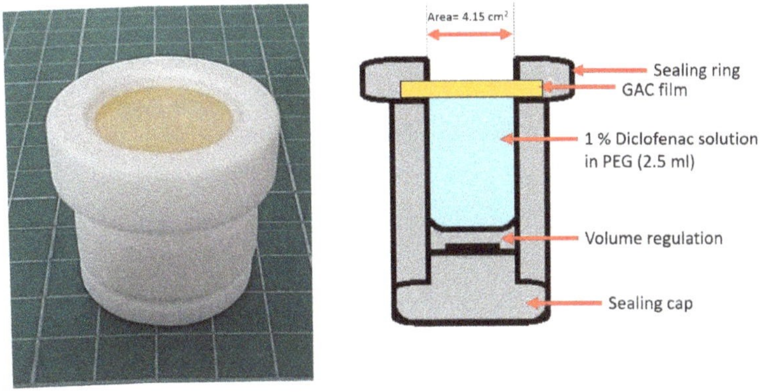

Figure 2. The enhancer cell with gelatin + Aquacoat + carrageenan (GAC) film.

3. Results

3.1. Microscopic Imaging of the Films and Capsules

At the first stage of the study, the prepared GA and GAC films were observed prior to and after submersion in HCl. Macroscopically it was visible that the samples after the acid treatment became opaque and swollen. Under the microscope, the untreated samples had a smooth surface with no structures visible [6]. As presented in Figure 3, the films after submersion in acid revealed a network-like structures, resembling scaffolds.

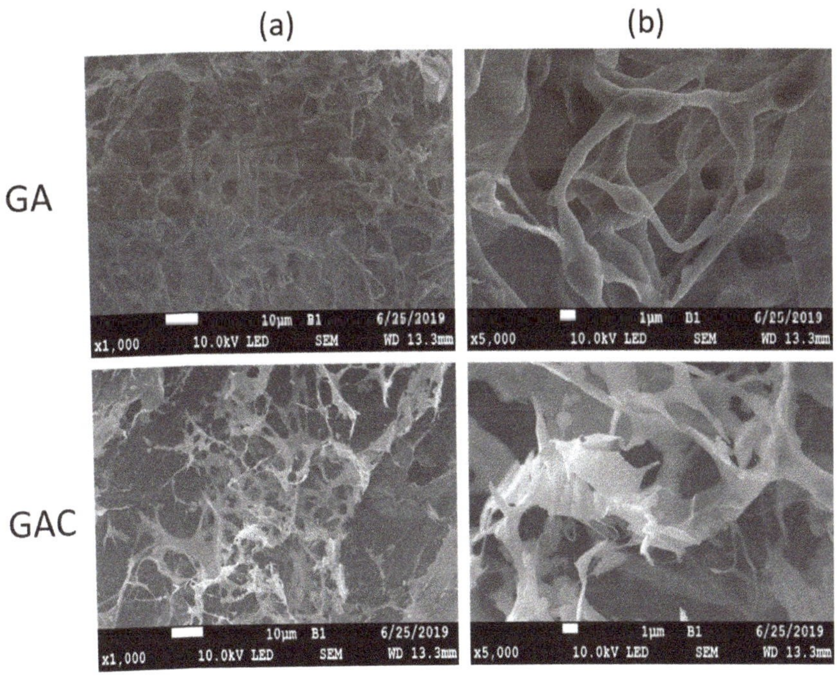

Figure 3. SEM image of GA and GAC films after 2 h in 0.1 M HCl. Scale bar: (**a**) 10 µm, (**b**) 1 µm.

There are clear differences between the images of a top and a middle layer of the sample (Figure 4). It appears that, after 2 h in acid, noticeably less solid material is left on the top of the film, than in the deeper part. The signals registered by CLSM can be potentially both from CAP and gelatin, due to very similar autofluorescence behavior. However, it is suspected that the outer layer consists mostly of CAP, while in the inner layer a swollen and undissolved gelatin can be present as well.

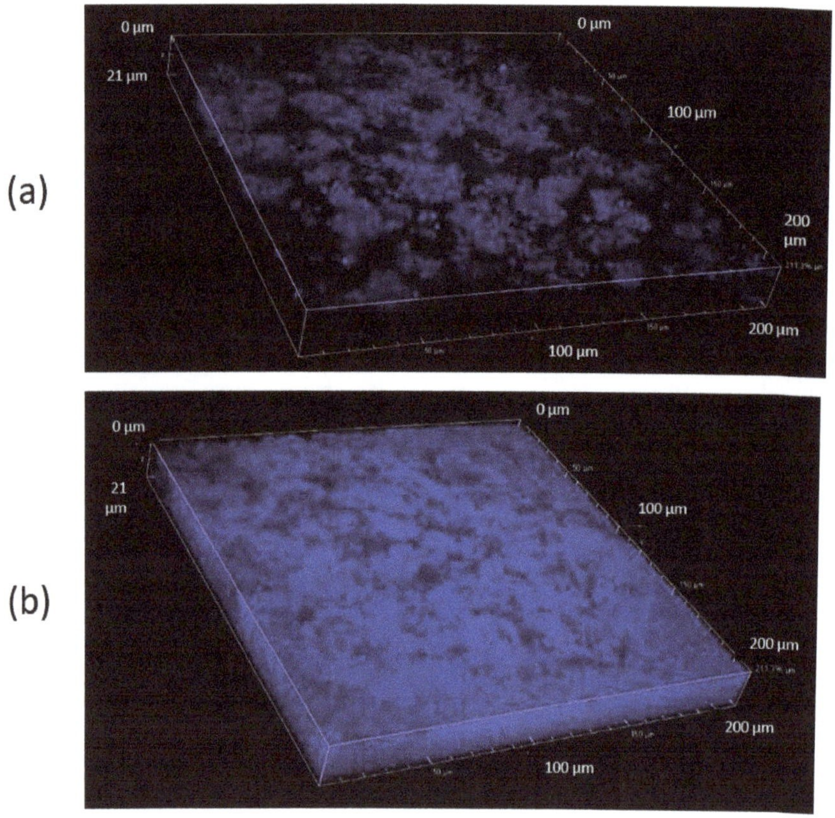

Figure 4. Confocal laser scanning microscopy (CLSM) images of GA film after 2 h immersion in 0.1M HCl: surface layer (**a**) and the inner central part (**b**) of the film.

Raman microscopy investigation was performed on GAC films before and after immersion in 0.1 M HCl. Several points have been scanned to obtain Raman spectra, which have been overlaid and compared. The spectra are shown in Figures 5 and 6.

As it can be seen from the spectra in Figure 5, the surface of the GAC sample is chemically uniform, without any phase separation visible. The acid-treated GAC samples display similar pattern in the spectrum as the untreated GAC. The spectra of the non-modified film (GEL) are not presented in the figure, but they were not different from the spectra of GAC. In Figure 6, different sets of spectra are overlaid. It appears that, in the untreated samples, the gelatin signals are overlapping with the peaks of CAP. After the acid-treatment, the signals from gelatin are weaker, but the signals from CAP are yet undetectable. This outcome can be explained by presence of a small amount of gelatin-rich phase residue undissolved in acid and covering the CAP scaffold. This corresponds well with SEM and CLSM results described above.

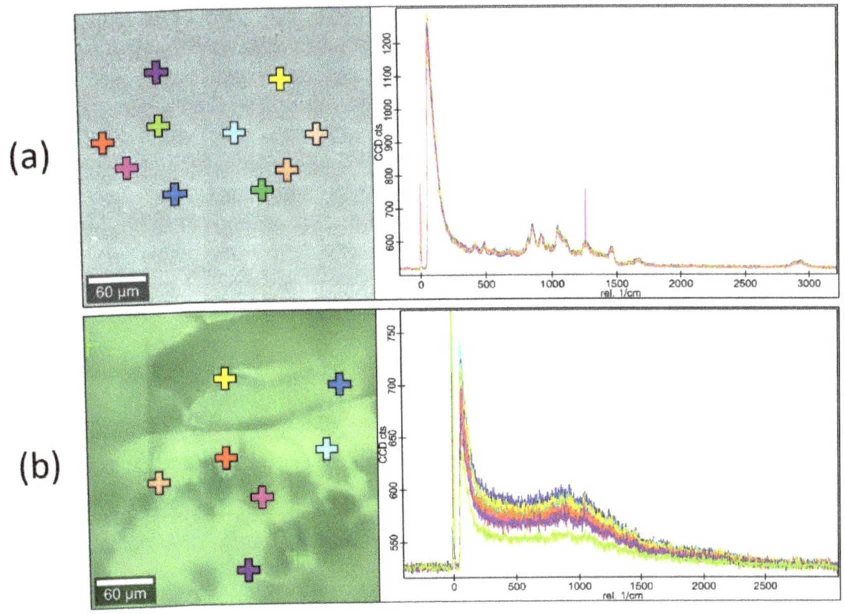

Figure 5. The Raman spectra of several points examined on the surface of GAC film: (**a**) before immersion in acid; (**b**) after immersion for 2 h in 0.1M HCl. Multiple overlaid spectra are presented on each graph.

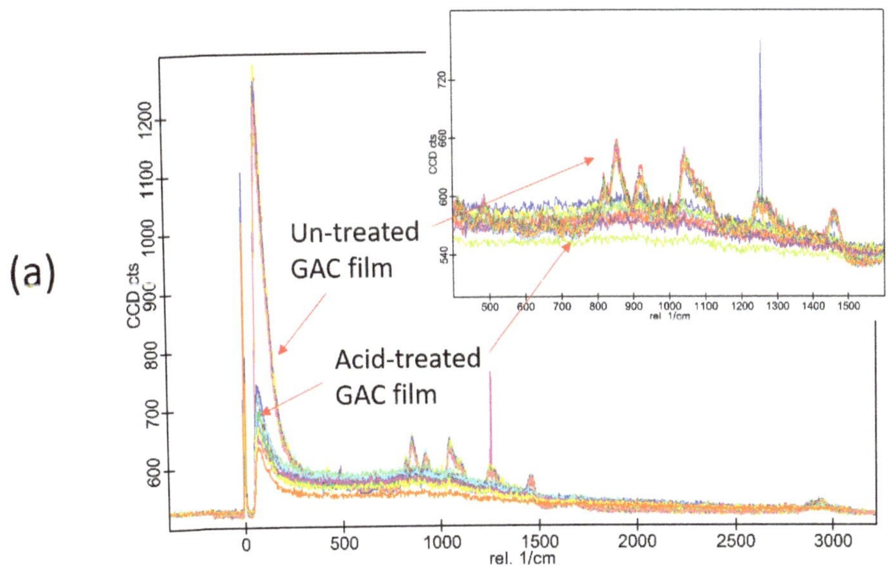

Figure 6. *Cont.*

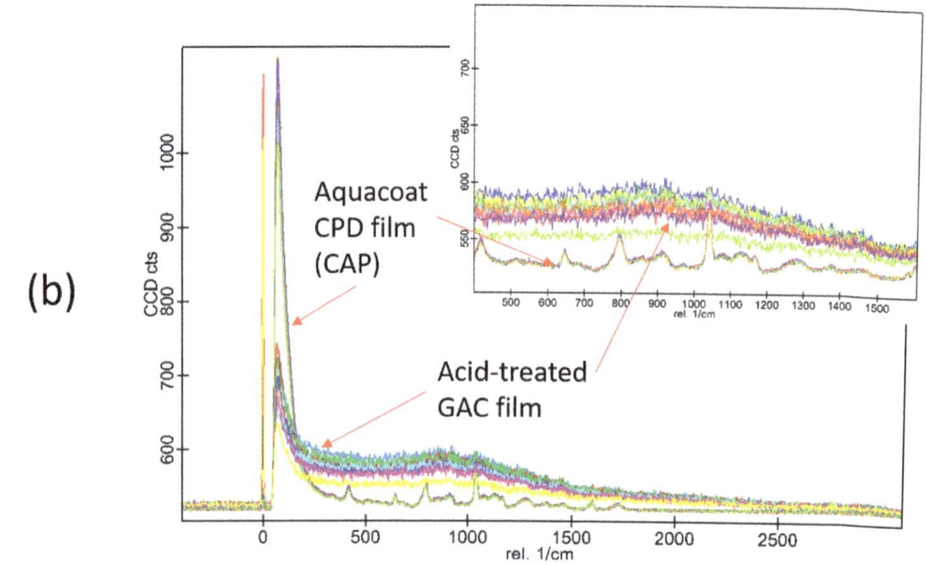

Figure 6. Comparison of the Raman spectra: (**a**) untreated and acid-treated GAC; (**b**) acid-treated GAC and cellulose acetate phthalate (CAP) film (without gelatin). Multiple spectra of each composition are presented.

3.2. QCM-D

A QCM-D study was performed to obtain supporting information on interactions between gelatin and CAP latex particles. Due to the fact that reliable results regarding particle deposition depend on the morphology of the used substrate, the films obtained in situ on the QCM sensors were investigated with AFM. It was confirmed that the films had uniform thickness and smooth surface, as shown in the Figure 7.

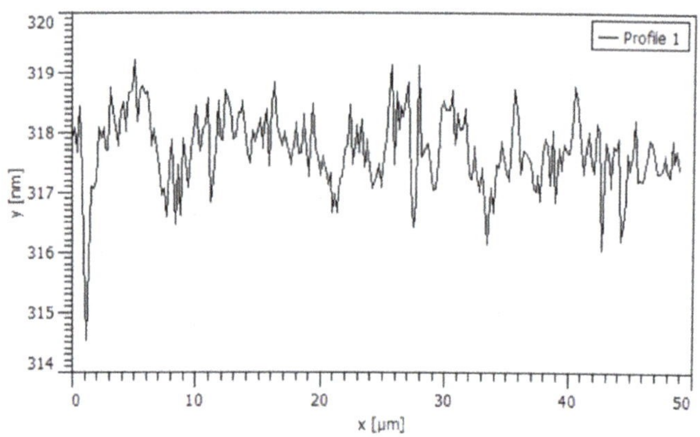

Figure 7. The surface morphology of the gelatin film on the quartz crystal microbalance with dissipation (QCM-D) sensor.

The QCM-D graph is shown in Figure 8. Although a large deposition of CAP particles on the gelatin film was detected, the very high extent of the frequency change (around 550 Hz) of all investigated overtones creates a risk of an error when calculating the mass deposition. Therefore, the obtained results were used only for the qualitative, and not for the quantitative analysis.

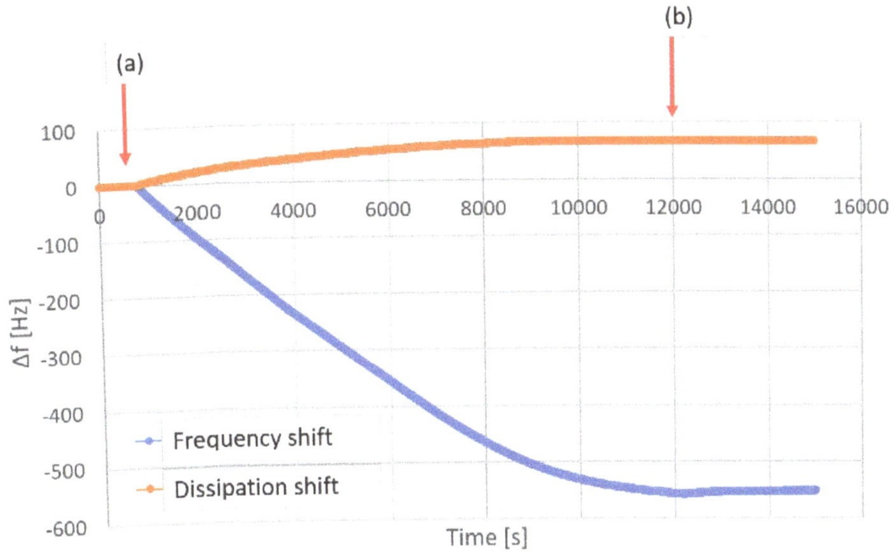

Figure 8. A QCM-D graph obtained at 5th overtone. The approx. 550 Hz drop in frequency carries risk of error on calculating the mass increase: (**a**) start of latex flow; (**b**) start of water flow (to remove particles that are not bound to the film).

The deposition of particles proceeded, until the full coverage of the QCM sensor occurred after approximately 12,000 s, which on the graph in Figure 8 is visible as a plateau in the frequency shift. Afterwards, the system was flushed with deionized water for 50 min, what did not cause any significant decrease in the amount of latex particles adsorbed on the gelatin.

3.3. Gas Permeability

Oxygen was a gas used for permeability test. Three compositions were investigated (GEL, GA, GAC). The thickness of investigated films was 650 ± 50 μm. The measured oxygen permeability ($cm^3/(m^2 \times 24\ h \times 0.1\ MPa)$) was 7.58 for GA sample, 3.13 for GAC and 3.43 for GEL samples. The test was performed twice for all the samples, and the same results were obtained.

3.4. Disintegration Time

The results of disintegration time measurements are shown in Table 2. The current pharmacopeial standards (European Pharmacopeia 10th) for disintegration time of gastro-resistant capsules state that the investigated sample should not disintegrate in 0.1 M HCl for 2 or 3 h (depending on the composition, however not less than 1 h), which should be followed by disintegration within 1 h at pH 6.8. In the investigated capsules, at the acid stage, no disruption of the capsule shell material was observed. However, the rupture of the capsule sealing was observed in several capsules.

Table 2. Disintegration time of GAC capsules with various fill. Three capsules from each batch were subjected to the test. The results are shown as mean ± standard deviation.

The GAC Capsule Filling Type	Batch No.	Tablet Disintegration Tester		Paddle Apparatus	
		0.1 M HCl	Phosphate Buffer pH 6.8	0.1 M HCl	Phosphate Buffer pH 6.8
MCT oil	1	86 ± 7 min	n/a	>2 h	6 ± 4.6 min
	2	>2 h	3.7 ± 1.5 min	>2 h	10.7 ± 7.4 min
	3	>2 h	2.5 ± 0.7 min	Single capsule leaking [1]	6.5 ± 7.8 min
PEG 400	1	77.7 ± 7.0 min	n/a	Single capsule leaking [1]	12.7 ± 4.2 min
	2	>2 h	7.3 ± 1.5 min	>2 h	39.0 ± 18.5 min
	3	>2 h	4.0 ± 2.0 min	>2 h	43.0 ± 13.2 min
Cetearyl alcohol	1	45.7 ± 7.4 min	n/a	79.3 ± 16.3 min	n/a
	2	48.7 ± 16.0 min	n/a	82.7 ± 23.3 min	n/a
	3	29.0 ± 33.0 min	n/a	81.3 ± 18.5 min	n/a

[1] the leakage was observed on the sealing of capsule.

The results of disintegration time measurements are not significantly different in regard to the method applied. The resistance of capsules to acid was similar when either MCT oil or PEG was used as a filling material. Surprisingly, the capsules filled with cetearyl alcohol disintegrated in acidic conditions within a relatively short time.

A careful observation in acid phase indicated that the shell did not disrupt in any other way but only through the sealing, while the walls of the capsules always retained their integrity. This indicates that the shell material itself is resistant to acid and the filling material does not change this property. The resistance of the capsules to acid, however, lacks reproducibility due to a variability of the seal quality. At pH 6.8, the capsules disintegrated through creation of a breach in the shell, caused by its thinning due to dissolution process. However, similarly to the test in acid, the disintegration always started at the sealing site. It was observed that the disintegration time at pH 6.8 was longer, when the paddle apparatus was used in the test, which can be attributed to different fluid dynamics that had impact on the rate of dissolution of the capsule shell.

The sealing sites of the investigated capsule shells and the reference commercial soft capsules were investigated microscopically. Although with an optical microscopy the image of the seals appeared similar to the commercial capsules, in SEM pictures, in some of the prepared capsules more sharp angle at the contact site of the fused films was observed. Such a defect is likely to induce formation of a rupture when capsules are swollen upon submersion in fluid. An evident difference between "commercial" sealing and the lab-scale sealing of the capsules is presented in Figure 9.

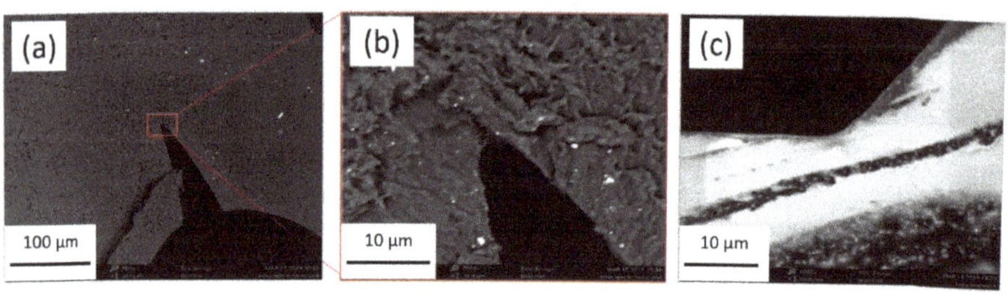

Figure 9. SEM images of the cross-sections of the capsule sealing site: (**a**) an apparently successful sealing; (**b**) close-up of the area; (**c**) a reference commercial soft gelatin capsule.

3.5. Drug Release Test

Due to the fact that during disintegration test, the disruption at the sealing zone appeared as a problem, the drug release test was performed in an enhancer cell, described in the Methods section.

Although this system can show potential differences in the kinetics of the drug release in comparison to a filled capsule, still the conclusions about acid-resistance and the rate of drug release can be drawn.

The pharmacopeial standards for drug release test from gastro-resistant forms require the release of less than 10% of the declared drug dose within 2 h in 0.1 M HCl, followed by at least 80% of the dose released at pH 6.8 within a specified time, usually not longer than 45 min.

The drug release test was performed for the diffusion cells filled with 1% m/v solution of diclofenac sodium in PEG (total dose 25 mg of API). The results obtained with different stirring rates are shown in Figure 10.

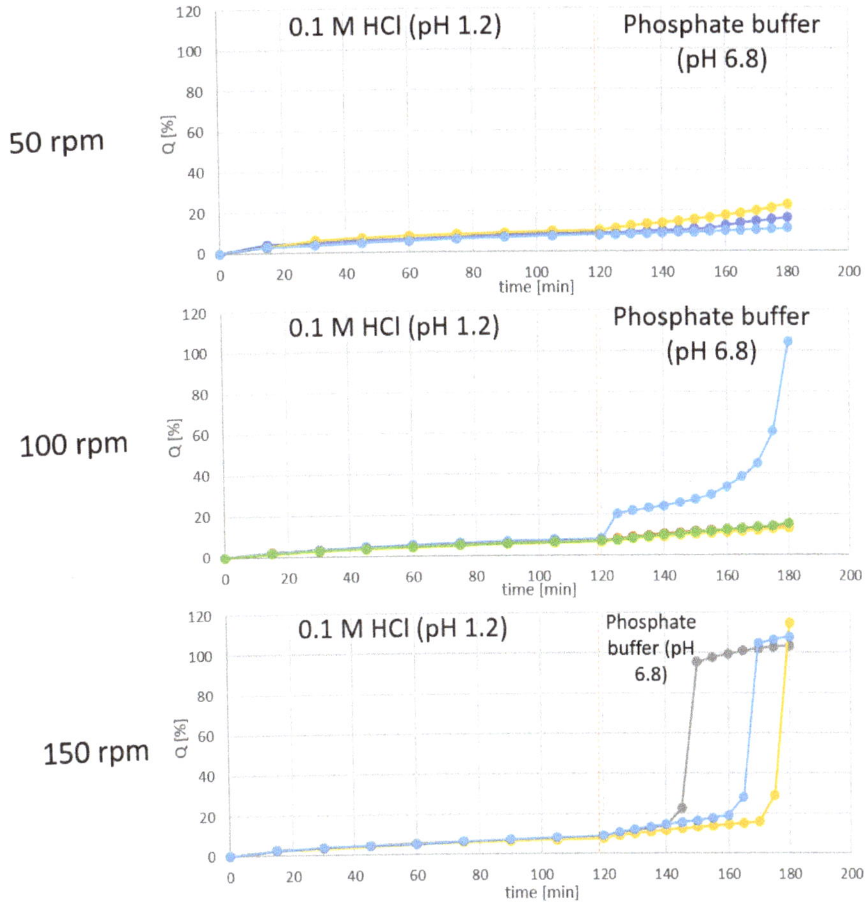

Figure 10. The effect of the stirring rate in a paddle apparatus on the release profiles of diclofenac sodium from the PEG 400 solution in a diffusion cell closed with GAC film.

The results show that at all of the used stirring rates, less than 10% of diclofenac was released during the acid phase of the test, which complies with the pharmacopeial requirements, and confirms the acid-resistance of the investigated films. On the other hand, at the buffer stage of the test, the release occurred in each sample only if they were tested at the high stirring rate, i.e., 150 rpm.

4. Discussion

The structures revealed by SEM in the films after they were treated with an acid (Figure 3) allow for conclusion that the formation of the modified films is based on a phase transformation in a gelatin-CAP mixture at a preparation stage. Due to the fact that pH of the utilized type B gelatin solution was around 4.5, one can expect that CAP should constitute a separate phase as this polymer is insoluble at low pH. The CAP phase can be considered continuous; due to the high temperature at the stage of mixture preparation, which is well above the glass transition temperature (Tg) of CAP, the particles can appear in a rubbery state and coagulate easily. Therefore, the two separate phases: gelatin gel and CAP phase, are physically mixed, forming a bi-continuous network with discreet separate microdomains. The structures revealed with SEM suggest that the phase separation proceeds with a spinodal decomposition mechanism, which is spontaneously initiated, and kinetically limited by increase in the viscosity of gelatin during the gelling process when the temperature drops at the casting stage. Similar "kinetic arrestation" of the phase separation process in the gelatin-containing mixtures was described by Lorén et al. [10] and by Tromp et al. [11].

To better explain the phase separation in the discussed systems, an additional experiment was performed. A premix of GA composition was placed in a glass vial and slowly heated to reach 80 °C. After 5 min at 80°C the temperature was lowered stepwise by 10 °C each 5 min. The appearance of turbidity indicated the phase separation process. After reaching 40 °C the sample was heated again to 80 °C and kept at that temperature for 48 h. The second heating revealed that the temperature-dependent phase separation is reversible (the sample became transparent again). The results are presented in Figure 11.

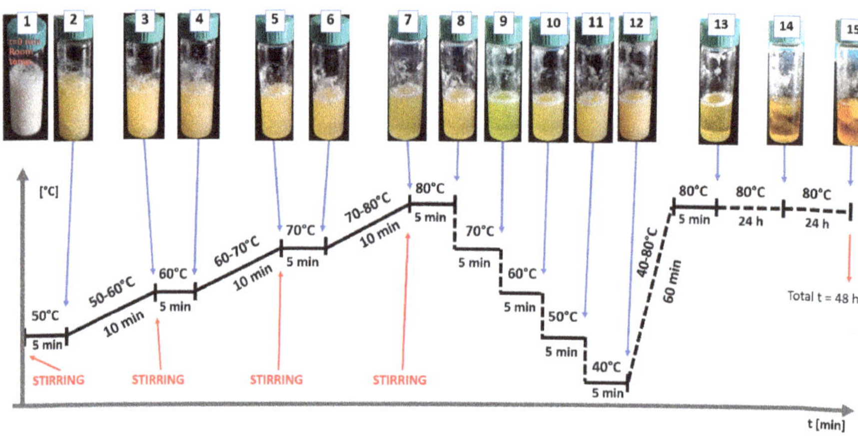

Figure 11. The visible phase-separation on lowering the temperature of the sample, and suspected spinodal decomposition process on storage at 80 °C for a prolonged time.

However, the storage at 80 °C for a prolonged time caused irreversible phase separation, with a fibrillar/sponge-like appearance of the precipitated phase within the liquid continuous phase. Additionally, the alteration of color of the sample and the fact that the gel was partially liquid at room temperature indicated gelatin degradation. The overall results of this experiment confirm that the phase separation process proceeds at high temperature and can be stopped by immobilization of the growing CAP structure in a gelatin gel when the temperature drops below approx. 50 °C. This supports the thesis that the separate phase of CAP acts like a "reinforcement" for the gelatin network, decreasing the rate of penetration of the acidic medium into the water-soluble phase, and explains the structure integrity of the films in acidic media. Furthermore, it appears that the temperature/time balance in the preparation procedure allows to exploit the natural imbalance between phases for the favor of functionality of the films.

The mechanism behind formation and acid-resistance of both binary (GA) and ternary (GAC) polymer systems appears the same. However, in comparison to the GA films, a more significant hindering of disintegration and dissolution can be observed in case of GAC films [5]. This can be explained by a possible interaction between gelatin and carrageenan, which is suspected to be a polyelectrolyte complex formation. This non-covalent interaction has already been reported and is widely described [6,12–16]. Even though carrageenan is present in the film in a small amount, it still can significantly impact the viscosity of the gel phase. Therefore, by increasing the density of the polymer network, it leads to formation of a gel diffusion layer more viscous than the gelatin alone. That causes higher swelling degree and slower dissolution rate of soluble ingredients present in the films, as it was previously described [5]. It appears also possible that the carrageenan in the film-forming mixture accumulates on the CAP-gelatin interface, supporting the CAP scaffold during the immersion in an acid. However, we believe that the more irregular appearance of the CAP scaffolds visible in Figure 4 are actually CAP coated with undissolved gelatin-carrageenan complex. This hypothesis appears to be supported by the results of Raman microscopy (see Figures 6 and 7), in which after immersion in acid, the structure shows a pattern in Raman spectra with the same features as observed in the spectra of the films before the acid treatment (both CAP and GAC), what suggests that during the immersion the gelatin was not fully dissolved and it is still present on the surface of the residual CAP scaffold structure.

The CLSM study (Figure 4) appears to correspond well with the SEM results. Additionally, it was discovered that there are clear differences in the density of the solid material left in structure inside the films and on its surface. Due to the fact that the soluble fraction of the film composition is supposedly gelatin-based gel and plasticizer, the mechanism of erosion of the film when placed in an acid is likely based on the diffusion of the medium through the gel layer. The erosion can be additionally limited by the presence of the insoluble CAP phase, which acts also as a scaffold. Therefore, it can be suspected that the penetration of the acidic medium into the membrane is delayed and can depend on both density of the CAP scaffold and viscosity of the gelatin-based phase.

The higher barrier properties of a ternary system (GAC) than the binary one (GA) also was demonstrated by the oxygen permeability test. However, in the literature the results of oxygen permeability can be found only for very thin gelatin films obtained from dilute gelatin solutions [17], where the values of oxygen permeability can be around 350–600 $cm^3/(m^2 \times 24\ h \times 0.1\ MPa)$. In the present study the permeability was measured for films with thickness around 650 μm, at which the measured values were between 3.5 and 7.5 $cm^3/(m^2 \times 24\ h \times 0.1\ MPa)$. Although, after addition of CAP, the oxygen permeability increased slightly, the differences between formulations were still very low and one can conclude that there is a lack of significant influence of the film ingredients on gas barrier properties.

During the formation of modified GA or GAC films, the CAP spherical particles (average size of 0.43 μm) are being incorporated in the gel structure. QCM-D is a surface sensitive technique which can be applied to analyze the interaction of the particles of CAP with gelatin. Quantitative values can be obtained with well-defined model systems. Saurebrey and Johannsman models [18] are often used to calculate the surface excess after adsorption. In our particular case, those models will not give reliable approximation because of the large size of the CAP particles. However, a qualitative information on interaction between CAP particles and gelatin films can be obtained. The large decrease in frequency of the vibration as soon as the CAP particles were introduced into the flow cell reflects the adsorption of CAP particles on the gelatin films. Due to the fact that the measurements were performed at 25 °C, a coalescence of the CAP is rather negligible, therefore such interaction should be based purely on surface charge of the particles. Although the test could not be performed at high temperature (80 °C), the confirmed high affinity of these two materials at 25 °C may be also relevant at higher temperature.

We assume that such type of interaction can potentially stabilize the CAP inside the gel matrix and allow formation of a network structure during the preparation of the film-forming mass.

Preparation of capsules on a lab scale with the proposed steel mold was a simple process, allowing for application of a liquid fill, and for obtaining visually sealed capsules. However, the results of the disintegration tests show large variability because of the significant tendency of capsules to disrupt at the sealing area. On the other hand, the results prove that formation of the capsules using GAC composition is generally possible, and the capsules can be filled with liquid oil, PEG or melted fatty alcohol. In addition, no case in which a capsule disintegrated in acid at other region than the sealing was observed. This confirms that the films being in contact with a filling, still retain their structural integrity when submersed in acid.

For the purpose of investigation, whether the filling formulation has an impact on acid-resistance of the capsule shell, the capsules were filled with three types of substances: PEG, MCT oil and cetearyl alcohol. The results show that the capsules filled with solid fatty alcohol show lower resistance of the sealing to disintegration in acid. On the other hand, there is no noticeable difference between the capsules filled with MCT oil or PEG. Overall, the mechanism of disintegration of the capsules appears to be related more to the capsule formation process, than to the filling composition. It was observed that the lab-manufactured capsules are prone to leakages on the sealing zone, especially when more intensive mechanical stress was involved, as in the tablet disintegration apparatus. We believe that the imperfect capsule sealing can be corrected when encapsulation process involves the conventional soft capsule manufacturing machines.

The imperfections in the sealing region did not allow to further test the capsules in the drug release test. This is why this study was performed in a vertical diffusion cell placed in a paddle dissolution apparatus. The test was performed to investigate whether the films display barrier properties against diffusion of diclofenac sodium at acidic pH. In addition, it was important that the films allow to release the API after switching the pH to neutral (6.8). In one of our previous articles, the barrier properties of the films towards radio-labeled water were described [5], and preliminary data on the diffusion-hindering by the modified gelatin-CAP compositions was obtained. The present investigation confirms appropriate barrier properties of GAC film, because no diffusion of diclofenac during 2 h in 0.1 M HCl was observed. However, the reproducibility of the diclofenac release at pH 6.8 is not very high. The release was initiated at different time points, what results from the mechanism of film rupture—not dissolving totally in a specified time, but forming a breach. Since at lower stirring rates the release of diclofenac did not occur or was accidental, the proposed model requires higher stirring rates, which shows the significance of the mechanical factor in the dissolution of the GAC film in the pH 6.8 buffer.

Although the performed experiment with diclofenac as a model drug demonstrates lack of the drug diffusion through the modified gelatin film immersed for 2 h in an acid, diffusion of an acid through the membrane was not measured in the course of this stage of the research. Impermeability of the new capsule-forming material to the acid is a condition for using it in the capsules filled with an acid-labile drugs.

5. Conclusions

In this work the discreet kinetically-limited phase separation was identified as the main factor influencing the resistance of the modified gelatin films to disintegration in the acidic environment. The imaging techniques (SEM, CLSM, Raman microscopy) provided the information on the mechanism of film partial dissolution in acid. The QCM-D analysis proved the affinity of the latex particles to the surface-wetted gelatin structures, which may be important in regard to the film formation process. The lab-scale soft capsule formation process was performed, and the tested filling materials were proved to be compatible with the films, however the obtained capsules showed the sealing area as a weak spot, limiting the acid-resistance of the capsules during the disintegration test. On the other hand, a modified dissolution test with a paddle apparatus and diffusion cell allowed to confirm that the films are hampering the drug release in acidic phase, while releasing the drug at pH 6.8. The drug release at pH 6.8 was possible, however, only when higher stirring rates (150 rpm) were applied.

Author Contributions: Conceptualization, B.M. and M.S.; methodology, B.M., M.S. and A.L.; validation, B.M., V.A.; investigation, B.M.; resources, M.S., A.L.; data curation, B.M., V.A.; writing—original draft preparation, B.M.; writing—review and editing, M.S., A.L.; supervision, M.S.; funding acquisition, B.M. All authors have read and agreed to the published version of the manuscript.

Funding: The work was supported by the National Science Centre Poland (Grant No. 2015/19/N/NZ7/03447).

Acknowledgments: The oxygen permeability test was performed by COBRO–Packaging Research Institute (Lukasiewicz Research Network, Warsaw, Poland). Katrina Logg and Archana Samanta from Chalmers University of Technology in Gothenburg are acknowledged for the support with the imaging techniques. Authors acknowledge funding from NordForsk for the Nordic University Hub project #85352 (Nordic POP, Patient Oriented Products).

Conflicts of Interest: The authors declare no conflict of interest.

References

1. Soni, H.; Patel, V.A. Gastro retentive drug delivery system. *Int. J. Pharm. Sci. Rev. Res.* **2015**, *31*, 81–85.
2. Benameur, H. Enteric capsule drug delivery technology—Achieving protection without coating. *Drug Dev. Deliv.* **2015**, *15*, 34–37.
3. Felton, L.A.; Haase, M.M.; Shah, N.H.; Zhang, G.; Infeld, M.H.; Malick, A.W.; Mcginity, J.W. Physical and enteric properties of soft gelatin capsules coated with Eutragit L30 D-55. *Int. J. Pharm.* **1995**, *113*, 17–24. [CrossRef]
4. Hassan, E.M.; Fatmi, A.A.; Chidambaram, N. Enteric Composition for the Manufacture of Soft Capsule Wall. U.S. Patent 8,685,445, 1 April 2014.
5. Maciejewski, B.; Ström, A.; Larsson, A.; Sznitowska, M. Soft gelatin films modified with cellulose acetate phthalate pseudolatex dispersion—structure and permeability. *Polymers (Basel)* **2018**, *10*, 981. [CrossRef] [PubMed]
6. Maciejewski, B.; Sznitowska, M. Gelatin films modified with acidic and polyelectrolyte polymers—material selection for soft gastroresistant capsules. *Polymers (Basel)* **2019**, *11*, 338. [CrossRef] [PubMed]
7. Reich, G. Formulation and physical properties of soft capsules. In *Pharmaceutical Capsules*; Podczeck, F., Jones, B., Eds.; Pharmaceutical Press: London, UK, 2004; pp. 201–212.
8. Ulrich, E.; Prosekov, A.; Petrov, A.; Dyshlyuk, L.; Kozlova, O. Properties of plant analogs of pharmaceutical gelatin for shells of soft capsules. *Biol. Med.* **2015**, *7*, 113–115.
9. Kamiya, S.; Nagae, K.; Hayashi, K.; Suzuki, N.; Hayakawa, E.; Kato, K.; Sonobe, T.; Nakashima, K. Development of a new evaluation method for gelatin film sheets. *Int. J. Pharm.* **2014**, *461*, 30–33. [CrossRef] [PubMed]
10. Lorén, N.; Langton, M.; Hermansson, A.-M. Confocal laser scanning microscopy and image analysis of kinetically trapped phase-separated gelatin/maltodextrin gels. *Food Hydrocoll.* **1999**, *13*, 185–198. [CrossRef]
11. Tromp, R.H.; van de Velde, F.; van Riel, J.; Paques, M. Confocal scanning light microscopy (CSLM) on mixtures of gelatine and polysaccharides. *Food Res. Int.* **2001**, *34*, 931–938. [CrossRef]
12. De Kruif, C.G.; Weinbreck, F.; de Vries, R. Complex coacervation of proteins and anionic polysaccharides. *Curr. Opin. Colloid Interface Sci.* **2004**, *9*, 340–349. [CrossRef]
13. McClements, D.J. Non-covalent interactions between proteins and polysaccharides. *Biotechnol. Adv.* **2006**, *24*, 621–625. [CrossRef] [PubMed]
14. De Kruif, C.G.; Tuinier, R. Polysaccharide protein interactions. *Food Hydrocoll.* **2001**, *15*, 555–563. [CrossRef]
15. Michon, C. Gelatin/iota-carrageenan interactions in non-gelling conditions. *Food Hydrocoll.* **2000**, *14*, 203–208. [CrossRef]
16. Derkach, S.R.; Ilyin, S.O.; Maklakova, A.A.; Kulichikhin, V.G.; Malkin, A.Y. The rheology of gelatin hydrogels modified by κ-carrageenan. *Lwt-Food Sci. Technol.* **2015**, *63*, 1–8. [CrossRef]

17. Avena-Bustillos, R.J.; Chiou, B.; Olsen, C.W.; Bechtel, P.J.; Olson, D.; Mchugh, T.H. Gelation, oxygen permeability, and mechanical properties of mammalian and fish gelatin films. *J. Food Sci.* **2011**, *76*, 519–524. [CrossRef] [PubMed]
18. Johannsmann, D.; Mathauer, K.; Wegner, G.; Knoll, W. Visco-elastic properties of thin films probed with a quartz crystal resonator. *Phys. Rev. B* **1991**, *46*, 7809–7815. [CrossRef]

materials

MDPI

Review

Hydrogels as Potential Nano-, Micro- and Macro-Scale Systems for Controlled Drug Delivery

Adam Chyzy [†], Monika Tomczykowa [†] and Marta E. Plonska-Brzezinska [*]

Department of Organic Chemistry, Faculty of Pharmacy with the Division of Laboratory Medicine, Medical University of Bialystok, Mickiewicza 2A, 15-222 Bialystok, Poland; adam.chyzy24@gmail.com (A.C.); monika.tomczyk@umb.edu.pl (M.T.)

* Correspondence: marta.plonska-brzezinska@umb.edu.pl; Tel.: +4885-748-5683

† These authors contributed equally to this work.

Received: 16 December 2019; Accepted: 27 December 2019; Published: 2 January 2020

Abstract: This review is an extensive evaluation and essential analysis of the design and formation of hydrogels (HGs) for drug delivery. We review the fundamental principles of HGs (their chemical structures, physicochemical properties, synthesis routes, different types, etc.) that influence their biological properties and medical and pharmaceutical applications. Strategies for fabricating HGs with different diameters (macro, micro, and nano) are also presented. The size of biocompatible HG materials determines their potential uses in medicine as drug carriers. Additionally, novel drug delivery methods for enhancing treatment are discussed. A critical review is performed based on the latest literature reports.

Keywords: hydrogel; drug delivery; polymer; biocompatibility; immobilization of drug

1. Introduction

Drug delivery systems grounded on hydrogels (HGs) are interesting because of their high biocompatibility and biodegradability. These properties are especially relevant for materials used for biomedical engineering applications, an example of which may be drug delivery or tissue engineering [1–3]. HGs are water-swollen polymeric networks containing chemical or physical cross-links able to absorb large quantities of water or biological fluids [4]. HGs have a variety of structures, architectures, sizes (from centimetres to sub-nanometres), and functions, and together with other properties, these variables determine HG use for drug delivery [1–3,5].

HGs can be prepared from one polymer (homo-polymeric HG), two or more polymers (multi-polymeric HG); they may also contain other nanostructures/nanoparticles in a polymeric network [6–9]. These polymeric networks can be chemically and physically modified imparting new and unusual properties [9–11]. Chemical structures, compositions, biodegradability, biological functions and different physicochemical properties (e.g., mechanical, rheological, spectral, thermosensitive, pH stability) can be modified [2,6,9,12]. These variations influence the performances of HGs and affect loading and releasing properties for drugs [7,9,11–15].

HGs can be used to form microparticles, nanoparticles, micelles and films [15,16]. For HG particles, the particle size (macro, micro and nano) determines the route by which HGs can be delivered into the human body [17–21]. For micro- or nano-sized HGs, the effects of various physical and chemical factors on drug release should be considered [15,21]. Therefore, drug immobilization in a polymer matrix should be considered in the context of controlled release at target sites. Various in vivo and in vitro drug application techniques have been developed with various therapeutic properties [12,22–24], including antifungal [25–27], antibacterial [28–32], antitumor [33–36], anti-inflammatory [37,38], immunomodulatory [39–41], anti-glycemic [42], antioxidant [32,37,43], tissue repair and regeneration [14,16,44].

The objective of this review is to explore the potential use of HG particles in drug delivery systems with respect to their size (macro, micro and nano). This review also attempts to identify the effects of HG particle size and physicochemical properties on biological performance and medical applications. Finally, novel drug delivery methods for enhancing treatments are discussed.

2. Types of Hydrogels

A gel is a liquid treated with gelling substances, including natural polymers (e.g., agar, alginates, and dextran), semi-synthetic polymers (cellulose derivatives) or synthetic polymers (acrylic and methacrylic acid derivatives) [45]. Lipophilic gels (oleogels) are obtained using oil as the dispersing phase. Hydrophilic gels (hydrogels) are obtained using water as the dispersing phase [46–48]. Due to their similarity to living human tissues, HGs with controlled drug release are widely used in pharmaceuticals. By modifying their compositions and physicochemical properties (e.g., to impart hydrophilic or hydrophobic character), HGs can be used as drug carriers for external or internal use [49].

HGs are classified using different criteria [50]. The simplest criterion is origin, i.e., natural or synthetic [50,51]. Natural HGs are biocompatible, biodegradable [1] and support cell activity. However, natural HGs have low mechanical strength and large inter-batch variety. Proteins such as collagen or polysaccharides (e.g., chitosan, dextran, and alginate) are examples of natural HGs [52–55]. Synthetic HG polymers are prepared from polymerizable monomers, including vinyl acetate, acrylamide, ethylene glycol and lactic acid (made from plants, mostly from corn and sugarcane) [56–59]. Synthetic HGs can be precisely controlled and tailored to achieve desired properties. However, synthetic HGs typically lack bioactivity and have low biodegradability. Hybrid HGs consist of chemically, functionally and morphologically different units [60–65]. Biologically active proteins, peptides, nano/microstructures are constituent parts of hybrid HGs and are connected with each other by physical or chemical forces [60–66]. Because of their construction, hybrid HGs derive their bioactivity from natural materials; furthermore, the easy control over physical and chemical properties of hybrid HGs are due to synthetic material properties [66].

2.1. Physical and Chemical Hydrogels

HGs can be classified into two groups based on the type of interactions involved in the creation of the network structure. The first group includes chemical solids [47,51,67,68], wherein HGs form three-dimensional (3D) networks with polymer chains are connected by permanent covalent bonds via cross-linking reactions [69]. Characteristic features of chemical HGs include their ability to swell resulting from interactions among the polymer network, water and the density of connections between polymer chains. Chemical HGs are not homogeneous due to the hydrophobic aggregation of cross-linking agents and high cross-link-density clusters [60].

The second group includes physical (reversible) HGs [51,67,68]. These HGs have chains connected by weak hydrogen bonds, ionic bonds and dipolar or hydrophobic interactions excluding those that dissolve before use [68]. These forces result in non-homogeneous HGs [60]. Examples of physical HG are solutions of agar, gelatine, and polyvinyl alcohol [70,71].

These two types of HGs encompass a wide variety of macromolecular structures formed from cross-linked and entangled linear homopolymers and linear, block or graft copolymers [72–74]. The HG networks can be stabilized by reactions of monovalent and polyvalent ions, multiple monovalent ions or complexes containing hydrogen bonds. The properties and applications of these HGs are closely related to cross-linking density, which determines swelling behaviour and the combined properties of solid and liquid phases [51,68,75].

2.2. Conventional and Stimuli-Responsive Hydrogels

HGs are also classified as conventional and stimuli-responsive. Conventional HGs comprise loosely connected hydrophilic, mostly non-ionic polymers with significant degrees of swelling in water without dissolution [76]. Stimuli-responsive HGs respond to various factors, such as small changes in temperature, ionic strength, pH, electric field, mechanical stress, light and selected substances (Figure 1a) [49,51,77–80]. Stimuli-responsive HGs can be tailored to react to various types of stimuli in the body, including ionic strength, pH and temperature, to act as potential drug carriers [49,51,77].

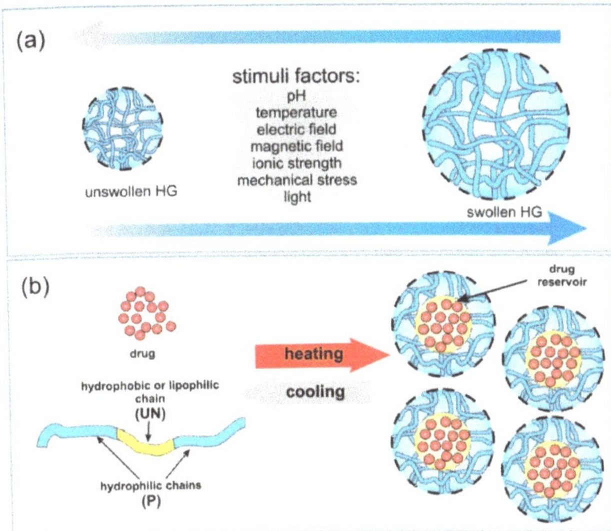

Figure 1. (**a**) Types of stimuli causing HG swelling. (**b**) Schematic illustration of a thermo-reactive HG formation loaded with a drug that uses temperature as a stimulus.

2.2.1. Thermosensitive Hydrogels

Thermosensitive HGs have a specific balance between a hydrophilic polymer gel and water molecules. In these systems, even small changes in temperature can disturb this balance and cause a phase change from sol to gel [50]. Such phase transitions often occur when the temperature changes from room temperature to physiological temperature of the human body [46]. This is because temperature influences interactions among water molecules and the hydrophilic and hydrophobic segments of the HG polymer [81].

HGs comprise hydrophilic (P) and hydrophobic (UN) segments arranged in the sequence: P-UN-P or UN-P-UN (Figure 1b). The hydrophilic segments used in HG formulations are characterized by high swelling properties, water solubility and biocompatibility [46]. The hydrophobic segments increase the loading capacity of other hydrophobic molecules, such as drugs, due to the micellization of nanoparticles (Figure 1b) [6,82].

For the proper design of thermosensitive HGs, the phase transition temperature should be identified [83]. The temperature at which the polymer solution separates into two phases is called the critical solution temperature (CST) [46] and is used to classify thermosensitive HGs as positive or negative temperature responsive systems [84,85]. The CST is typically lower for HG polymers with higher molecular weight or hydrogen-bonding interactions. Thermo-responsive HG polymers with lower CST (LCST) remain liquid at low temperature and are called negative temperature HGs. At temperatures lower than the LCST, these HGs swell [86]. When the temperature increases these HGs undergo a sol-gel transition and become insoluble upon heating. The opposite phenomenon is observed for HG polymers with upper CST (UCST, positive temperature HGs), which become soluble upon heating [46]. At temperatures lower than the UCST, positive temperature HGs dehydrate [87]. LCST HGs are used in the medical field because they form gels in situ at physiological temperatures, i.e., 30–37 °C (Table 1) [88]. Examples of synthetic polymers used to develop LCST HGs include copolymer blocks of poly(ethylene oxide) and poly(pentapeptide), poly(*N*-isopropylacrylamide) (PNIPAM), and poly(*N*,*N*-diethyl acrylamide) (PDEAM) [88]. Natural LCST polymers include cellulose derivatives, chitosan and gellan gum (Table 1) [88–90].

Table 1. Phase transition temperatures for selected polymers with LCST and UCST.

Polymer Name	Abbreviation	Transition Temperature (in Water)	Refs.
Poly(*N*-isopropylacrylamide)	PNIPAM	30–34 °C	[88]
Poly(*N*,*N*-diethylacrylamide)	PDEAM	32–34 °C	[88]
Poly(methyl vinyl ether)	PMVE	37 °C	[91]
Polyvinyl chloride	PVC	30–50 °C	[92,93]
Gellan gum	-	50–60 °C	[88]
Methylcellulose	-	40 °C	[94,95]
Acrylamide and acrylic acid	AAm and AAc	15–25 °C	[96]

2.2.2. Photo-Responsive Hydrogels

Light-responsive HGs are subjected to sol-gel changes under the influence of light at different wavelengths: infrared (IR), ultraviolet (UV), or visible (VIS) [50]. The initiation of sol-gel phase transitions in photo-responsive HGs using light is straightforward and non-invasive making these HGs a potential source of functional material in biomedicine, e.g., for drug delivery [50,97,98]. To prepare a photo-responsive HG, a light-responsive chromophore group is attached to HG polymers [99,100]. Incident light initiates swelling/deswelling processes or sol-gel phase transition to release drugs at desired locations. UV-light-responsive HGs undergo photopolymerization or photocleavage depending on the exposure length [101,102]. Light wavelength is also extremely important because it affects the drug release control process [103].

2.2.3. pH-, Electric- and Magnetic-Responsive Hydrogels

pH-responsive HGs have specific pH-dependent physicochemical properties [104]. These HGs have acidic and alkaline groups associated with HG polymer chains [50,105]. The human body is a dynamic environment with different tissues having different pH ranges, which provides delivery means for in vivo pH-responsive HGs in pharmaceutical and biomedical applications [106–108].

Ionizable groups embedded in the polymer network of pH-responsive HGs accept or donate protons in response to pH change, resulting in changes to the structure and solubility of the HGs and consequently, swelling or deswelling (Figure 1a). The most frequently used synthetic monomer with acidic characteristics are acrylic acid, methacrylic acid, maleic anhydride, *N*,*N*-dimethylaminoethyl methacrylate and sulphonamide-containing polymers [2,18,20,91,109]. Weak synthetic polybases, such as aromatics 4-vinylpyridine, 2-vinylpyridine, poly(vinyl imidazole), poly(*N*,*N*-dimethyl aminoethyl methacrylate), and poly(*N*,*N*-diethyl aminoethyl methacrylate), accept protons at low pH [50,110,111]. There is also a large group of natural polymers used in pH-responsive HGs that has advantages over synthetic polymers due to their degradability within the human body, making them excellent biocompatible components in drug delivery [9,33,88,112].

There are also HGs that respond to external physical stimuli, like electric and magnetic fields [52,105]. Magnetic or electrically responsive HG networks have been developed for drug-delivery applications. The influence of an electric field causes electro-responsive HGs to swell or de-swell, which can be adapted for drug delivery systems [113]. Similar applications have been determined for magnetic field-based stimuli, especially high frequency fields, which are much safer for humans than electric fields or UV light [114]. Magnetic and electrical stimuli-responsive HGs allow for the study of cell behaviour by altering hydrogel properties in situ. These stimuli-responsive HGs can serve in potential drug delivery systems for targeting and localizing to delivery sites for controlled drug release upon application of electric or magnetic fields [115–117].

2.3. Other Classifications for Hydrogels

HGs could also be classified by the substrates used for their production. Polymer networks may be synthesized from linking monomers to form high-molecular weight polymers. Hydrophilic monomers, such as polyphenylene oxide (PPO) or polyethylene glycol (PEG), are copolymerised with crosslinkers to form networks for drug delivery [118,119]. When prepolymers (oligomers) are used as synthetic substrates, different types of HGs are obtained. A series of model networks consisting of PEG, tetra-PEG gels, have been prepared by Sakai et al. [119,120]. Small-angle neutron scattering measurements confirmed, that network structure is extremely uniform, even in the presence of different functional groups at the ends, i.e., amine group and succinimidyl ester group [121,122]. Additionally, the modified with ionic liquid form of tetra-PEG gels possesses high ion conductivity and high mechanical properties [123,124]. For example, polyurethane-based HGs are used in the fabrication of flexible medical devices, prosthetics for example [125]. Other substrate-specific HGs are those with network structures made by crosslinking hydrophilic polymer chains, e.g., chitosan crosslinked with glutaraldehyde [75].

Due to large differences in their chemical compositions, HGs are sometimes broadly classified [51,126]. For example, in terms of the types of monomers used to prepare HGs (polymeric composition), three groups of HGs are recognized: (i) homopolymeric HGs (consisting of one type of hydrophilic monomer); (ii) copolymeric HGs (two or more monomer types of which at least one is hydrophilic), and (iii) multipolymer HGs (obtained from more than one type of polymer).

Multipolymer interpenetrating polymeric (IPN) HGs represent a class of HGs made of two independent cross-linked synthetic and/or natural polymers [51,75,127–129]. HGs there are two polymer networks, one of which is polymerized around and within the on the second polymer network, and what is very important, there are no covalent linkages between those two polymeric networks.

HGs may take significantly different morphological forms, which can be a criterion for their division depending on the state of aggregation (physical form) [17,105,129]: (i) solid (amorphous, semicrystalline or crystalline); (ii) semisolid (for example shear-thinning HGs), and (iii) liquid [17,51,75,130]. Solid HGs can be useful in formation of functional tissues [75,131]. At room temperature they are solid in nature with strong cross-linked network structure [17]. Solid HGs swell in the presence of water or other hydrophilic solutions like buffers or biological fluids. Semisolid HG consists of two types components where at least one possess biological nature, e.g., plant resins or gums [17]. The shear-thinning HGs also belong to this group [17,132,133]. These HGs can be pre-gelled outside of the body and injected by applying shear stress and flow like low-viscosity fluids. The shear-thinning behaviour is a result of reversible physical cross-links. That group of HGs is often called bio- or muco-adhesive HGs according to their specific adhesive properties [16,75,134]. Liquid HGs are injected in liquid form and a sol-gel transition is performed inside the human body [19,109,135–137]. The resulting HGs takes the shape of space available at the injection site which allows to achieve the sol–gel transition with variety of strategies. Slow-gelling systems based on gelation mechanisms (charge interaction, stereocomplexation, click chemistry, etc.), termo- or pH-responsive systems are good representatives of this group.

It is also possible to classify HG based on ionic charges into the following four groups: (i) neutral (no charge, e.g., dextran); (ii) anionic (negative charge) like carrageenan; (iii) cationic (with positive charge) such as chitosan, and (iv) amphiphilic (having the ability to strongly interact with both polar and nonpolar solvent molecules—such as collagen) [17,138–143].

3. Types of Physical Appearance of Hydrogels

HGs have a number of properties, such as oxygen and nutrient permeability and the ability to bind, transport and control drug delivery and release, attractive for use in biological applications. Different tissues have different requirements and would be better served using different types of HGs [144]. HGs can be formed in almost any shape and size [145], e.g., as a matrix, a film, microparticles, nanoparticles, macro-beads, membranes, or coatings, depending on the polymerization technique [46,51,82,146]. HGs are altered to improve their compatibility with hydrophobic or hydrophilic compounds or drugs with different properties [147,148].

Polymeric HGs are classified as macro-, micro- and nanogels with respect to particle size [17,149]. Microgels, cross-linked structure is large, usually from millimetres to centimetres [150]. A group of gels called microgels, whose particles are smaller, separate and cross-linked, having a size between 1 µm and 100 nm [151]. Nanogels are those gels whose particles have a size smaller than 100 nm [49,106,152]. It should be noted that gels with particle sizes only slightly larger than 100 nm were called quasi-nanogels [153].

Microgel is a submicron- or micron-sized network polymer which is highly swellable but insoluble in water [154]. Cross-linked microgel particles possess special kind of control over particle shape and size, which is probably due to specific extensive swelling pattern [155]. Microgels are obtained in two ways. One of them is molecular assembly of existing polymer molecules in aqueous solutions. The other one, particle-forming polymerization, includes two types of polymerization: precipitation and inverse emulsion polymerization [154,156].

Nanogels are prepared in water by self-aggregation of polymers, mostly of natural origin; natural polysaccharides (dextran, pullulan) or cholesterol-containing polysaccharide [14,157]. The dimensions of these HGs are usually of 20–30 nm. Such a small size causes the use of nanogels for cell targeting, because swelling caused by pH changes in the surrounding environment is the reason to release the entrapped drug [158]. Nanogels react to the external stimuli changes much quicker than macrogels. The reason for this is probably their small size and short relaxation time [159,160]. Swelling and deswelling properties and small size of biocompatible and biodegradable nanogel particles are the cause of easy crossing the blood brain barrier [21,149,153].

Obtaining a specific shape of HG at a nanoscale is difficult. HG nanofibers, for example, can be obtained by the by electrostatic spinning method [161]. HGs are also given the form of microcapsules and microspheres, whose size is 1–1000 µm. This allows more protection of substances enclosed in the interior of microcapsules and microspheres. HGs can form micelles in an aqueous environment, as a result of aggregation of amphiphilically connected blocks or end-modified polymers. It is often difficult to differentiate micelles from reversible HGs, because HGs can form micelles above a appropriate concentration called the micelle gel concentration [162]. Similarly, biodegradable nanoparticles of HGs are classified as nanocapsules or nanospheres having drug molecules adsorbed on the surface or closed inside [163,164].

There also exist the group of in situ-gelling HGs, which undergo a sol-gel transition inside the human body [165]. That group of HG always adapts to the free space with its shape. An example of such HGs are shear-thinning HGs. Because of reversible physical cross-links shear-thinning hydrogels pre-gelled outside of the human body, after shear stress injection can regain its initial stiffness.

Despite of many nano- and microgels advantages, it should remember about the possibility of using macroporous HGs and the benefits resulting from it. Macroporous HGs can mechanically collapse at up to 90% and recover reversibly almost immediately after injection in the human body. Such macroporous HGs allow for the preparation of highly defined shapes for drug delivery [151].

4. Immobilization of Drugs in Hydrogel

4.1. HG Structure and Physical Properties

A physical and chemical properties, including structural parameters, swelling behaviour, diffusive characteristics, and surface properties, determine the biomedical and pharmaceutical applications of HGs [5]. After the creation of porous HG structures, different drug molecules may be incorporated into the scaffolding (post-loading) or in the network during formation (in situ loading); these drugs are subsequently released from the hydrophilic matrix to biological media. Thus, network mesh size (or pore size) is a major parameter which determines the sizes of the drug system that can be immobilized in the porous matrix of the HG. The network mesh size (ξ) indicates the distance between adjacent junctions, cross-links and tie points. Cross-links may be chemical (mainly covalent bonds) or physical (e.g., electrostatic, hydrophobic and dipole-dipole) [5].

The cross-linking density (ρ_x) of HGs, which refers to the number of cross-links in a given volume, is a parameter that defines the structural properties of HGs [5,166,167]. Relationships between cross-linking density and HG properties (e.g., shear modulus, equilibrium swelling ratio, drug diffusivity) are presented in Figure 2 [5]. These parameters allow for the prediction of HG release behaviour.

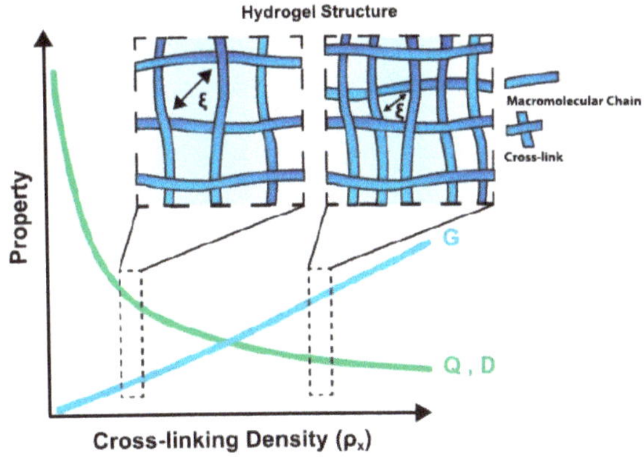

Figure 2. Relationship between the cross-linking density and HG properties. Two network structures representative of low and high cross-linking densities are depicted to demonstrate the relationship between the cross-linking density and basic HG properties of highly swollen, non-ionic gels, including shear modulus (G), equilibrium volumetric swelling ratio (Q) and diffusivity (D). As the cross-linking density increases, the mesh size (ξ) decreases, which is a measure of the space available between macromolecular chains for the diffusion. Reprinted with permission from Ref. [5].

From experimentally measured volumetric swelling ratios of HGs (Q), the average molecular weight can be calculated between cross-links ($\overline{M}_c$) and the equilibrium polymer volume fraction (v_2) of a given network [167]. There are many equations describing the relationship between the network structure and its physical properties, which allow the calculation and prediction of transport properties of homogenous porous materials:

$$\xi = v_2^{-\frac{1}{3}} \left(\overline{r_0^2}\right)^{\frac{1}{2}} \tag{1}$$

$$\frac{D_g}{D_0} = \left(1 - \frac{r_s}{\xi}\right) \exp\left(-Y\left(\frac{v_2}{1 - v_2}\right)\right) \tag{2}$$

$$Q \sim \left(\overline{M_c}\right)^{3/5} \tag{3}$$

where ξ is the mesh size; $\left(\overline{r_0^2}\right)^{\frac{1}{2}}$ describes the root-mean-squared end-to-end distance of the polymer chains in the unperturbated state; D_g means the solute diffusivity in e swollen gel; D_0 means the diffusivity of the solute in the swelling solvent; r_s is a sign of the radius of the solute; Y is the proportion of the critical volume needed for a successful translational motion of the solute molecule to a medium free volume attributable per molecule of the liquid [167–169]. After some simplification, these equations can be used to find the correlation between all mentioned parameters (Equation (4)) for highly swollen, non-ionic HGs, and scale with $\overline{M}_c$:

$$1 - \frac{D_g}{D_0} = \frac{r_s}{\xi} \sim \left(\overline{M_c}\right)^{-\frac{7}{10}} \tag{4}$$

The HG network structure is also important in determining the mechanical properties of porous material. The network structure controls swelling behaviour and drug release in biological environment. Swelling behaviour is defined as the ratio of the volume of the water-swollen gel to the volume of dry polymer and is indicative of the water content of the swollen HG. However, higher water content resulting from high network porosity is more beneficial for immobilizing active substances and can simultaneously lead to polymer degradation and prevent the controlled delivery of drugs. In this context, the cross-linking density must be high enough to allow for the immobilization of the active substance in the polymer matrix. Beyond the structural properties, the chemical composition of HGs regulates the final biochemical properties, e.g., charge, hydrophilicity and bioactivity.

In this brief discussion, we highlighted the structural properties of macroscale HGs. These HGs have consistent physicochemical properties throughout, and drugs can be distributed homogeneously in "bulk HGs" (Figure 3a). Macroscopic HGs have dimensions of millimetres to centimetres and are typically surgically implanted or arranged in contact with the patient's body to achieve transepithelial drug delivery [17,137].

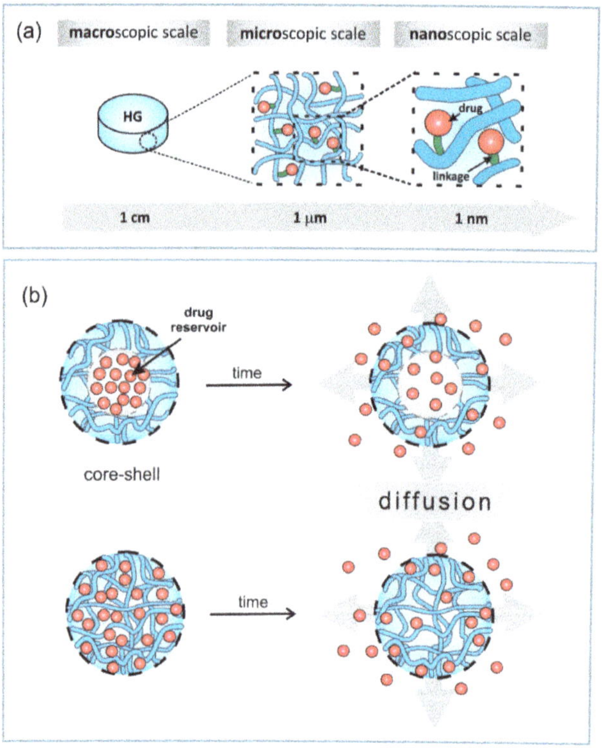

Figure 3. (**a**) Macroscopic design of HGs including the size and porous structure. (**b**) Schematic structures of micro- and nanogels showing the release of drugs.

Other structures, e.g., micro- and nanogels, draw attention to their use as injection materials (Figures 3 and 4). Their relative scale (1 μm to 100 nm microgels and 10–100 nm nanogels) determines the course by which HGs can be delivered to the human body [17–20]. Small HG particles are needle-injectable, provide large surfaces for bioconjugation and enhance penetration through tissue barriers [135,136,170,171].

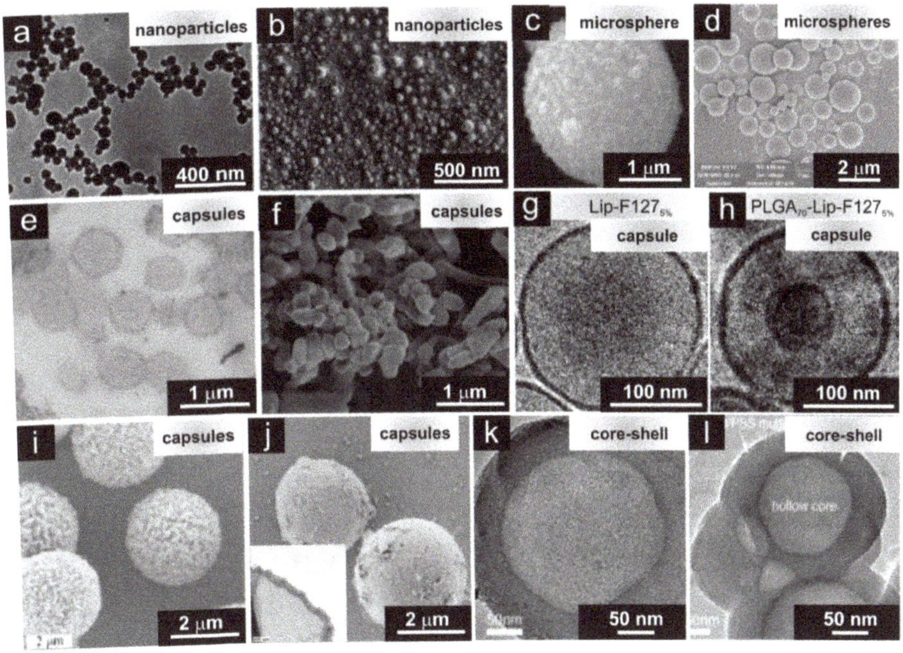

Figure 4. Morphology (structure) of HGs observed by (**a**,**c**,**e**,**g**,**h**) TEM and (**b**,**d**,**f**,**i**,**j**) SEM at different magnifications (the scale bars are marked on the images) obtained by post-loading or in situ loading drugs. (**a**,**b**) Piperlongumine-loaded poly(ethylene glycol)-poly(dehydroabietic ethyl methacrylate) HG nanoparticles [172]. (**c**) Silver alginate HG microspheres [173]. (**d**) Thermosensitive HG microspheres loaded with tenofovir [174]. (**e**,**f**) Polyacrylamide-poly(*N*-isopropylacrylamide) HG capsules loaded with doxorubicin hydrochloride [175]. (**g**) Unloaded lipid capsule (Lip) with F127$_{5\%}$ (5 wt% content of Pluronic F127) [176]. (**h**) Doxorubicin-loaded poly(lactic-co-glycolic acid) (PLGA) core-lipid shell (Lip) with F127$_{5\%}$ structure [176]. Smart organic/inorganic HG polyelectrolyte capsules (poly(allylamine hydrochloride): (**i**) hollow and (**j**) hydroxyapatite [177]. (**k**,**l**) Core-shell mesoporous silica spheres (**l**) loaded with ibuprofen [178]. Reprinted with permission from Refs. [172–178]. The Royal Society of Chemistry, Hindawi, Nature Group, Springer and Wiley-VCH Verlag GmbH & Co.

Drug delivery based on micro- and nanogels are superior when compared with macrosystems (Figure 3b) because: (i) the smaller sizes of these HGs allows for active and passive drug targeting; (ii) controlled drug release leads to improved therapeutic impact and reduced side effects; (iii) drug loading takes place with no chemical reaction (environmental stimuli); (iv) the smaller HGs may infiltrate into tissues via paracellular and transcellular routes; and (v) the smaller HGs are biocompatible and biodegradable [15,179]. Examples of different structures/morphology of HGs are presented in Figure 4. These systems were observed using TEM (transmission electron microscopy) or SEM (scanning electron microscopy). We can distinguish two types of HGs: (i) "core-shell" systems and (ii) homogenously dispersed drugs (Figures 3b and 4). In the core-shell system, a shell is formed by HGs around a central core is drug reservoir (Figure 4e–h,l,k). This structure can be produced in the form of capsules, spheres or slabs, having a high concentration of the drug in the core of the systems, which facilitate the constant release of the drug in the body [126,174,175,180,181]. In the second system, the drug is dispersed homogeneously in a polymer matrix (Figure 4a–d), which in contact with a bio-fluid usually begins swelling, and takes forms of nanoparticles, polymeric micelles, microspheres, etc. [68,172,180,181]. For both types of HGs, releasing of drugs is time-depended process and it is mainly diffusion controlled. Drug releasing from the HG network will be discussed in detail in the next paragraph.

4.2. Immobilization of Drugs

HGs due to their 3D structures may be applied as carriers for drugs, proteins, lipids or cells [15,17,68,182,183]. Additionally, a large amount of water within their structures makes a convenient environment for the immobilization of drugs. However, 3D structures of HG allow the immobilization of drugs in their matrix, in this regard, when immobilization process of active substances is performed, two aspects must be taken into account. The large pore size of HG (large mesh size), as well as the high water content causes that water-soluble drugs with small particles quickly escape from the network, thanks to which they have a short release time [182]. Drug release is much slower when the drug particle size is comparable to the mesh size [17]. When the drug particle size exceeds the mesh size, then the drug is physically trapped in the network [17]. To minimize these problems, the pore size in HG (mesh size) should be "matched" to the size of the immobilized drug.

There are many methods leading to loading of drugs in HG networks/spheres/capsules, and the choice of method depends mainly on the place where the drug should be delivered. Briefly, drugs can be loaded into HG matrices in two ways: post-loading and in situ loading (Table 2 and Figure 5). HGs have proven to be very efficient for local drug delivery, which could lead to prolonged and faster release. Some examples of HG applications are presented in Table 2; different structures are presented in Figure 5.

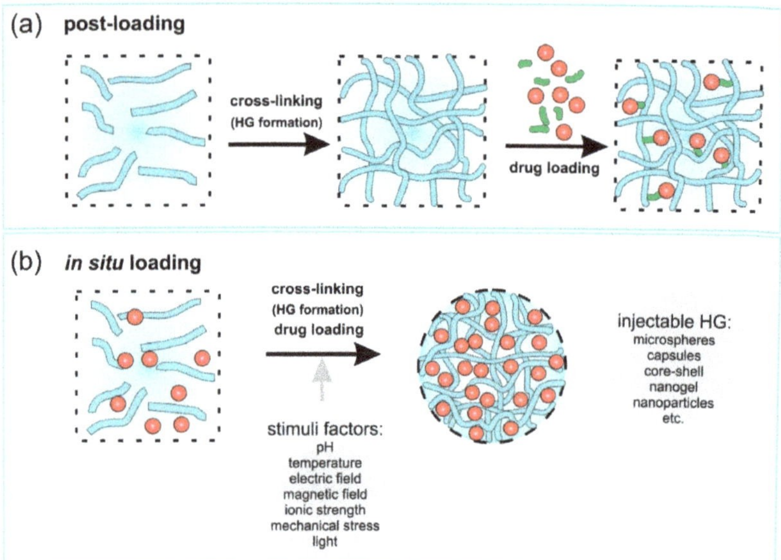

Figure 5. Schematic representation of drug loading in a HG network using (**a**) post-loading and (**b**) in situ loading mechanisms.

Briefly, the post-loading drug process consists of forming a polymer HG matrix (a preformed HG) and immobilizing the drug in the polymeric network (Figure 5a). HGs swell in drug solutions till equilibrium. Depending on the HG macro- or microscopic size, HGs are administered by more or less invasive procedures. Preformed HG gels outside the human body, while being quite simple films or sticky solutions that retain their rheological or mechanical properties after administration [184].

Table 2. Examples of HGs with different routes drug immobilization of and their structures and applications.

Type of Loading Drugs	Hydrogel Precursors	Drug	Structure	Applications	Refs.
Post-loading	AA [17] –BA [18] –DEAP [19]	17-DMAPG [15]	Core-shell [16]	Antitumor activity	[152]
	Carbopol 940 [20]	Vor [11]	Core-shell LPN	Dermal applications	[185]
	Carbopol/stearic acid	Vor	Core-shell	Ophthalmic Application	[186]
	PNIPAAm-*b*-PLA-*b*-PEG-*b*-PLA [7] -*b*-PNIPAAm	Riluzole	-	Neuroprotective drug	[183]
	PAA-PNIPAAm	DOX	Core-shell capsule	Antitumor activity	[175]
	PEG-*b*-PDAEMA [8]	PLGM [9]	Nanoparticle	Antitumor activity	[172]
	Chitosan/Polysaccharide	Tenofovir	Microsphere	Vaginal drug	[174]
	Chitosan/β-GB	DOX-loaded in LTSL [29]	PP [33] HG	Antitumor activity	[187]
	PLGA [30] MicroC PEG derivatives (HG)	DOX-loaded in MicroC [31] Fu [32]-loaded in HG	PP HG	Antitumor activity	[188]
In situ loading	Polysaccharide	Ibuprofen	Core-shell capsule	Oral drug	[126]
	Carbohydrate-NIPAM [22]	Bupivacaine	HG-microgel composite	Anaesthetic drug	[189]
	PNIPAAm-co-AIA [21]	Lopinavir	Microspheres	Antiretroviral drug	[190]
	CmetCel [25] -Dextran	AmB [23]	MacroHG	Antifungal therapy	[191]
	NiPAAm [26]-NtBAAm [27]	Fluvastatin	PP HG	HMG-CoA [25]	[192]
	PPZ [34]	Silibinin	Microspheres	Anticancer and antiangiogenic activity	[193]
	PEG–PCL [35]–PEG	PTX micelles [36]	PP HG	Antitumor activity	[194]

Abbreviations: [1] PNIPAAm: poly(*N*-isopropylacrylamide); [2] PAA: polyacrylamide; [3] PHEMA: poly(hydroxylethyl methacrylate); [4] PVP: poly(vinylpyrrolidone); [5] PEG: poly(ethylene glycol); [6] LPN: lipid nanoparticles; [7] PLA: poly(lactide); [8] PDAEMA: poly(dehydroabietic ethyl methacrylate); [9] PLGM: piperlongumine; [10] DOX: Doxorubicin hydrochloride; [11] Vor: Voriconazole; [12] EGDMA: 2-(2-methyl-acryloyloxy)ethyl 2-methyl-acrylate; [13] HEMA: 2-hydroxethyl 2-methylprop-2-enoate; [14] Indo: Indomethacin; [15] 17-DMAPG: geldanamycin derivative (aminated form, which readily protonates at low pH; [16] core-shell structure of HG-in-LPN; [17] AA: acrylic acid; [18] BA: *N,N'*-methylenebis(acrylamide); [19] DEAP: 2,2-diethoxyacetophenone; [20] Carbopol 940: HG composed of Precirol ATO 5, Labrafil 1944 CS, and Tween 80; [21] AIA: allylamine; [22] NIPAM: poly(*N*-isopropylacrylamide); [23] AmB: Amphotericin B; [24] CmetCel: carboxymethylcellulose-hydrazide; [25] HMG-CoA: reductase inhibitor (statin); [26] NiPAAm: *N*-isopropylacrylamide; [27] NtBAAm: *N-tert*-butylacrylamide; [28] β-GB: β-glycerophosphate; [29] LTSL: lysolipid thermally sensitive liposomes; [30] PLGA: poly(lactide-co-glycolide); [31] MicroC: microcapsule; [32] Fu: 5-fluorouracil; [33] PP: "plum padding"; [34] PPZ: poly(organophosphazene); [35] PCL: poly(ε-caprolactone); [36] PTX: Paclitaxel.

Drugs can be physically or chemically immobilized in HGs and can also be loaded into other species (secondary delivery vehicles) to provide the appropriate environment for targeting. Preformed HGs are solid even when injected [184].

In situ loading of drugs is associated with the simultaneous formation of injectable HGs in the body and the encapsulation of drugs (Figure 5b). Briefly, the whole process shows a sol-to-gel transition in situ that form after injection in vivo.

The resulting HG take the shape of the available space. In situ gel formation usually includes the subsequent steps: (i) gelation as a response to the changes of temperature or pH changes (thermoresponsive or chemical responsive polymers; i.e., "smart" polymers), (ii) ionic or covalent cross-linking, (iii) solvent exchange or crystallization, and/or (iv) thickening upon removal of the injection shear. These external stimuli are shortly summarized in Figure 1a (stimuli factors).

Taking into account the lifetime of HGs, their formation mechanism can be categorized as: (i) pre-gelation (polymer precursor in solution), (ii) therapeutic window (after injection and gelation containing drug), and (iii) degradation (HG degradation products) [18]. These systems may be homopolymeric or multipolymeric or may also be composed of different components, including other systems (e.g., capsule, microgels, and nanoparticles) serving as drug carriers. This latter mentioned gel systems are characterized in the next section.

Immobilization of Drugs in Hydrogel Composites

Some systems containing drugs in matrix/capsules are poorly water-soluble or insoluble. In this sense, soluble therapeutic agents are poorly retained in HGs due to their hydrophilic nature. In this case, there are many problems with loading process of drugs due to their tendency to aggregate resulting in high local concentrations causing toxicity [195]. To overcome these problems, HG composites have been created to exploit the hydrophilic-hydrophobic nature of various components [192,196]. HG composites contain polymeric networks (hydrophilic) swollen with water and nanostructures/microstructures with different physico-chemical properties [192,197–199]. These composites represent a new class of materials with new properties.

The inclusion of nanomaterials in the HG polymer network is an interesting way to adjust the mechanical properties of HG and/or to provide the composite with sensitivity to external stimuli [192,196,197,199]. Different nanomaterials have been immobilized in polymeric networks, including inorganic nanoparticles [200,201], carbon nanomaterials [202], and lipids [99]. Nanoparticle systems have gained considerable attention by being one of the most interesting and promising biomedical materials with the exceptional physicochemical properties, controlled shapes, nano-sized characteristics, comprehensive modification options and well-defined multi-functionality [83]. The preparation of HG composites may be performed using physical and chemical forces.

To overcome the incompatibility of hydrophobic drugs and hydrophilic HG networks, lipid nanoparticles (LPNs) are frequently used to promote good solubility. LPNs have been used for dermal, mucosal, transdermal and intramuscular applications [176,182]. Three types of LPNs have been used in drug delivery systems: (i) lipid nano emulsions (LNEs, where the core is composed of liquid lipids), (ii) solid lipid nanoparticles (SLNs, where the core has lipids in a solid state at room and body temperatures), and (iii) nanostructured lipid carriers (NLCs, where the lipid core is a heterogeneous mixture of solid and liquid lipids) [182]. The lipophilic core of LNPs entraps active ingredients, whereas the surfactant membrane (consisting of phospholipids) ensures the stability of LNPs in hydrophilic environments. In pure lipophilic form, these systems have unsuitable rheological properties and therefore, require structural modifications. These modifications result in the formation of LNP-HG composites [187] as shown in Figure 6.

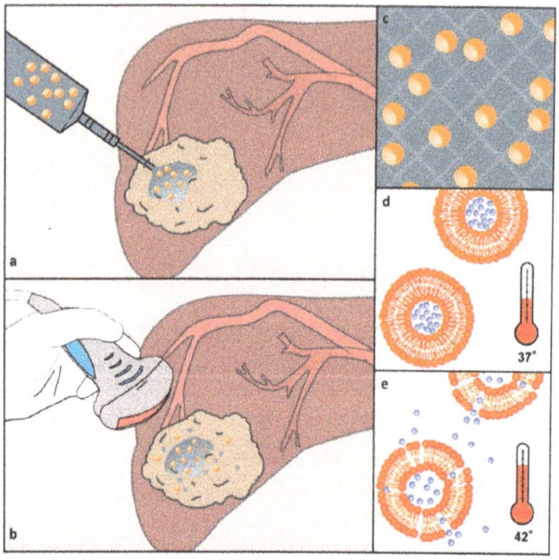

Figure 6. (**a**) Fully syringe-injectable Lipogel consisting of a chitosan/β-GP thermoresponsive gel containing a suspension of DOX-loaded thermosensitive liposomes. (**b**) Release from the gel in situ is controlled using minimally invasive hyperthermia, which is achievable using high intensity-focused ultrasound. A small portion of the drug-loaded liposomes is released from the Lipogel, maximizing the drug delivery distance from the gel implant. (**c**) The majority of liposomes are locked into the gel upon initiation of cross-linking during thermogelation. (**d**,**e**) Liposomes sequester the majority of the drug at body temperature but rapidly become more permeable upon mild hyperthermia and release their drug payload. Reprinted with permission from Ref. [187].

An in situ gelation of chitosan/β-glycerophosphate (GP) and thermoresponsive liposomes was performed [187]. This biocompatible and biodegradable HG was used as a matrix for lysolipid thermally sensitive liposomes (LTSL) loaded with doxorubicin (DOX). LTSLs are bi-layered spherical vesicles that rapidly change structure upon mild hyperthermia (41–43 °C), creating openings in the liposome. DOX was loaded into LTSL by changing solution pH. DOX delivery was also based on the pH-sensitivity of liposomes to acidic pH. A schematic of the controlled-delivery of chitosan/β-GP/DOX-loaded LTSLs and DOX release is presented in Figure 6.

The next very interesting group of composite HGs are "plum pudding gels" (PP gels), which are schematically shown in Figure 7 [179,187,188,192,194]. This type of the composite HG contains microgels or nanogels inside a bulk HG network, which improves the loading and release of drug. In the PP gel structure, the microgels or nanogels act as reservoirs for drugs and can be incorporated into a conventional macroscale HG at the different concentrations to provide two-component gel matrices.

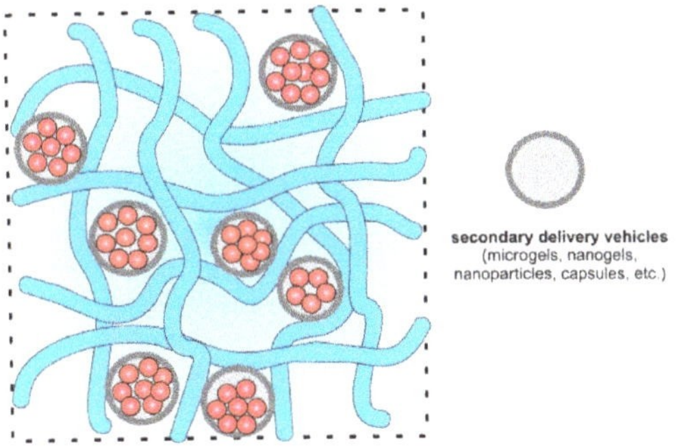

Figure 7. Structure of "plum pudding gels" (PP gels). Composite HG containing a drug embedded in a secondary controlled delivery vehicle (microgels, nanogels, nanoparticles, capsules, etc.).

From a drug delivery perspective, these systems are particularly interesting because the drug delivery limitations of microgels and HGs are minimized, whereas their synergistic effects are observed as soft nanocomposites [179]. The presented chitosan/β-GP/DOX-loaded LTSL system belongs to this group of composites [187]. Other similar systems are listed in Table 2.

4.3. Release Mechanism of Drug from Hydrogel Matrices

The physical and chemical properties of HGs affect their delivery properties, including controlled release. HG structure, diameter, and cross-linking agent density affect the rate of drug diffusion. The kinetics of drug release from HGs is connected to the chemical structure and crosslinking density of the materials, including the HG matrix monomers and coating (Figure 8) [203].

The use of HG networks increases the local concentrations of pharmaceuticals and their slow release at delivery sites. Suitable controlled-release mechanisms include (Figure 8): diffusion [17], swelling [68], chemical and environmental stimuli (degradation or deformation) [67]. Drug release from HGs in response to environmental stimuli may occur due to changes in pH [110,204], temperature [110,205–207], electric field [109,111,208] or ionic strength [209,210]. Some ideas are schematically presented in Figure 8 [17].

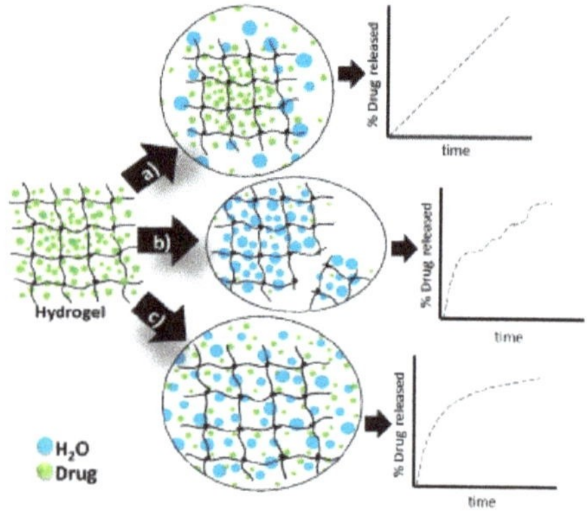

Figure 8. Hydrogels drug release mechanisms and their respective kinetic profiles: (**a**) drug diffusion, (**b**) degradation of the polymeric matrix and (**c**) swelling. Reprinted with permission from Ref. [203].

5. The Use of Hydrogels in Modern Pharmacy

Interest in using HGs has not decreased over the years. The development of specially modified HGs has provided new pathways for drug delivery and offers advantages as a vehicle for active substances [191,192,211–213]. One of the most important advantages is administration route versatility. HGs offer several routes: oral, injection (both intramuscular, into bone and subcutaneous), rectal, vaginal, through wounds, and ophthalmic (Table 3). Drug delivery is strongly connected to drug molecule size and is dependent on administration route. As mentioned previously, HG delivery systems can be classified into three main categories based on their size: macro-, micro- and nanosized.

Table 3. Summary of HGs that can be used as drug carriers.

Hydrogel	Structure/Size	Active Substance	Route of Delivery	Refs.
CTS-g-poly(AA-co-AAm)PVP/MMT	Nanogel	Clarithromycin	Oral	[214]
Salecan/PAMPS	Microgel	Insulin	Oral	[215]
HA	Nanogel	Insulin	Oral	[216]
CTS	Nanogel	Curcumin	Oral	[217]
CTS	-	Camptothecin	Oral	[218]
SS/PVA	Microgel	Gentamycin sulphate	Dermal	[219]
TG/SA/PVA	Microgel	Moxifloxacin	Dermal	[220]
P407/CMCs	Microgel	Cortex Moutan extract	Dermal	[221]
CMCTS/PAD	Macrogel	Voriconazole	Ocular	[12]
HECTS	Macrogel	-	Ocular	[222]
P407	Nanogel	Theaflavin/Nifeviroc	Vaginal	[2]
CTS	Microgel	Tenofovir	Vaginal	[174]
CHC	-	Naringenin	Topical oral	[223]
CTS	Nanogel	Thymol	Topical oral	[224]
GCTS	-	Paclitaxel	Injection	[225]
PF127/HA	Nanogel	Paclitaxel and doxorubicin	Injection	[226]
CTS	-	Curcumin	Injection	[227]
CMCTS	Nanogel	Curcumin	Injection	[228]
CTS/GP	-	Docetaxel	Injection	[229]
Gelatine/laponite (+chitosan or PNIPAM-co-AA)	Nanogel	Rhodamine B	Injection	[230]
Salecan	-	Doxorubicin	Injection	[231]
Agarose	-	Doxorubicin	Injection	[232]
SA	Nanogel	Bevacizumab	Injection	[233]

Abbreviations: CTS: chitosan; AA: acrylic acid; AAm: acrylamide; PVP: polyvinylopyrrolidone; MMT: montmorillonite; PAMPS: poly(2-acryloamido-2-methyl-1-propanesulfonic acid); HA: hyaluronic acid; SS: sericin; PVA: polyvinyl alcohol; TG: tragacanth gum; SA: sodium alginate; P407: poloxamer 407; CMCs: sodium carboxymethyl cellulose; PAD: poly aldehyde dextran; HECTS: hydroxyethylated chitosan; CHC: carboxymethyl-hexanoyl chitosan; GCTS: glycol chitosan; CTS/GP: chitosan/β-glycerophosphate; PNIPAM: poly(*N*-isoproplacrylamide)-co-acrylic acid.

We present some examples of HGs applications in drug delivery. Most informative reports from the last couple of years regarding HGs in health care have been well summarized; however, the literature on HGs is constantly increasing, as well as interest in HG materials. Herein, we highlight only a few studies that only begins to hint at the wide applications of HGs (Table 3).

5.1. *Application of Hydrogels for Oral Administration*

Oral administrative routes are classic and accepted means for delivering drugs. The vast majority of medicines available on the market are taken through oral routes. Several pills are often taken daily at appropriate intervals to achieve effective therapy. This reduces associated risks of skipping or missing doses and thereby lowering the effectiveness of the therapy. The development of delayed release capsules that prolong the delivery of active substances has been a breakthrough and has allowed for increased compliance. With these delayed release capsules, the amount of tablets consumed is reduced, usually to only one per day. Acid-sensitive drugs require protection against harmful effects of gastric juice typically encountered when delivering drugs vial oral routes. Protection can be offered by using special tablets coated with polymers soluble only at basic pH, such as in the intestines. As a result, drugs survive transit through the stomach and are released only in the intestines thereby promoting higher absorption into the bloodstream.

HGs have also been explored for oral delivery applications. The appropriate selection of gel or the addition of pH-dependent coatings enable the controlled release of active substances from HGs.

The first modification was presented by Panahi et al. [214] (Table 3). Chitosan (CTS)-based gels with acrylic acid (AA), acrylamide (AAm) and polyvinylpyrrolidone (PVP) were prepared [214]. During HG formation, a mineral (montmorillonite, MMT), which has ability to absorb water, was used. Clarithromycin (CAM), a macrolide antibiotic, was used to eradicate *H. Pylori* from the gastrointestinal tract was immobilized in the HG network. It was important to achieve a prolonged release of the antibiotic to increase the chances of an effective therapy. MMT increased the pore diameter in the gel structure, which increased the immobilization of the active substance. However, this also prevented solvent from readily reaching CAM through more intricate pathways in the network, which complicated drug release.

The dependence of the release of the drug on the pH level was explored by Qi et al. [215] (Table 3). An HG system was prepared based on salecan (beta glucan) with pH-sensitive poly(2-acrylamido-2-methyl-1-propanesulfonic acid) (PAMPS). The system had the ability to take or donate protons depending on the pH of the environment, while reducing or increasing the network volume. The authors showed that salecan in combination with PAMPS had the ability to release the active substance depending on the pH of the environment. Using insulin as an exemplary drug, at acidic pH insulin was released at a lower level than at neutral or slightly alkaline pH.

Insulin release from HGs was more extensively studied [216]. Special nanocarriers based on methacrylic acid were synthesized. In a neutral environment, the methacrylic polymer chains started to repel each other, thereby loosening the nanocarrier network and allowing the release of insulin. This property was exploited in preparing HA-doped HG. In the intestines, where the pH is above 7, insulin release increased dramatically compared with release in the stomach. By using HGs as a drug carrier, the release of insulin was extended over time.

In another study, a unique biodegradable, super porous, swellable and pH-sensitive nanocellulose reinforced CTS HG was prepared for the oral administration of curcumin [217]. The in vitro degradation of HG was dependent on the swelling ratio and the number of cellulose nanocrystals (CNCs) in the HG. All HGs showed maximum swelling ratios greater than 300%. The drug release occurred in simulated gastric media; the drug maintained its chemical activity after in vitro release. According to this study, CNC-reinforced CTS HGs can be used to improve the bioavailability of curcumin for absorption from the stomach and upper intestinal tract.

Finally, tetrakis(hydroxymethyl)phosphonium chloride was used as a crosslinking agent in a Mannich reaction to obtain chitosan-based HGs [218]. These pH-sensitive HGs showed low toxicity, high biocompatibility, and allowed for the modified release of encapsulated drugs, namely camptothecin, for 48 h. According to the obtained results, the oral administration of camptothecin through HGs provided low concentrations of the drug at the absorption site, avoiding carrier saturation and reducing intestinal toxicity [218].

5.2. Hydrogels for Dermal Applications

Due to their fairly compact consistency, HGs can also be applied to wounds or other skin issues. Commercially available HG-based dressings used for exudative wounds, pressure sores or burns, exploit the adhesive properties and ability of HGs to absorb liquids. The addition of antibacterial substances may further improve the applications of HG dressings.

Researchers from Southwest University in China developed a sericin (SS) and polyvinyl alcohol (PVA)-based HG (Table 3) [219]. This gel showed good biocompatibility, humidity and self-healing properties, i.e., ideal for dressings. Gentamycin sulphate, a known antibiotic substance and aminoglycoside, was added to this HG. To determine the properties of the obtained antibacterial HG, a number of in vitro tests were carried out, which measured wettability, swelling, microbiological activity, drug substance release, cytotoxicity, and immunotoxicity. The studies showed that the antibacterial HG ensured a prolonged release. For a deeper analysis of the HG properties, a model infected tissue was prepared. Tests confirmed that the HG was cytocompatible with mammalian cells and did not affect the growth of healthy cells.

The material used in wound dressing should fulfil many requirements [220], including isolation from harmful external factors, such as secondary wound infection, but should also provide sufficient water vapour permeability and oxygen availability. HGs based on tragacanth gum (TG), sodium alginate (SA) and PVA meet these properties and have potential applications [220]. In addition to good permeability for water vapour and oxygen, these HGs also provide barriers against secondary wound infection. These properties were confirmed by in vitro studies. Additionally, the HGs had haemolytic and mucoadhesive properties. Dressings based on these HGs were able to release antimicrobial active substances (moxifloxacin, an antibiotic from the fluoroquinolone group used to treat a wide spectrum of microorganisms). The results showed prolonged release of up to 24 h without an initial burst release.

Patients struggling with atopic skin changes very often use strong steroid drugs to alleviate emerging inflammation. Unfortunately, topical steroid therapy also leads to skin dryness and irritation, which patients with atopic dermatitis should avoid. Therefore, adequate skin hydration in atopic dermatitis therapy is maintained through the systematic and frequent use of appropriate emollients. Wang et al. combined the administration of an anti-inflammatory substance and maintenance of proper skin hydration (Figure 9) [221].

A HG prepared from a mixture of poloxamer 407 (P407) and sodium carboxymethyl cellulose (CMC) was used to achieve this goal (Table 3) [221]. A special nonwoven fabric was covered with the HG mixture. The addition of CMCs resulted in an increase in hydrophilicity of the resulting gel structure and also significantly reduced the sol-gel transition temperature, which advantageously promoted the fabrication of the coated nonwoven fabric. The HG exhibited moisture retaining properties, and the nonwoven fabric material prevented excessive water transpiration. The anti-inflammatory effect was provided by the addition of a Cortex Moutan extract, a well-known plant popular in Chinese medicine, during HG preparation. To determine the drug release profile, in vitro and ex vivo studies were performed. The in vitro studies showed that the release of the anti-inflammatory substance occurred in a prolonged manner up to several dozen hours, ensuring prolonged and elevated levels of the anti-inflammatory substance. Ex vivo tests on pork ear confirmed the results of the in vitro tests. The nonwoven fabric coated with a mixture of HGs composed of P407 and CMCs and Cortex Moutan extract showed potential for treating and caring for atopic lesions in patients with active skin inflammation.

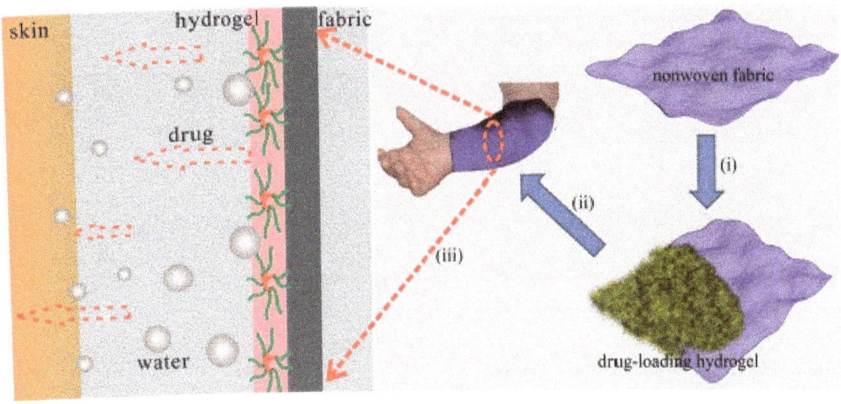

Figure 9. Diagram of a dual-functional fabric; (i) representation showing how the fabric is covered with HG, (ii) representation showing how the HG-coated fabric is applied to the patient's skin, and (iii) representation showing how the drug diffuses across the skin. Reprinted with permission from Ref. [221].

5.3. *Hydrogels for Ocular Applications*

The administration of a slightly soluble substance in the eye can be quite challenging [12,222]. Chitosan-based HGs showed promise as polymeric carriers for both hydrophilic and lyophilic drugs for ocular applications [234]. HGs based on carboxymethyl chitosan (CMCTS) and polyaldehyde dextran (PAD) were prepared [12]. The poor water solubility of the active substance (voriconazole), was addressed by encapsulating in cyclodextrin, which significantly increased the bioavailability of voriconazole and facilitated its application. The HG has the ability to gel in situ, allowing for easier intraocular injection. However, the prolonged release of voriconazole from the gel remained an issue.

Scientists from the Ocean University of China focused on the gelling process [222]. HGs based on hydroxyethylated chitosan (HECTS) with a special azide group showed the ability to polymerize under UV radiation. The polymerization lasted up to 30 s and was carried out in vivo. While the experiments, which were performed on New Zealand rabbits, should be confirmed on the human body, they already gave promising results and were thriving to recognize this HG as a good carrier for the intraocular administration of anti-glaucoma medications.

5.4. *Hydrogels for Vaginal Applications*

A HG made of P407 with nanosized layered double hydroxides allowed to include both hydrophilic and hydrophobic substances in the gel structure, which significantly widens the spectrum of administered drugs to patients [2]. Additionally, the HG has an important ability to gel transition at body temperature, so that during application it is a solution and in contact with the human body temperature it solidifies and stays on the tissues. As sample drugs, water-soluble theaflavin and Nile red dye were used to determine release profiles from the HG in rabbit vagina in independent and simultaneous states [2]. The antiviral properties of theaflavin were preserved and was able to limit the entry of HIV into the immune system. These tests were repeated against nifeviroc as a hydrophobic active and the effect was amplified. Thus, this HG acts as a potential carrier for the delivery of a broad spectrum of antiviral drugs to the human body.

HGs have also been applied in the vaginal administration of antiviral drug in the treatment of HIV [174]. HGs based on CTS with glycerol phosphate gelled upon contact with the human body, at a temperature of approximately 37 °C. Additionally, by enclosing active substances (e.g., tenofovir) in microspheres and immobilizing them in the HG structure (PP gels, Figure 7), a much longer release profile was obtained when compared with the release profile from the microspheres or HG alone. This provides an opportunity to improve on the effectiveness of immunotherapy.

5.5. Hydrogels for Topical Oral Applications

In the use of HG structures in the treatment of inflammatory periodontal bent, Chang et al. prepared a gel based on carboxymethyl-hexanoyl chitosan (CHC) [223]. Through the addition of glycerol phosphate into the HG network, the HG was modified to solidify at human body temperatures, which provided significant advantages over the use of other substances in oral inflammation therapy. Anti-inflammatory properties were carried out using naringenin. Experiments on rats showed that the anti-inflammatory substance was more rapidly released from the HG when the pH of the environment was acidic (~5.5), thereby rapidly treating the periodontal inflammation.

CTS HG can also be used to treat oral inflammation when used with thymol [224]. In vitro studies showed that the resulting structure allowed for the prolonged release and high antimicrobial activity of thymol against microorganisms up to 72 h. Furthermore, in the first two days, the active substance was increasingly released. This dual action HG demonstrated anti-inflammatory properties and the treatment of bacterial biofilms in the oral cavity.

5.6. Injection of Hydrogels

Therapeutic injections are usually least favoured by the patient and requires specialized HGs. The key challenge is to prepare a structure that will be fluid enough to squeeze through the needle but also sufficiently rigid that when applied does not spread and remains at the injection site [17,135,136]. Therefore, an important property of these structures is shear thinning, i.e., the ability to decrease viscosity of fluid under shear strain.

CTS-based in situ HGs are frequently adopted in the treatment for age-related macular degeneration (AMD), glaucoma and mucosal allergic diseases [234]. The HGs offer a convenient matrix for in situ gelling systems containing other nanoparticles, such as micelles. Injectable CTS-based in situ gels have been applied as in situ forming implants and as drug carriers in nasal and ocular delivery due to good biocompatibility, simple manufacture and sustained-release properties. These systems have been utilized to deliver several chemotherapeutic agents, including camptothecin [218,235], paclitaxel (PTX) [225,226], DOX [226], curcumin [217,227,228], and docetaxel (DTX) [229,236].

An injectable drug delivery system based on a visible light-cured glycol chitosan (GCS) HG containing PTX-complexed beta-cyclodextrin (β-CD) (GCS/CD/PTX) was tested for ovarian cancer therapy using a tumour-bearing mouse model [225]. The swollen GCS HG affected the release of PTX and CD/PTX. These systems exhibited a controlled release of PTX for 7 days. Additionally, GC/CD/PTX resulted in a superior antitumor effect against ovarian cancer.

An interesting application of HGs as dual carriers for PTX and DOX was proposed by Rezazadeh et al. [226]. An injectable thermosensitive HG for simultaneous intra-tumoral administration of PTX and DOX was prepared using mixed micelles composed of Pluronic F127 and α-tocopheryl polyethylene glycol 1000 succinate (TPGS) and a thermosensitive Pluronic F127/hyaluronic acid (PF127/HA) HG containing fixed amount of DOX [226]. PP gel formation temperature, rheological properties, injectability, degradation rates of the HG, and the release rate of DOX and PTX from the HG were examined. The HG containing PTX-loaded micelles and DOX converted to a semisolid with increasing temperature to 35 °C. DOX and PTX were released from the HG at different rates within 12 h and 3 days, respectively. This novel thermosensitive HG could be used for the co-delivery of PTX and DOX in solid tumours.

A series of injectable in situ-forming chitosan-based HGs were prepared by the chemical cross-linking of CTS and genipin in the presence of different inorganic salts [227]. In situ HG formation, with curcumin as an active substance, was detected after subcutaneous injections in rats. In vitro curcumin release profiles exhibited sustained release properties with an initial burst release with approximately three to six times higher cumulative release than other gel controls.

CTS-based HGs may also show magnetic properties, when magnetic Fe_3O_4 and $MnFe_2O_4$ nanoparticles are incorporated into their networks [228]. The Fe_3O_4 and $MnFe_2O_4$ on CMCTS HGs indicated a pH-sensitive behaviour; their release performances were investigated by curcumin as a model drug. The effects of applied magnetic fields on drug release for composites containing Fe_3O_4 were higher compared with $MnFe_2O_4$. The results showed that HGs containing Fe_3O_4 and $MnFe_2O_4$ can be applied for novel drug delivery systems.

DTX is an anticancer drug used for treatment of various solid tumours [229]. But it's low water solubility and a lack of specification has limited its clinical use. Therefore, new solutions have been explored to use this drug in targeted therapy. A thermosensitive chitosan/β-glycerophosphate (CTS/GP) DTX-loaded HG for intratumoral delivery was studied. The results of an in vitro release study demonstrated that DTX-C/GP was released over a period of 3 weeks. The tumour volume was minimized by the intratumoral injection of DTX-CTS/GP (at 20 mg/kg in H22 tumour-bearing mice). Further, the in vivo pharmacokinetic characteristics of DTX-CTS/GP correlated well with the in vitro release. DTX-CTS/GP significantly prolonged DTX retention, supported a high DTX concentration in tumours, and toxicity was effectively reduced.

Carbon nanostructures were also successfully applied in hydrophilic, porous materials as a component of HGs for DTX carriers [236]. A functionalized graphene oxide (GO)-based thermosensitive HG loaded with DTX for intratumoral delivery was designed. First, GO was functionalized by using chitosan to achieve high stability in physiological solutions. Next, HGs containing few components were formed: GO-chitosan, Poloxamer 407 and Poloxamer 188 and DTX. GO/chitosan HG was stable in physiological solution; DTX released much more slowly from this gel compared with free DTX with a pH-responsive feature. The DTX–GO/chitosan gel released higher concentrations and longer resident times in the tumour tissues of mice in vivo giving nontoxic effects to normal organs. Additionally, the combination of near-infrared laser irradiation at 808 nm significantly enhanced tumour inhibition in vitro and in vivo.

Other composite-based HGs were also evaluated. Gharaie et al. used gelatin and laponite in various proportions to obtain appropriate shear thinning properties (Figure 10) [230]. Additionally, by loading with CTS or poly(*N*-isopropylacrylamide)-co-acrylic acid (PNIPAM-co-AA) to the gel structure, pH-responsive abilities were gained, resulting in a different release profiles. As a model substance, rhodamine B was used to test the release. The results of these experiments showed that the gel could be easily and accessibly applied, whereas the release of the active substance was dependent on the pH level (in acidic and neutral environments, the release was negligible, whereas release increased at basic pH.

The key point of interest of many scientists is the fight against cancer. Also in this case HG can also be used to treat cancer [36]. Hu et al. combined salecan, a beta-glucan polysaccharide, with Fe_3O_4 nanoparticles and agarose [231]. The resulting HG gained magnetic properties and was used as a carrier for DOX (an anthracycline cytostatic used in anti-cancer therapy). The salecan-based HG with iron oxide nanoparticles had a pH-dependent drug release profile. In an acidic environment, unlike alkaline environments, a larger amount of DOX was released. Due to the acidic pH of cancer cells, this result indicated significant promise. Additionally, due to the presence of Fe_3O_4 nanoparticles, the amount of active drug released was also manipulated. With an external electromagnetic field, inorganic particles can be vibrated, thereby loosening the gel structure enabling easier release of DOX.

The technical aspects of DOX release from HGs were examined by a group from Drexel University [232]. In this study, an agarose-based HG was used wherein DOX-dextran sulphate complexes were provided. The structure was doped with divalent metallic ions, Ca^{2+} from $CaCl_2$ and Mg^{2+} from $MgCl_2$, at different concentrations. The divalent ions interfered with the binding between DOX and dextran sulphate, promoting prolonged release of the active substance from the HG and thereby providing a sustained chemotherapeutic effect for destroying cancer cells. Alginate (sodium alginate, SA)-based HGs with immobilized bevacizumab were also examined [233]. Anti-angiogenic active ingredients were released at prolonged and elevated levels to cancerous and neighbouring cells. As a result, tumour growth was inhibited.

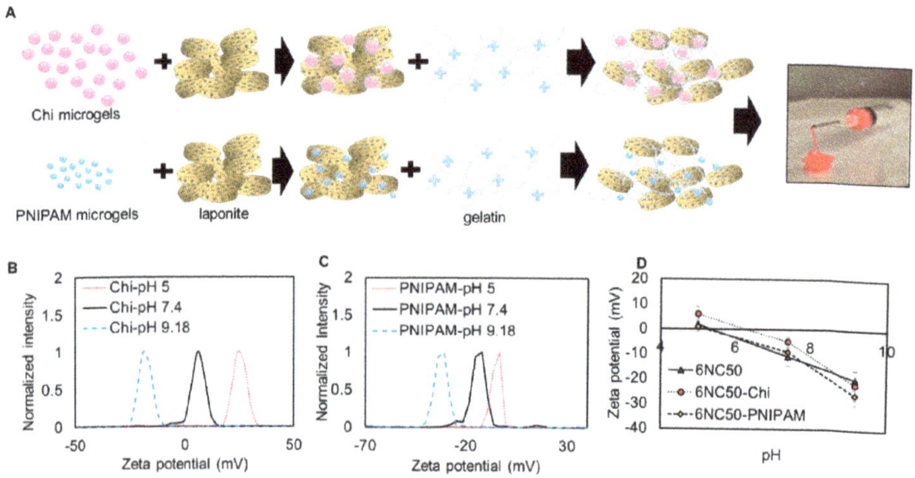

Figure 10. Electrostatic interactions between gelatin and laponite resulted in formation of pH responsive nanocomposite. (**A**) Schematic preparation of shear-thinning HG with laponite. Zeta potential at different pH of (**B**) CTS particles, (**C**) PNIPAM-*co*-Acrylic acid particles and (**D**) PNIPAM-*co*-Acrylic acid and laponite particles. Reprinted with permission from Ref. [230].

6. Future Outlook

Recent developments of HGs as drug delivery carriers for biological and biomedical applications were reviewed. Major synthetic strategies for the preparation of HGs, their classification, the physicochemical properties and their applications were described in detail. The wide variety of recently reported work has demonstrated the application of HG materials as drug delivery systems with different dimensions (macro, micro and nano) in various administration routes.

There are advantages in using HGs as vehicles for active substances. First, HGs can be modified for prolonged or rapid release. This feature is very important in planning therapies for patients and improves compliance with drug regimens. Additionally, by using modified HG materials, drug release can be tailored on demand. Therefore, greater control over health and treatment process can be achieved. Another advantage of functional modifications of HGs is the ability to adjust the controlled release of drugs. By adding specific enzymes or ionisable groups, the entire HG network structure acquires the ability to react to pH changes or to the presence of specific ions.

Technological challenges and problems in the production of HGs are primarily observed in maintaining a balance between chemical structure, composition, drug release and biocompatibility. HGs appear to be very promising materials and forward-looking and have yet to be fully realized. Due to a relatively simple manufacturing process, HGs can be readily modified and functionalized for use in targeted therapy. By using HG structures in medicine, it is easier to achieve better compliance, which is desirable in therapy.

However, the studies reviewed herein are mostly proof-of-concept. There remains limited information on HG in vivo experiments and HG biocompatibility. Most of these studies have focused on in vitro studies to show the non-cytotoxicity of the materials studied. Research on the biocompatibility and biodegradability of these materials, combined with in vivo research, remains a niche topic and has many unresolved questions. The future design and development of effective HG-based drug carriers requires a high degree of control over their properties in vitro and in vivo. These properties include controlled stability for prolonged circulation and biodegradability for facile removal after drug release. One future goal should be the improved design of HG with specific targeting residues to enable highly selective uptake in particular cells or organs. In this way, careful control over size, biodegradability, stability, functionality, and bioconjugation will guide the development of next generation HGs.

Author Contributions: Conceptualization, M.E.P.-B.; writing—original draft preparation, M.E.P.-B.; M.T.; A.C.; writing—review and editing, M.E.P.-B., M.T.; visualization, supervision and project administration, M.E.P.-B.; funding acquisition, M.E.P.-B. All authors have read and agreed to the published version of the manuscript.

Funding: This research was funded by the National Science Centre, Poland, grant number 2017/25/B/ST5/01414 to M.E.P-B.

Conflicts of Interest: The authors declare no conflict of interest.

References

1. Korde, J.M.; Kandasubramanian, B. Naturally biomimicked smart shape memory hydrogels for biomedical functions. *Chem. Eng. J.* **2020**, *379*, 122430. [CrossRef]
2. Tian, W.; Han, S.; Huang, X.; Han, M.; Cao, J.; Liang, Y.; Sun, Y. LDH hybrid thermosensitive hydrogel for intravaginal delivery of anti-HIV drugs. *Artif. Cells Nanomed. Biotechnol.* **2019**, *47*, 1234–1240. [CrossRef] [PubMed]
3. Tibbitt, M.W.; Dahlman, J.E.; Langer, R. Emerging frontiers in drug delivery. *J. Am. Chem. Soc.* **2016**, *138*, 704–717. [CrossRef] [PubMed]
4. Fujiyabu, T.; Yoshikawa, Y.; Chung, U.; Sakai, T. Structure-property relationship of a model network containing solvent. *Sci. Technol. Adv. Mater.* **2019**, *20*, 608–621. [CrossRef] [PubMed]
5. Kirschner, C.M.; Anseth, K.S. Hydrogels in healthcare: From static to dynamic material microenvironments. *Acta Mater.* **2013**, *61*, 931–944. [CrossRef] [PubMed]
6. Chatterjee, S.; Chi-leung Hui, P. Stimuli-responsive hydrogels: An interdisciplinary overview. In *Hydrogels—Smart Materials for Biomedical Applications*; Popa, L., Violeta Ghica, M., Dinu-Pîrvu, C.-E., Eds.; IntechOpen: London, UK, 2019; ISBN 978-1-78985-875-4.
7. Shafiee, S.; Ahangar, H.A.; Saffar, A. Taguchi method optimization for synthesis of Fe_3O_4 @chitosan/Tragacanth Gum nanocomposite as a drug delivery system. *Carbohydr. Polym.* **2019**, *222*, 114982. [CrossRef] [PubMed]
8. Ullah, K.; Sohail, M.; Buabeid, M.A.; Murtaza, G.; Ullah, A.; Rashid, H.; Khan, M.A.; Khan, S.A. Pectin-based (LA-co-MAA) semi-IPNS as a potential biomaterial for colonic delivery of oxaliplatin. *Int. J. Pharm.* **2019**, *569*, 118557. [CrossRef]
9. Nazlı, A.B.; Açıkel, Y.S. Loading of cancer drug resveratrol to pH-Sensitive, smart, alginate-chitosan hydrogels and investigation of controlled release kinetics. *J. Drug Deliv. Sci. Technol.* **2019**, *53*, 101199. [CrossRef]
10. Kim, C.; Jeong, D.; Kim, S.; Kim, Y.; Jung, S. Cyclodextrin functionalized agarose gel with low gelling temperature for controlled drug delivery systems. *Carbohydr. Polym.* **2019**, *222*, 115011. [CrossRef]
11. Cocarta, A.-I.; Hobzova, R.; Sirc, J.; Cerna, T.; Hrabeta, J.; Svojgr, K.; Pochop, P.; Kodetova, M.; Jedelska, J.; Bakowsky, U.; et al. Hydrogel implants for transscleral drug delivery for retinoblastoma treatment. *Mater. Sci. Eng. C* **2019**, *103*, 109799. [CrossRef]
12. Yang, C.; Gao, L.; Liu, X.; Yang, T.; Yin, G.; Chen, J.; Guo, H.; Yu, B.; Cong, H. Injectable Schiff base polysaccharide hydrogels for intraocular drug loading and release. *J. Biomed. Mater. Res.* **2019**, *107*. [CrossRef] [PubMed]
13. Alvarez-Rivera, F.; Serro, A.P.; Silva, D.; Concheiro, A.; Alvarez-Lorenzo, C. Hydrogels for diabetic eyes: Naltrexone loading, release profiles and cornea penetration. *Mater. Sci. Eng. C* **2019**, *105*, 110092. [CrossRef] [PubMed]
14. Dragan, E.S.; Dinu, M.V. Polysaccharides constructed hydrogels as vehicles for proteins and peptides. A review. *Carbohydr. Polym.* **2019**, *225*, 115210. [CrossRef] [PubMed]
15. Duwa, R.; Emami, F.; Lee, S.; Jeong, J.-H.; Yook, S. Polymeric and lipid-based drug delivery systems for treatment of glioblastoma multiforme. *J. Ind. Eng. Chem.* **2019**, *79*, 261–273. [CrossRef]
16. Riley, L.; Schirmer, L.; Segura, T. Granular hydrogels: Emergent properties of jammed hydrogel microparticles and their applications in tissue repair and regeneration. *Curr. Opin. Biotechnol.* **2019**, *60*, 1–8. [CrossRef]

17. Li, J.; Mooney, D.J. Designing hydrogels for controlled drug delivery. *Nat. Rev. Mater.* **2016**, *1*, 10671. [CrossRef]
18. Bakaic, E.; Smeets, N.M.B.; Hoare, T. Injectable hydrogels based on poly (ethylene glycol) and derivatives as functional biomaterials. *RSC Adv.* **2015**, *5*, 35469–35486. [CrossRef]
19. Norouzi, M.; Nazari, B.; Miller, D.W. Injectable hydrogel-based drug delivery systems for local cancer therapy. *Drug Dis. Today* **2016**, *21*, 1835–1849. [CrossRef]
20. Nguyen, Q.V.; Huynh, D.P.; Park, J.H.; Lee, D.S. Injectable polymeric hydrogels for the delivery of therapeutic agents: A review. *Eur. Polym. J.* **2015**, *72*, 602–619. [CrossRef]
21. Soni, K.S.; Desale, S.S.; Bronich, T.K. Nanogels: An overview of properties, biomedical applications and obstacles to clinical translation. *J. Control. Release* **2016**, *240*, 109–126. [CrossRef]
22. Nita, L.E.; Chiriac, A.P.; Rusu, A.G.; Bercea, M.; Ghilan, A.; Dumitriu, R.P.; Mititelu-Tartau, L. New self-healing hydrogels based on reversible physical interactions and their potential applications. *Eur. Polym. J.* **2019**, *118*, 176–185. [CrossRef]
23. Wang, J.; Liu, L.; Chen, J.; Deng, M.; Feng, X.; Chen, L. Supramolecular nanoplatforms via cyclodextrin host-guest recognition for synergistic gene-photodynamic therapy. *Eur. Polym. J.* **2019**, *118*, 222–230. [CrossRef]
24. Marafon, P.; Fachel, F.N.S.; Dal Prá, M.; Bassani, V.L.; Koester, L.S.; Henriques, A.T.; Braganhol, E.; Teixeira, H.F. Development, physico-chemical characterization and in-vitro studies of hydrogels containing rosmarinic acid-loaded nanoemulsion for topical application. *J. Pharm. Pharmacol.* **2019**, *71*, 1199–1208. [CrossRef] [PubMed]
25. Miranda, M.; Cruz, M.T.; Vitorino, C.; Cabral, C. Nanostructuring lipid carriers using *Ridolfia segetum* (L.) Moris essential oil. *Mater. Sci. Eng. C* **2019**, *103*, 109804. [CrossRef] [PubMed]
26. Sharaf, S.; El-Naggar, M.E. Wound dressing properties of cationized cotton fabric treated with carrageenan/cyclodextrin hydrogel loaded with honey bee propolis extract. *Int. J. Biol. Macromol.* **2019**, *133*, 583–591. [CrossRef]
27. Liu, Z.; Zheng, Y.; Dang, J.; Zhang, J.; Dong, F.; Wang, K.; Zhang, J. A novel antifungal plasma-activated hydrogel. *ACS Appl. Mater. Interfaces* **2019**, *11*, 22941–22949. [CrossRef]
28. Wang, K.; Wang, J.; Li, L.; Xu, L.; Feng, N.; Wang, Y.; Fei, X.; Tian, J.; Li, Y. Synthesis of a novel anti-freezing, non-drying antibacterial hydrogel dressing by one-pot method. *Chem. Eng. J.* **2019**, *372*, 216–225. [CrossRef]
29. Chen, W.; Zhu, Y.; Zhang, Z.; Gao, Y.; Liu, W.; Borjihan, Q.; Qu, H.; Zhang, Y.; Zhang, Y.; Wang, Y.-J.; et al. Engineering a multifunctional N-halamine-based antibacterial hydrogel using a super-convenient strategy for infected skin defect therapy. *Chem. Eng. J.* **2020**, *379*, 122238. [CrossRef]
30. Xue, H.; Hu, L.; Xiong, Y.; Zhu, X.; Wei, C.; Cao, F.; Zhou, W.; Sun, Y.; Endo, Y.; Liu, M.; et al. Quaternized chitosan-Matrigel-polyacrylamide hydrogels as wound dressing for wound repair and regeneration. *Carbohydr. Polym.* **2019**, *226*, 115302. [CrossRef]
31. Huang, L.; Zhu, Z.; Wu, D.; Gan, W.; Zhu, S.; Li, W.; Tian, J.; Li, L.; Zhou, C.; Lu, L. Antibacterial poly (ethylene glycol) diacrylate/chitosan hydrogels enhance mechanical adhesiveness and promote skin regeneration. *Carbohydr. Polym.* **2019**, *225*, 115110. [CrossRef]
32. Liang, Y.; Zhao, X.; Hu, T.; Han, Y.; Guo, B. Mussel-inspired, antibacterial, conductive, antioxidant, injectable composite hydrogel wound dressing to promote the regeneration of infected skin. *J. Coll. Interface Sci.* **2019**, *556*, 514–528. [CrossRef] [PubMed]
33. Solomevich, S.O.; Bychkovsky, P.M.; Yurkshtovich, T.L.; Golub, N.V.; Mirchuk, P.Y.; Revtovich, M.Y.; Shmak, A.I. Biodegradable pH-sensitive prospidine-loaded dextran phosphate based hydrogels for local tumor therapy. *Carbohydr. Polym.* **2019**, *226*, 115308. [CrossRef] [PubMed]
34. Chen, H.; Wei, X.; Chen, H.; Wei, H.; Wang, Y.; Nan, W.; Zhang, Q.; Wen, X. The study of establishment of an in vivo tumor model by three-dimensional cells culture systems methods and evaluation of antitumor effect of biotin-conjugated pullulan acetate nanoparticles. *Artif. Cells Nanomed. Biotechnol.* **2019**, *47*, 123–131. [CrossRef] [PubMed]
35. Cao, D.; Zhang, X.; Akabar, M.D.; Luo, Y.; Wu, H.; Ke, X.; Ci, T. Liposomal doxorubicin loaded PLGA-PEG-PLGA based thermogel for sustained local drug delivery for the treatment of breast cancer. *Artif. Cells Nanomed. Biotechnol.* **2019**, *47*, 181–191. [CrossRef]

36. Wu, Q.; He, Z.; Wang, X.; Zhang, Q.; Wei, Q.; Ma, S.; Ma, C.; Li, J.; Wang, Q. Cascade enzymes within self-assembled hybrid nanogel mimicked neutrophil lysosomes for singlet oxygen elevated cancer therapy. *Nat. Commun.* **2019**, *10*, 240. [CrossRef]
37. Puertas-Bartolomé, M.; Benito-Garzón, L.; Fung, S.; Kohn, J.; Vázquez-Lasa, B.; San Román, J. Bioadhesive functional hydrogels: Controlled release of catechol species with antioxidant and antiinflammatory behavior. *Mater. Sci. Eng. C* **2019**, *105*, 110040. [CrossRef]
38. Zheng, J.; Fan, R.; Wu, H.; Yao, H.; Yan, Y.; Liu, J.; Ran, L.; Sun, Z.; Yi, L.; Dang, L.; et al. Directed self-assembly of herbal small molecules into sustained release hydrogels for treating neural inflammation. *Nat. Commun.* **2019**, *10*, 1604. [CrossRef]
39. Uehara, M.; Li, X.; Sheikhi, A.; Zandi, N.; Walker, B.; Saleh, B.; Banouni, N.; Jiang, L.; Ordikhani, F.; Dai, L.; et al. Anti-IL-6 eluting immunomodulatory biomaterials prolong skin allograft survival. *Sci. Rep.* **2019**, *9*, 6535. [CrossRef]
40. García, J.R.; Quirós, M.; Han, W.M.; O'Leary, M.N.; Cox, G.N.; Nusrat, A.; García, A.J. IFN-γ-tethered hydrogels enhance mesenchymal stem cell-based immunomodulation and promote tissue repair. *Biomaterials* **2019**, *220*, 119403. [CrossRef]
41. Hertegård, S.; Nagubothu, S.R.; Malmström, E.; Ström, C.E.; Tolf, A.; Davies, L.C.; Le Blanc, K. Hyaluronan hydrogels for the local delivery of mesenchymal stromal cells to the injured vocal fold. *Stem Cells Dev.* **2019**, *28*, 1177–1190. [CrossRef]
42. Grigoras, A.G. Drug delivery systems using pullulan, a biocompatible polysaccharide produced by fungal fermentation of starch. *Environ. Chem. Lett.* **2019**, *17*, 1209–1223. [CrossRef]
43. Ge, W.; Cao, S.; Shen, F.; Wang, Y.; Ren, J.; Wang, X. Rapid self-healing, stretchable, moldable, antioxidant and antibacterial tannic acid-cellulose nanofibril composite hydrogels. *Carbohydr. Polym.* **2019**, *224*, 115147. [CrossRef] [PubMed]
44. Clegg, J.R.; Wagner, A.M.; Shin, S.R.; Hassan, S.; Khademhosseini, A.; Peppas, N.A. Modular fabrication of intelligent material-tissue interfaces for bioinspired and biomimetic devices. *Prog. Mater. Sci.* **2019**, *106*, 100589. [CrossRef]
45. Bezerra, U.T.; Ferreira, H.S.; Barbosa, N.P. Hydrogels: Types, structure, properties, and applications. In *Frontiers in Biomaterials*; Razavi, M., Ed.; Bentham Science Publishers Ltd.: Sharjah, UAE, 2017; Volume 4, pp. 143–169. ISBN 978-1-68108-536-4.
46. Chatterjee, S.; Hui, P.; Kan, C. Thermoresponsive hydrogels and their biomedical applications: Special insight into their applications in textile based transdermal therapy. *Polymers* **2018**, *10*, 480. [CrossRef]
47. Akhtar, M.F.; Hanif, M.; Ranjha, N.M. Methods of synthesis of hydrogels … A review. *Saudi Pharm. J.* **2016**, *24*, 554–559. [CrossRef]
48. Buwalda, S.J.; Boere, K.W.M.; Dijkstra, P.J.; Feijen, J.; Vermonden, T.; Hennink, W.E. Hydrogels in a historical perspective: From simple networks to smart materials. *J. Control. Release* **2014**, *190*, 254–273. [CrossRef]
49. Zhang, Y.S.; Khademhosseini, A. Advances in engineering hydrogels. *Science* **2017**, *356*, eaaf3627. [CrossRef]
50. Chai, Q.; Jiao, Y.; Yu, X. Hydrogels for biomedical applications: Their characteristics and the mechanisms behind them. *Gels* **2017**, *3*, 6. [CrossRef]
51. Ahmed, E.M. Hydrogel: Preparation, characterization, and applications: A review. *J. Adv. Res.* **2015**, *6*, 105–121. [CrossRef]
52. Straccia, M.; d'Ayala, G.; Romano, I.; Oliva, A.; Laurienzo, P. Alginate hydrogels coated with chitosan for wound dressing. *Mar. Drugs* **2015**, *13*, 2890–2908. [CrossRef]
53. Yuan, L.; Wu, Y.; Fang, J.; Wei, X.; Gu, Q.; El-Hamshary, H.; Al-Deyab, S.S.; Morsi, Y.; Mo, X. Modified alginate and gelatin cross-linked hydrogels for soft tissue adhesive. *Artif. Cells Nanomed. Biotechnol.* **2017**, *45*, 76–83. [CrossRef] [PubMed]
54. Kaczmarek-Pawelska, A. Alginate-based hydrogels in regenerative medicine. In *Alginates*; IntechOpen: London, UK, 2019.
55. Hu, Y.; Zheng, M.; Dong, X.; Zhao, D.; Cheng, H.; Xiao, X. Preparation and characterization of alginate-hyaluronic acid-chitosan based composite gel beads. *J. Wuhan Univ. Technol.-Mater. Sci. Ed.* **2015**, *30*, 1297–1303. [CrossRef]
56. Parlato, M.; Reichert, S.; Barney, N.; Murphy, W.L. Poly(ethylene glycol) hydrogels with adaptable mechanical and degradation properties for use in biomedical applications. *Macromol. Biosci.* **2014**, *14*, 687–698. [CrossRef] [PubMed]

57. Zustiak, S.P.; Leach, J.B. Hydrolytically degradable poly(ethylene glycol) hydrogel scaffolds with tunable degradation and mechanical properties. *Biomacromolecules* **2010**, *11*, 1348–1357. [CrossRef]
58. Basu, A.; Kunduru, K.R.; Doppalapudi, S.; Domb, A.J.; Khan, W. Poly(lactic acid) based hydrogels. *Adv. Drug Deliv. Rev.* **2016**, *107*, 192–205. [CrossRef]
59. Korzhikov-Vlakh, V.; Krylova, M.; Sinitsyna, E.; Ivankova, E.; Averianov, I.; Tennikova, T. Hydrogel layers on the surface of polyester-based materials for improvement of their biointeractions and controlled release of proteins. *Polymers* **2016**, *8*, 418. [CrossRef]
60. Palmese, L.L.; Thapa, R.K.; Sullivan, M.O.; Kiick, K.L. Hybrid hydrogels for biomedical applications. *Curr. Opin. Chem. Eng.* **2019**, *24*, 143–157. [CrossRef]
61. Stagnaro, P.; Schizzi, I.; Utzeri, R.; Marsano, E.; Castellano, M. Alginate-polymethacrylate hybrid hydrogels for potential osteochondral tissue regeneration. *Carbohydr. Polym.* **2018**, *185*, 56–62. [CrossRef]
62. Jia, X.; Kiick, K.L. Hybrid multicomponent hydrogels for tissue engineering. *Macromol. Biosci.* **2009**, *9*, 140–156. [CrossRef]
63. Vieira, V.M.P.; Hay, L.L.; Smith, D.K. Multi-component hybrid hydrogels—Understanding the extent of orthogonal assembly and its impact on controlled release. *Chem. Sci.* **2017**, *8*, 6981–6990. [CrossRef]
64. Hu, S.; Zhou, L.; Tu, L.; Dai, C.; Fan, L.; Zhang, K.; Yao, T.; Chen, J.; Wang, Z.; Xing, J.; et al. Elastomeric conductive hybrid hydrogels with continuous conductive networks. *J. Mater. Chem. B* **2019**, *7*, 2389–2397. [CrossRef]
65. Radvar, E.; Azevedo, H.S. Supramolecular peptide/polymer hybrid hydrogels for biomedical applications. *Macromol. Biosci.* **2019**, *19*, 1800221. [CrossRef] [PubMed]
66. Li, J.; Illeperuma, W.R.K.; Suo, Z.; Vlassak, J.J. Hybrid hydrogels with extremely high stiffness and toughness. *ACS Macro Lett.* **2014**, *3*, 520–523. [CrossRef]
67. Narayanaswamy, R.; Torchilin, V.P. Hydrogels and their applications in targeted drug delivery. *Molecules* **2019**, *24*, 603. [CrossRef] [PubMed]
68. Caccavo, D.; Cascone, S.; Lamberti, G.; Barba, A.A.; Larsson, A. Swellable hydrogel-based systems for controlled drug delivery. In *Smart Drug Delivery System*; Sezer, A.D., Ed.; InTech: London, UK, 2016; ISBN 978-953-51-2247-0.
69. Parhi, R. Cross-linked hydrogel for pharmaceutical applications: A review. *Adv. Pharm. Bull.* **2017**, *7*, 515–530. [CrossRef] [PubMed]
70. Tonda-Turo, C.; Gnavi, S.; Ruini, F.; Gambarotta, G.; Gioffredi, E.; Chiono, V.; Perroteau, I.; Ciardelli, G. Development and characterization of novel agar and gelatin injectable hydrogel as filler for peripheral nerve guidance channels: Novel agar and gelatin injectable hydrogel for peripheral nerve guidance channels. *J. Tissue Eng. Regen. Med.* **2017**, *11*, 197–208. [CrossRef]
71. Wakhet, S.; Singh, V.K.; Sahoo, S.; Sagiri, S.S.; Kulanthaivel, S.; Bhattacharya, M.K.; Kumar, N.; Banerjee, I.; Pal, K. Characterization of gelatin–agar based phase separated hydrogel, emulgel and bigel: A comparative study. *J. Mater. Sci. Mater. Med.* **2015**, *26*, 118. [CrossRef]
72. Sadeghi, M.; Heidari, B. Crosslinked graft copolymer of methacrylic acid and gelatin as a novel hydrogel with pH-responsiveness properties. *Materials* **2011**, *4*, 543–552. [CrossRef]
73. Lanzalaco, S.; Turon, P.; Weis, C.; Alemán, C.; Armelin, E. The mechanism of adhesion and graft polymerization of a PNIPAAm thermoresponsive hydrogel to polypropylene meshes. *Soft Matter* **2019**, *15*, 3432–3442. [CrossRef]
74. Singh, R.; Mahto, V. Synthesis, characterization and evaluation of polyacrylamide graft starch/clay nanocomposite hydrogel system for enhanced oil recovery. *Pet. Sci.* **2017**, *14*, 765–779. [CrossRef]
75. Varaprasad, K.; Raghavendra, G.M.; Jayaramudu, T.; Yallapu, M.M.; Sadiku, R. A mini review on hydrogels classification and recent developments in miscellaneous applications. *Mater. Sci. Eng. C* **2017**, *79*, 958–971. [CrossRef] [PubMed]
76. Mohammadi, S.; Vafaie Sefti, M.; Baghban Salehi, M.; Mousavi Moghadam, A.; Rajaee, S.; Naderi, H. Hydrogel swelling properties: Comparison between conventional and nanocomposite hydrogels for water shutoff treatment: Comparing Conventional hydrogels with Nanocomposite. *Asia Pac. J. Chem. Eng.* **2015**, *10*, 743–753. [CrossRef]
77. Monir, T.S.B.; Afroz, S.; Khan, R.A.; Miah, M.Y.; Takafuji, M.; Alam, M.A. pH-sensitive hydrogel from polyethylene oxide and acrylic acid by gamma radiation. *J. Compos. Sci.* **2019**, *3*, 58. [CrossRef]

78. Hennink, W.E.; van Nostrum, C.F. Novel crosslinking methods to design hydrogels. *Adv. Drug Deliv. Rev.* **2012**, *64*, 223–236. [CrossRef]
79. Bauri, K.; Nandi, M.; De, P. Amino acid-derived stimuli-responsive polymers and their applications. *Polym. Chem.* **2018**, *9*, 1257–1287. [CrossRef]
80. Kahn, J.S.; Hu, Y.; Willner, I. Stimuli-responsive dna-based hydrogels: From basic principles to applications. *Acc. Chem. Res.* **2017**, *50*, 680–690. [CrossRef] [PubMed]
81. Alexander, A.; Ajazuddin; Khan, J.; Saraf, S.; Saraf, S. Poly(ethylene glycol)–poly(lactic-co-glycolic acid) based thermosensitive injectable hydrogels for biomedical applications. *J. Control. Release* **2013**, *172*, 715–729. [CrossRef]
82. Larrañeta, E.; Stewart, S.; Ervine, M.; Al-Kasasbeh, R.; Donnelly, R. Hydrogels for hydrophobic drug delivery. Classification, synthesis and applications. *J. Funct. Biomater.* **2018**, *9*, 13. [CrossRef]
83. Zhao, F.; Yao, D.; Guo, R.; Deng, L.; Dong, A.; Zhang, J. Composites of polymer hydrogels and nanoparticulate systems for biomedical and pharmaceutical applications. *Nanomaterials* **2015**, *5*, 2054–2130. [CrossRef]
84. Laftah, W.A.; Hashim, S.; Ibrahim, A.N. Polymer hydrogels: A review. *Polym. Plast. Technol. Eng.* **2011**, *50*, 1475–1486. [CrossRef]
85. Purushotham, S.; Ramanujan, R.V. Thermoresponsive magnetic composite nanomaterials for multimodal cancer therapy. *Acta Biomater.* **2010**, *6*, 502–510. [CrossRef] [PubMed]
86. Qiu, Y.; Park, K. Environment-sensitive hydrogels for drug delivery. *Adv. Drug Deliv. Rev.* **2012**, *64*, 49–60. [CrossRef]
87. Gong, C.; Qi, T.; Wei, X.; Qu, Y.; Wu, Q.; Luo, F.; Qian, Z. Thermosensitive polymeric hydrogels as drug delivery systems. *Curr. Med. Chem.* **2012**, *20*, 79–94. [CrossRef]
88. Ferreira, N.N.; Ferreira, L.M.B.; Cardoso, V.M.O.; Boni, F.I.; Souza, A.L.R.; Gremião, M.P.D. Recent advances in smart hydrogels for biomedical applications: From self-assembly to functional approaches. *Eur. Polym. J.* **2018**, *99*, 117–133. [CrossRef]
89. Cochis, A.; Bonetti, L.; Sorrentino, R.; Contessi Negrini, N.; Grassi, F.; Leigheb, M.; Rimondini, L.; Farè, S. 3D printing of thermo-responsive methylcellulose hydrogels for cell-sheet engineering. *Materials* **2018**, *11*, 579. [CrossRef]
90. Contessi Negrini, N.; Bonetti, L.; Contili, L.; Farè, S. 3D printing of methylcellulose-based hydrogels. *Bioprinting* **2018**, *10*, e00024. [CrossRef]
91. Larrañeta, E.; Barturen, L.; Ervine, M.; Donnelly, R.F. Hydrogels based on poly(methyl vinyl ether-co-maleic acid) and Tween 85 for sustained delivery of hydrophobic drugs. *Int. J. Pharm.* **2018**, *538*, 147–158. [CrossRef]
92. Hwang, T.; Frank, Z.; Neubauer, J.; Kim, K.J. High-performance polyvinyl chloride gel artificial muscle actuator with graphene oxide and plasticizer. *Sci. Rep.* **2019**, *9*, 1–9. [CrossRef]
93. Bae, J.W.; Shin, E.-J.; Jeong, J.; Choi, D.-S.; Lee, J.E.; Nam, B.U.; Lin, L.; Kim, S.-Y. High-performance PVC gel for adaptive micro-lenses with variable focal length. *Sci. Rep.* **2017**, *7*, 2068. [CrossRef]
94. Kim, M.H.; Kim, B.S.; Park, H.; Lee, J.; Park, W.H. Injectable methylcellulose hydrogel containing calcium phosphate nanoparticles for bone regeneration. *Int. J. Biol. Macromol.* **2018**, *109*, 57–64. [CrossRef]
95. Contessi, N.; Altomare, L.; Filipponi, A.; Farè, S. Thermo-responsive properties of methylcellulose hydrogels for cell sheet engineering. *Mater. Lett.* **2017**, *207*, 157–160. [CrossRef]
96. Safronov, A.P.; Terziyan, T.V. Formation of chemical networks of acrylamide and acrylic acid hydrogels initiated by ammonium persulfate. *Polym. Sci. Ser. B* **2015**, *57*, 481–487. [CrossRef]
97. Jalani, G.; Naccache, R.; Rosenzweig, D.H.; Haglund, L.; Vetrone, F.; Cerruti, M. Photocleavable hydrogel-coated upconverting nanoparticles: A multifunctional theranostic platform for nir imaging and on-demand macromolecular delivery. *J. Am. Chem. Soc.* **2016**, *138*, 1078–1083. [CrossRef]
98. Li, L.; Scheiger, J.M.; Levkin, P.A. Design and applications of photoresponsive hydrogels. *Adv. Mater.* **2019**, *31*, 1807333. [CrossRef]
99. Makhlouf, A.S.H.; Abu-Thabit, N.Y. (Eds.) *Stimuli Responsive Polymeric Nanocarriers for Drug Delivery Applications*; Elsevierz: Amsterdam, The Netherlands, 2018; Volume 1, ISBN 978-0-08-101997-9.
100. Tomatsu, I.; Peng, K.; Kros, A. Photoresponsive hydrogels for biomedical applications. *Adv. Drug Deliv. Rev.* **2011**, *63*, 1257–1266. [CrossRef]
101. Truong, V.X.; Li, F.; Ercole, F.; Forsythe, J.S. Wavelength-selective coupling and decoupling of polymer chains via reversible [2 + 2] photocycloaddition of styrylpyrene for construction of cytocompatible photodynamic hydrogels. *ACS Macro Lett.* **2018**, *7*, 464–469. [CrossRef]

102. Rosales, A.M.; Vega, S.L.; DelRio, F.W.; Burdick, J.A.; Anseth, K.S. Hydrogels with reversible mechanics to probe dynamic cell microenvironments. *Angew. Chem. Int. Ed.* **2017**, *56*, 12132–12136. [CrossRef]
103. Yan, B.; Boyer, J.-C.; Habault, D.; Branda, N.R.; Zhao, Y. near infrared light triggered release of biomacromolecules from hydrogels loaded with upconversion nanoparticles. *J. Am. Chem. Soc.* **2012**, *134*, 16558–16561. [CrossRef]
104. Ariffin, A.; Musa, M.S.; Othman, M.B.H.; Razali, M.A.A.; Yunus, F. Effects of various fillers on anionic polyacrylamide systems for treating kaolin suspensions. *Coll. Surf. A Physicochem. Eng. Asp.* **2014**, *441*, 306–311. [CrossRef]
105. Ullah, F.; Othman, M.B.H.; Javed, F.; Ahmad, Z.; Akil, H.M. Classification, processing and application of hydrogels: A review. *Mater. Sci. Eng. C* **2015**, *57*, 414–433. [CrossRef]
106. Steichen, S.D.; Caldorera-Moore, M.; Peppas, N.A. A review of current nanoparticle and targeting moieties for the delivery of cancer therapeutics. *Eur. J. Pharm. Sci.* **2013**, *48*, 416–427. [CrossRef] [PubMed]
107. Sharpe, L.A.; Daily, A.M.; Horava, S.D.; Peppas, N.A. Therapeutic applications of hydrogels in oral drug delivery. *Exp. Opin. Drug Deliv.* **2014**, *11*, 901–915. [CrossRef] [PubMed]
108. Liu, L.; Yao, W.; Rao, Y.; Lu, X.; Gao, J. pH-Responsive carriers for oral drug delivery: Challenges and opportunities of current platforms. *Drug Deliv.* **2017**, *24*, 569–581. [CrossRef] [PubMed]
109. Qu, J.; Zhao, X.; Ma, P.X.; Guo, B. Injectable antibacterial conductive hydrogels with dual response to an electric field and pH for localized "smart" drug release. *Acta Biomater.* **2018**, *72*, 55–69. [CrossRef] [PubMed]
110. Chen, Y.; Cheng, G.; Hu, R.; Chen, S.; Lu, W.; Gao, S.; Xia, H.; Wang, B.; Sun, C.; Nie, X.; et al. A nasal temperature and pH dual-responsive in situ gel delivery system based on microemulsion of huperzine A: Formulation, evaluation, and in vivo pharmacokinetic study. *AAPS PharmSciTech* **2019**, *20*, 301. [CrossRef] [PubMed]
111. Kiaee, G.; Mostafalu, P.; Samandari, M.; Sonkusale, S. A pH-mediated electronic wound dressing for controlled drug delivery. *Adv. Healthcare Mater.* **2018**, *7*, 1800396. [CrossRef]
112. Anirudhan, T.S.; Mohan, A.M. Novel pH switchable gelatin based hydrogel for the controlled delivery of the anti cancer drug 5-fluorouracil. *RSC Adv.* **2014**, *4*, 12109. [CrossRef]
113. Indermun, S.; Choonara, Y.E.; Kumar, P.; du Toit, L.C.; Modi, G.; Luttge, R.; Pillay, V. An interfacially plasticized electro-responsive hydrogel for transdermal electro-activated and modulated (TEAM) drug delivery. *Int. J. Pharm.* **2014**, *462*, 52–65. [CrossRef]
114. Reddy, N.N.; Mohan, Y.M.; Varaprasad, K.; Ravindra, S.; Joy, P.A.; Raju, K.M. Magnetic and electric responsive hydrogel-magnetic nanocomposites for drug-delivery application. *J. Appl. Polym. Sci.* **2011**, *122*, 1364–1375. [CrossRef]
115. Siangsanoh, C.; Ummartyotin, S.; Sathirakul, K.; Rojanapanthu, P.; Treesuppharat, W. Fabrication and characterization of triple-responsive composite hydrogel for targeted and controlled drug delivery system. *J. Mol. Liq.* **2018**, *256*, 90–99. [CrossRef]
116. Uva, M.; Pasqui, D.; Mencuccini, L.; Fedi, S.; Barbucci, R. Influence of alternating and static magnetic fields on drug release from hybrid hydrogels containing magnetic nanoparticles. *J. Biomater. Nanobiotechnol.* **2014**, *5*, 116–127. [CrossRef]
117. Crippa, F.; Moore, T.L.; Mortato, M.; Geers, C.; Haeni, L.; Hirt, A.M.; Rothen-Rutishauser, B.; Petri-Fink, A. Dynamic and biocompatible thermo-responsive magnetic hydrogels that respond to an alternating magnetic field. *J. Magn. Magn. Mater.* **2017**, *427*, 212–219. [CrossRef]
118. Jain, E.; Hill, L.; Canning, E.; Sell, S.A.; Zustiak, S.P. Control of gelation, degradation and physical properties of polyethylene glycol hydrogels through the chemical and physical identity of the crosslinker. *J. Mater. Chem. B* **2017**, *5*, 2679–2691. [CrossRef]
119. Deng, Y.; Yang, L. Preparation and characterization of polyethylene glycol (PEG) hydrogel as shape-stabilized phase change material. *Appl. Therm. Eng.* **2017**, *114*, 1014–1017. [CrossRef]
120. Fukasawa, M.; Sakai, T.; Chung, U.; Haraguchi, K. Synthesis and mechanical properties of a nanocomposite gel consisting of a tetra-PEG/clay network. *Macromolecules* **2010**, *43*, 4370–4378. [CrossRef]
121. Matsunaga, T.; Sakai, T.; Akagi, Y.; Chung, U.; Shibayama, M. SANS and SLS studies on tetra-arm PEG gels in as-prepared and swollen states. *Macromolecules* **2009**, *42*, 6245–6252. [CrossRef]
122. Matsunaga, T.; Sakai, T.; Akagi, Y.; Chung, U.; Shibayama, M. Structure characterization of tetra-PEG gel by small-angle neutron scattering. *Macromolecules* **2009**, *42*, 1344–1351. [CrossRef]

123. Fujii, K.; Asai, H.; Ueki, T.; Sakai, T.; Imaizumi, S.; Chung, U.; Watanabe, M.; Shibayama, M. High-performance ion gel with tetra-PEG network. *Soft Matter* **2012**, *8*, 1756–1759. [CrossRef]
124. Asai, H.; Fujii, K.; Ueki, T.; Sakai, T.; Chung, U.; Watanabe, M.; Han, Y.-S.; Kim, T.-H.; Shibayama, M. Structural analysis of high performance ion-gel comprising tetra-PEG network. *Macromolecules* **2012**, *45*, 3902–3909. [CrossRef]
125. Oveissi, F.; Naficy, S.; Le, T.Y.L.; Fletcher, D.F.; Dehghani, F. Tough hydrophilic polyurethane-based hydrogels with mechanical properties similar to human soft tissues. *J. Mater. Chem. B* **2019**, *7*, 3512–3519. [CrossRef]
126. Chen, Z.; Wang, T.; Yan, Q. Building a polysaccharide hydrogel capsule delivery system for control release of ibuprofen. *J. Biomater. Sci. Polym. Ed.* **2018**, *29*, 309–324. [CrossRef] [PubMed]
127. Jana, S.; Saha, A.; Nayak, A.K.; Sen, K.K.; Basu, S.K. Aceclofenac-loaded chitosan-tamarind seed polysaccharide interpenetrating polymeric network microparticles. *Coll. Surf. B Biointerfaces* **2013**, *105*, 303–309. [CrossRef] [PubMed]
128. Liou, G.-S.; Lin, P.-H.; Yen, H.-J.; Yu, Y.-Y.; Tsai, T.-W.; Chen, W.-C. Highly flexible and optical transparent 6F-PI/TiO2 optical hybrid films with tunable refractive index and excellent thermal stability. *J. Mater. Chem.* **2010**, *20*, 531–536. [CrossRef]
129. Koetting, M.C.; Peters, J.T.; Steichen, S.D.; Peppas, N.A. Stimulus-responsive hydrogels: Theory, modern advances, and applications. *Mater. Sci. Eng. R Rep.* **2015**, *93*, 1–49. [CrossRef]
130. Jayaramudu, T.; Raghavendra, G.M.; Varaprasad, K.; Sadiku, R.; Raju, K.M. Development of novel biodegradable Au nanocomposite hydrogels based on wheat: For inactivation of bacteria. *Carbohydr. Polym.* **2013**, *92*, 2193–2200. [CrossRef]
131. Gaharwar, A.K.; Peppas, N.A.; Khademhosseini, A. Nanocomposite hydrogels for biomedical applications: Nanocomposite hydrogels. *Biotechnol. Bioeng.* **2014**, *111*, 441–453. [CrossRef]
132. Chen, M.H.; Wang, L.L.; Chung, J.J.; Kim, Y.-H.; Atluri, P.; Burdick, J.A. Methods to assess shear-thinning hydrogels for application as injectable biomaterials. *ACS Biomater. Sci. Eng.* **2017**, *3*, 3146–3160. [CrossRef]
133. Rodell, C.B.; Lee, M.E.; Wang, H.; Takebayashi, S.; Takayama, T.; Kawamura, T.; Arkles, J.S.; Dusaj, N.N.; Dorsey, S.M.; Witschey, W.R.T.; et al. Injectable shear-thinning hydrogels for minimally invasive delivery to infarcted myocardium to limit left ventricular remodeling. *Circ. Cardiovasc. Interv.* **2016**, *9*, e004058. [CrossRef]
134. Nep, E.I.; Conway, B. Grewia gum 2: Mucoadhesive properties of compacts and gels. *Trop. J. Pharm. Res.* **2011**, *10*, 393–401. [CrossRef]
135. Tsintou, M.; Dalamagkas, K.; Seifalian, A. Injectable hydrogel versus plastically compressed collagen scaffold for central nervous system applications. *Int. J. Biomater.* **2018**, *2018*, 1–10. [CrossRef]
136. Bae, K.H.; Wang, L.-S.; Kurisawa, M. Injectable biodegradable hydrogels: Progress and challenges. *J. Mater. Chem. B* **2013**, *1*, 5371. [CrossRef]
137. Mathew, A.P.; Uthaman, S.; Cho, K.-H.; Cho, C.-S.; Park, I.-K. Injectable hydrogels for delivering biotherapeutic molecules. *Int. J. Biol. Macromol.* **2018**, *110*, 17–29. [CrossRef] [PubMed]
138. Tong, R.; Chen, G.; Pan, D.; Qi, H.; Li, R.; Tian, J.; Lu, F.; He, M. Highly stretchable and compressible cellulose ionic hydrogels for flexible strain sensors. *Biomacromolecules* **2019**, *20*, 2096–2104. [CrossRef]
139. Zhou, Y.; Wan, C.; Yang, Y.; Yang, H.; Wang, S.; Dai, Z.; Ji, K.; Jiang, H.; Chen, X.; Long, Y. Highly Stretchable, Elastic, and Ionic Conductive Hydrogel for Artificial Soft Electronics. *Adv. Funct. Mater.* **2019**, *29*, 1806220. [CrossRef]
140. Zhang, X.; Sheng, N.; Wang, L.; Tan, Y.; Liu, C.; Xia, Y.; Nie, Z.; Sui, K. Supramolecular nanofibrillar hydrogels as highly stretchable, elastic and sensitive ionic sensors. *Mater. Horiz.* **2019**, *6*, 326–333. [CrossRef]
141. Liu, S.; Oderinde, O.; Hussain, I.; Yao, F.; Fu, G. Dual ionic cross-linked double network hydrogel with self-healing, conductive, and force sensitive properties. *Polymer* **2018**, *144*, 111–120. [CrossRef]
142. Franzén, H.; Draget, K.; Langebäck, J.; Nilsen-Nygaard, J. Characterization and properties of hydrogels made from neutral soluble chitosans. *Polymers* **2015**, *7*, 373–389. [CrossRef]
143. Shams Es-haghi, S.; Weiss, R.A. Fabrication of tough hydrogels from chemically cross-linked multiple neutral networks. *Macromolecules* **2016**, *49*, 8980–8987. [CrossRef]
144. Weller, C.; Weller, C.; Team, V. Interactive dressings and their role in moist wound management. In *Advanced Textiles for Wound Care*; Elsevier: Amsterdam, The Netherlands, 2019; pp. 105–134. ISBN 978-0-08-102192-7.
145. Seliktar, D. Designing cell-compatible hydrogels for biomedical applications. *Science* **2012**, *336*, 1124–1128. [CrossRef]

146. Kumar, A.; Kumar, A. Antimicrobial polymeric gels. In *Polymeric Gels*; Elsevier: Amsterdam, The Netherlands, 2018; pp. 357–371. ISBN 978-0-08-102179-8.
147. McKenzie, M.; Betts, D.; Suh, A.; Bui, K.; Kim, L.; Cho, H. Hydrogel-based drug delivery systems for poorly water-soluble drugs. *Molecules* **2015**, *20*, 20397–20408. [CrossRef]
148. Gu, D.; O'Connor, A.J.; Qiao, G.G.H.; Ladewig, K. Hydrogels with smart systems for delivery of hydrophobic drugs. *Exp. Opin. Drug Deliv.* **2017**, *14*, 879–895. [CrossRef] [PubMed]
149. Jiang, Y.; Chen, J.; Deng, C.; Suuronen, E.J.; Zhong, Z. Click hydrogels, microgels and nanogels: Emerging platforms for drug delivery and tissue engineering. *Biomaterials* **2014**, *35*, 4969–4985. [CrossRef] [PubMed]
150. Barthelat, F.; Yin, Z.; Buehler, M.J. Structure and mechanics of interfaces in biological materials. *Nat. Rev. Mater.* **2016**, *1*, 16007. [CrossRef]
151. Thorne, J.B.; Vine, G.J.; Snowden, M.J. Microgel applications and commercial considerations. *Coll. Polym. Sci.* **2011**, *289*, 625–646. [CrossRef]
152. Wang, Y.; Tu, S.; Pinchuk, A.N.; Xiong, M.P. Active drug encapsulation and release kinetics from hydrogel-in-liposome nanoparticles. *J. Coll. Interface Sci.* **2013**, *406*, 247–255. [CrossRef]
153. Nayak, A.K.; Das, B. Introduction to polymeric gels. In *Polymeric Gels*; Elsevier: Amsterdam, The Netherlands, 2018; pp. 3–27. ISBN 978-0-08-102179-8.
154. Kawaguchi, H. Micro hydrogels: Preparation, properties, and applications. *J. Oleo Sci.* **2013**, *62*, 865–871. [CrossRef]
155. Vinogradov, S. Colloidal microgels in drug delivery applications. *Curr. Pharm. Des.* **2006**, *12*, 4703–4712. [CrossRef]
156. Panda, J.J.; Mishra, A.; Basu, A.; Chauhan, V.S. Stimuli responsive self-assembled hydrogel of a low molecular weight free dipeptide with potential for tunable drug delivery. *Biomacromolecules* **2008**, *9*, 2244–2250. [CrossRef]
157. Kim, B.; Peppas, N.A. In vitro release behavior and stability of insulin in complexation hydrogels as oral drug delivery carriers. *Int. J. Pharm.* **2003**, *266*, 29–37. [CrossRef]
158. Na, K.; Lee, K.H.; Bae, Y.H. pH-sensitivity and pH-dependent interior structural change of self-assembled hydrogel nanoparticles of pullulan acetate/oligo-sulfonamide conjugate. *J. Control. Release* **2004**, *97*, 513–525. [CrossRef]
159. Fleige, E.; Quadir, M.A.; Haag, R. Stimuli-responsive polymeric nanocarriers for the controlled transport of active compounds: Concepts and applications. *Adv. Drug Deliv. Rev.* **2012**, *64*, 866–884. [CrossRef] [PubMed]
160. Setia, A.; Ahuja, P. Nanohydrogels. In *Organic Materials as Smart Nanocarriers for Drug Delivery*; Elsevier: Amsterdam, The Netherlands, 2018; pp. 293–368. ISBN 978-0-12-813663-8.
161. Kim, S.H.; Kim, S.-H.; Nair, S.; Moore, E. Reactive electrospinning of cross-linked poly(2-hydroxyethyl methacrylate) nanofibers and elastic properties of individual hydrogel nanofibers in aqueous solutions. *Macromolecules* **2005**, *38*, 3719–3723. [CrossRef]
162. Gil, E.; Hudson, S. Stimuli-reponsive polymers and their bioconjugates. *Prog. Polym. Sci.* **2004**, *29*, 1173–1222. [CrossRef]
163. Konwar, R.; Ahmed, A.B. Nanoparticle: An overview of preparation, characterization and application. *Int. Res. J. Pharm.* **2016**, *4*, 47–57. [CrossRef]
164. Kumari, A.; Yadav, S.K.; Yadav, S.C. Biodegradable polymeric nanoparticles based drug delivery systems. *Coll. Surf. B Biointerfaces* **2010**, *75*, 1–18. [CrossRef]
165. Guvendiren, M.; Lu, H.D.; Burdick, J.A. Shear-thinning hydrogels for biomedical applications. *Soft Matter* **2012**, *8*, 260–272. [CrossRef]
166. Metters, A.T.; Bowman, C.N.; Anseth, K.S. Verification of scaling laws for degrading PLA-*b*-PEG-*b*-PLA hydrogels. *AIChE J.* **2001**, *47*, 1432–1437. [CrossRef]
167. Mason, M.N.; Metters, A.T.; Bowman, C.N.; Anseth, K.S. Predicting controlled-release behavior of degradable PLA-*b*-PEG-*b*-PLA hydrogels. *Macromolecules* **2001**, *34*, 4630–4635. [CrossRef]
168. Lustig, S.R.; Peppas, N.A. Solute diffusion in swollen membranes. IX. Scaling laws for solute diffusion in gels. *J. Appl. Polym. Sci.* **1988**, *36*, 735–747. [CrossRef]
169. Metters, A.T.; Anseth, K.S.; Bowman, C.N. A statistical kinetic model for the bulk degradation of PLA-*b*-PEG-*b*-PLA hydrogel networks: Incorporating network non-idealities. *J. Phys. Chem. B* **2001**, *105*, 8069–8076. [CrossRef]

170. Tan, H.; Marra, K.G. Injectable, biodegradable hydrogels for tissue engineering applications. *Materials* **2010**, *3*, 1746–1767. [CrossRef]
171. Lee, J.H. Injectable hydrogels delivering therapeutic agents for disease treatment and tissue engineering. *Biomater. Res.* **2018**, *22*, 1–14. [CrossRef] [PubMed]
172. Wang, J.; Yao, K.; Wang, C.; Tang, C.; Jiang, X. Synthesis and drug delivery of novel amphiphilic block copolymers containing hydrophobic dehydroabietic moiety. *J. Mater. Chem. B* **2013**, *1*, 2324. [CrossRef]
173. Lengert, E.; Yashchenok, A.M.; Atkin, V.; Lapanje, A.; Gorin, D.A.; Sukhorukov, G.B.; Parakhonskiy, B.V. Hollow silver alginate microspheres for drug delivery and surface enhanced Raman scattering detection. *RSC Adv.* **2016**, *6*, 20447–20452. [CrossRef]
174. Yang, T.-T.; Cheng, Y.-Z.; Qin, M.; Wang, Y.-H.; Yu, H.-L.; Wang, A.-L.; Zhang, W.-F. Thermosensitive chitosan hydrogels containing polymeric microspheres for vaginal drug delivery. *BioMed Res. Int.* **2017**, *2017*, 1–12. [CrossRef]
175. Nan, J.; Chen, Y.; Li, R.; Wang, J.; Liu, M.; Wang, C.; Chu, F. Polymeric hydrogel nanocapsules: A thermo and pH dual-responsive carrier for sustained drug release. *Nano-Micro Lett.* **2014**, *6*, 200–208. [CrossRef]
176. Yu, M.; Xu, L.; Tian, F.; Su, Q.; Zheng, N.; Yang, Y.; Wang, J.; Wang, A.; Zhu, C.; Guo, S.; et al. Rapid transport of deformation-tuned nanoparticles across biological hydrogels and cellular barriers. *Nat. Commun.* **2018**, *9*, 2607. [CrossRef]
177. Shchukin, D.G.; Sukhorukov, G.B.; Möhwald, H. Smart inorganic/organic nanocomposite hollow microcapsules. *Angew. Chem. Int. Ed.* **2003**, *42*, 4472–4475. [CrossRef]
178. Zhu, Y.; Shi, J.; Shen, W.; Dong, X.; Feng, J.; Ruan, M.; Li, Y. Stimuli-responsive controlled drug release from a hollow mesoporous silica sphere/polyelectrolyte multilayer core-shell structure. *Angew. Chem. Int. Ed.* **2005**, *44*, 5083–5087. [CrossRef]
179. Bayat, M.; Nasri, S. Injectable microgel–hydrogel composites "plum pudding gels": New system for prolonged drug delivery. In *Nanomaterials for Drug Delivery and Therapy*; Elsevier: Amsterdam, The Netherlands, 2019; pp. 343–372. ISBN 978-0-12-816505-8.
180. Basso, J.; Miranda, A.; Nunes, S.; Cova, T.; Sousa, J.; Vitorino, C.; Pais, A. Hydrogel-based drug delivery nanosystems for the treatment of brain tumors. *Gels* **2018**, *4*, 62. [CrossRef]
181. Motornov, M.; Roiter, Y.; Tokarev, I.; Minko, S. Stimuli-responsive nanoparticles, nanogels and capsules for integrated multifunctional intelligent systems. *Prog. Polym. Sci.* **2010**, *35*, 174–211. [CrossRef]
182. Desfrançois, C.; Auzély, R.; Texier, I. Lipid nanoparticles and their hydrogel composites for drug delivery: A review. *Pharmaceuticals* **2018**, *11*, 118. [CrossRef] [PubMed]
183. Pertici, V.; Pin-Barre, C.; Rivera, C.; Pellegrino, C.; Laurin, J.; Gigmes, D.; Trimaille, T. Degradable and injectable hydrogel for drug delivery in soft tissues. *Biomacromolecules* **2019**, *20*, 149–163. [CrossRef] [PubMed]
184. Sosnik, A.; Seremeta, K. Polymeric hydrogels as technology platform for drug delivery applications. *Gels* **2017**, *3*, 25. [CrossRef]
185. Song, S.H.; Lee, K.M.; Kang, J.B.; Lee, S.G.; Kang, M.J.; Choi, Y.W. Improved skin delivery of voriconazole with a nanostructured lipid carrier-based hydrogel formulation. *Chem. Pharm. Bull.* **2014**, *62*, 793–798. [CrossRef]
186. Khare, A.; Singh, I.; Pawar, P.; Grover, K. Design and evaluation of voriconazole loaded solid lipid nanoparticles for ophthalmic application. *J. Drug Deliv.* **2016**, *2016*, 1–11. [CrossRef]
187. López-Noriega, A.; Hastings, C.L.; Ozbakir, B.; O'Donnell, K.E.; O'Brien, F.J.; Storm, G.; Hennink, W.E.; Duffy, G.P.; Ruiz-Hernández, E. Hyperthermia-induced drug delivery from thermosensitive liposomes encapsulated in an injectable hydrogel for local chemotherapy. *Adv. Healthcare Mater.* **2014**, *3*, 854–859. [CrossRef]
188. Kim, D.Y.; Kwon, D.Y.; Kwon, J.S.; Park, J.H.; Park, S.H.; Oh, H.J.; Kim, J.H.; Min, B.H.; Park, K.; Kim, M.S. Synergistic anti-tumor activity through combinational intratumoral injection of an in-situ injectable drug depot. *Biomaterials* **2016**, *85*, 232–245. [CrossRef]
189. Sivakumaran, D.; Maitland, D.; Hoare, T. Injectable microgel-hydrogel composites for prolonged small-molecule drug delivery. *Biomacromolecules* **2011**, *12*, 4112–4120. [CrossRef]
190. Town, A.R.; Giardiello, M.; Gurjar, R.; Siccardi, M.; Briggs, M.E.; Akhtar, R.; McDonald, T.O. Dual-stimuli responsive injectable microgel/solid drug nanoparticle nanocomposites for release of poorly soluble drugs. *Nanoscale* **2017**, *9*, 6302–6314. [CrossRef]

191. Hudson, S.P.; Langer, R.; Fink, G.R.; Kohane, D.S. Injectable in situ cross-linking hydrogels for local antifungal therapy. *Biomaterials* **2010**, *31*, 1444–1452. [CrossRef] [PubMed]
192. McGillicuddy, F.C.; Lynch, I.; Rochev, Y.A.; Burke, M.; Dawson, K.A.; Gallagher, W.M.; Keenan, A.K. Novel "plum pudding" gels as potential drug-eluting stent coatings: Controlled release of fluvastatin. *J. Biomed. Mater. Res. Part A* **2006**, *79*, 923–933. [CrossRef] [PubMed]
193. Cho, J.-K.; Park, J.W.; Song, S.-C. Injectable and biodegradable poly(organophosphazene) gel containing silibinin: Its physicochemical properties and anticancer activity. *J. Pharm. Sci.* **2012**, *101*, 2382–2391. [CrossRef]
194. Lei, N.; Gong, C.; Qian, Z.; Luo, F.; Wang, C.; Wang, H.; Wei, Y. Therapeutic application of injectable thermosensitive hydrogel in preventing local breast cancer recurrence and improving incision wound healing in a mouse model. *Nanoscale* **2012**, *4*, 5686. [CrossRef]
195. Chen, M.-C.; Tsai, H.-W.; Liu, C.-T.; Peng, S.-F.; Lai, W.-Y.; Chen, S.-J.; Chang, Y.; Sung, H.-W. A nanoscale drug-entrapment strategy for hydrogel-based systems for the delivery of poorly soluble drugs. *Biomaterials* **2009**, *30*, 2102–2111. [CrossRef]
196. Biondi, M.; Borzacchiello, A.; Mayol, L.; Ambrosio, L. Nanoparticle-integrated hydrogels as multifunctional composite materials for biomedical applications. *Gels* **2015**, *1*, 162–178. [CrossRef]
197. Li, Y.; Huang, G.; Zhang, X.; Li, B.; Chen, Y.; Lu, T.; Lu, T.J.; Xu, F. Magnetic hydrogels and their potential biomedical applications. *Adv. Funct. Mater.* **2013**, *23*, 660–672. [CrossRef]
198. Meid, J.; Friedrich, T.; Tieke, B.; Lindner, P.; Richtering, W. Composite hydrogels with temperature sensitive heterogeneities: Influence of gel matrix on the volume phase transition of embedded poly-(N-isopropylacrylamide) microgels. *Phys. Chem. Chem. Phys.* **2011**, *13*, 3039–3047. [CrossRef]
199. Salvati, A.; Söderman, O.; Lynch, I. Plum-pudding gels as a platform for drug delivery: Understanding the effects of the different components on the diffusion behavior of solutes. *J. Phys. Chem. B* **2007**, *111*, 7367–7376. [CrossRef]
200. Li, X.; Rombouts, W.; van der Gucht, J.; de Vries, R.; Dijksman, J.A. Mechanics of composite hydrogels approaching phase separation. *PLoS ONE* **2019**, *14*, e0211059. [CrossRef]
201. Tan, H.-L.; Teow, S.-Y.; Pushpamalar, J. Application of metal nanoparticle–hydrogel composites in tissue regeneration. *Bioengineering* **2019**, *6*, 17. [CrossRef] [PubMed]
202. Alam, A.; Zhang, Y.; Kuan, H.-C.; Lee, S.-H.; Ma, J. Polymer composite hydrogels containing carbon nanomaterials—Morphology and mechanical and functional performance. *Prog. Polym. Sci.* **2018**, *77*, 1–18. [CrossRef]
203. Rocha-García, D.; Guerra-Contreras, A.; Rosales-Mendoza, S.; Palestino, G. Role of porous silicon/hydrogel composites on drug delivery. *Open Mater. Sci.* **2016**, *3*, 93–101. [CrossRef]
204. Ayub, A.D.; Chiu, H.I.; Mat Yusuf, S.N.A.; Abd Kadir, E.; Ngalim, S.H.; Lim, V. Biocompatible disulphide cross-linked sodium alginate derivative nanoparticles for oral colon-targeted drug delivery. *Artif. Cells Nanomed. Biotechnol.* **2019**, *47*, 353–369. [CrossRef] [PubMed]
205. Havanur, S.; Batish, I.; Cheruku, S.P.; Gourishetti, K.; JagadeeshBabu, P.E.; Kumar, N. Poly(N,N-diethyl acrylamide)/functionalized graphene quantum dots hydrogels loaded with doxorubicin as a nano-drug carrier for metastatic lung cancer in mice. *Mater. Sci. Eng. C* **2019**, *105*, 110094. [CrossRef]
206. Hou, F.; Xi, B.; Wang, X.; Yang, Y.; Zhao, H.; Li, W.; Qin, J.; He, Y. Self-healing hydrogel with cross-linking induced thermo-response regulated light emission property. *Coll. Surf. B Biointerfaces* **2019**, *183*, 110441. [CrossRef]
207. Zarrintaj, P.; Jouyandeh, M.; Ganjali, M.R.; Hadavand, B.S.; Mozafari, M.; Sheiko, S.S.; Vatankhah-Varnoosfaderani, M.; Gutiérrez, T.J.; Saeb, M.R. Thermo-sensitive polymers in medicine: A review. *Eur. Polym. J.* **2019**, *117*, 402–423. [CrossRef]
208. Zhang, Y.; An, Q.; Tong, W.; Li, H.; Ma, Z.; Zhou, Y.; Huang, T.; Zhang, Y. A new way to promote molecular drug release during medical treatment: A polyelectrolyte matrix on a piezoelectric-dielectric energy conversion substrate. *Small* **2018**, *14*, 1802136. [CrossRef]
209. Dadou, S.M.; El-Barghouthi, M.I.; Antonijevic, M.D.; Chowdhry, B.Z.; Badwan, A.A. Elucidation of the controlled-release behavior of metoprolol succinate from directly compressed xanthan gum/chitosan polymers: Computational and experimental studies. *ACS Biomater. Sci. Eng.* **2019**. [CrossRef]

210. Gu, S.; Yang, L.; Li, S.; Yang, J.; Zhang, B.; Yang, J. Thermo- and glucose-sensitive microgels with improved salt tolerance for controlled insulin release in a physiological environment: Thermo- and glucose-sensitive microgels with improved salt tolerance. *Polym. Int.* **2018**, *67*, 1256–1265. [CrossRef]
211. Wang, X.; Wang, C.; Zhang, Q.; Cheng, Y. Near infrared light-responsive and injectable supramolecular hydrogels for on-demand drug delivery. *Chem. Commun.* **2016**, *52*, 978–981. [CrossRef] [PubMed]
212. Sattari, S.; Dadkhah Tehrani, A.; Adeli, M. pH-responsive hybrid hydrogels as antibacterial and drug delivery systems. *Polymers* **2018**, *10*, 660. [CrossRef] [PubMed]
213. Li, X.; Fu, M.; Wu, J.; Zhang, C.; Deng, X.; Dhinakar, A.; Huang, W.; Qian, H.; Ge, L. pH-sensitive peptide hydrogel for glucose-responsive insulin delivery. *Acta Biomater.* **2017**, *51*, 294–303. [CrossRef]
214. Panahi, Y.; Gharekhani, A.; Hamishehkar, H.; Zakeri-Milani, P.; Gharekhani, H. Stomach-specific drug delivery of clarithromycin using asemi interpenetrating polymeric network hydrogel made ofmontmorillonite and chitosan: Synthesis, characterization and in vitro drug release study. *Adv. Pharm. Bull.* **2019**, *9*, 159–173. [CrossRef]
215. Qi, X.; Wei, W.; Li, J.; Zuo, G.; Pan, X.; Su, T.; Zhang, J.; Dong, W. Salecan-based pH-sensitive hydrogels for insulin delivery. *Mol. Pharm.* **2017**, *14*, 431–440. [CrossRef]
216. Li, L.; Jiang, G.; Yu, W.; Liu, D.; Chen, H.; Liu, Y.; Huang, Q.; Tong, Z.; Yao, J.; Kong, X. A composite hydrogel system containing glucose-responsive nanocarriers for oral delivery of insulin. *Mater. Sci. Eng. C* **2016**, *69*, 37–45. [CrossRef]
217. Udeni Gunathilake, T.; Ching, Y.; Chuah, C. Enhancement of curcumin bioavailability using nanocellulose reinforced chitosan hydrogel. *Polymers* **2017**, *9*, 64. [CrossRef]
218. Martínez-Martínez, M.; Rodríguez-Berna, G.; Bermejo, M.; Gonzalez-Alvarez, I.; Gonzalez-Alvarez, M.; Merino, V. Covalently crosslinked organophosphorous derivatives-chitosan hydrogel as a drug delivery system for oral administration of camptothecin. *Eur. J. Pharm. Biopharm.* **2019**, *136*, 174–183. [CrossRef]
219. Tao, G.; Wang, Y.; Cai, R.; Chang, H.; Song, K.; Zuo, H.; Zhao, P.; Xia, Q.; He, H. Design and performance of sericin/poly(vinyl alcohol) hydrogel as a drug delivery carrier for potential wound dressing application. *Mater. Sci. Eng. C* **2019**, *101*, 341–351. [CrossRef]
220. Singh, B.; Varshney, L.; Francis, S.; Rajneesh. Designing tragacanth gum based sterile hydrogel by radiation method for use in drug delivery and wound dressing applications. *Int. J. Biol. Macromol.* **2016**, *88*, 586–602. [CrossRef]
221. Wang, W.; Wat, E.; Hui, P.C.L.; Chan, B.; Ng, F.S.F.; Kan, C.-W.; Wang, X.; Hu, H.; Wong, E.C.W.; Lau, C.B.S.; et al. Dual-functional transdermal drug delivery system with controllable drug loading based on thermosensitive poloxamer hydrogel for atopic dermatitis treatment. *Sci. Rep.* **2016**, *6*, 24112. [CrossRef] [PubMed]
222. Qiao, X.; Peng, X.; Qiao, J.; Jiang, Z.; Han, B.; Yang, C.; Liu, W. Evaluation of a photocrosslinkable hydroxyethyl chitosan hydrogel as a potential drug release system for glaucoma surgery. *J. Mater. Sci. Mater. Med.* **2017**, *28*, 149. [CrossRef] [PubMed]
223. Chang, P.-C.; Chao, Y.-C.; Hsiao, M.-H.; Chou, H.-S.; Jheng, Y.-H.; Yu, X.-H.; Lee, N.; Yang, C.; Liu, D.-M. Inhibition of periodontitis induction using a stimuli-responsive hydrogel carrying naringin. *J. Periodontol.* **2017**, *88*, 190–196. [CrossRef] [PubMed]
224. Alvarez Echazú, M.I.; Olivetti, C.E.; Anesini, C.; Perez, C.J.; Alvarez, G.S.; Desimone, M.F. Development and evaluation of thymol-chitosan hydrogels with antimicrobial-antioxidant activity for oral local delivery. *Mater. Sci. Eng. C* **2017**, *81*, 588–596. [CrossRef] [PubMed]
225. Hyun, H.; Park, M.; Jo, G.; Kim, S.; Chun, H.; Yang, D. Photo-cured glycol chitosan hydrogel for ovarian cancer drug delivery. *Mar. Drugs* **2019**, *17*, 41. [CrossRef]
226. Rezazadeh, M.; Akbari, V.; Amuaghae, E.; Emami, J. Preparation and characterization of an injectable thermosensitive hydrogel for simultaneous delivery of paclitaxel and doxorubicin. *Res. Pharm. Sci.* **2018**, *13*, 181.
227. Songkroh, T.; Xie, H.; Yu, W.; Liu, X.; Sun, G.; Xu, X.; Ma, X. Injectable in situ forming chitosan-based hydrogels for curcumin delivery. *Macromol. Res.* **2015**, *23*, 53–59. [CrossRef]
228. Naderi, Z.; Azizian, J. Synthesis and characterization of carboxymethyl chitosan/Fe3O4 and MnFe2O4 nanocomposites hydrogels for loading and release of curcumin. *J. Photochem. Photobiol. B Biol.* **2018**, *185*, 206–214. [CrossRef]

229. Li, C.; Ren, S.; Dai, Y.; Tian, F.; Wang, X.; Zhou, S.; Deng, S.; Liu, Q.; Zhao, J.; Chen, X. Efficacy, pharmacokinetics, and biodistribution of thermosensitive chitosan/β-glycerophosphate hydrogel loaded with docetaxel. *AAPS PharmSciTech* **2014**, *15*, 417–424. [CrossRef]
230. Samimi Gharaie, S.; Dabiri, S.; Akbari, M. Smart shear-thinning hydrogels as injectable drug delivery systems. *Polymers* **2018**, *10*, 1317. [CrossRef]
231. Hu, X.; Wang, Y.; Zhang, L.; Xu, M.; Zhang, J.; Dong, W. Magnetic field-driven drug release from modified iron oxide-integrated polysaccharide hydrogel. *Int. J. Biol. Macromol.* **2018**, *108*, 558–567. [CrossRef] [PubMed]
232. Niu, X.; Zhang, Z.; Zhong, Y. Hydrogel loaded with self-assembled dextran sulfate-doxorubicin complexes as a delivery system for chemotherapy. *Mater. Sci. Eng. C* **2017**, *77*, 888–894. [CrossRef] [PubMed]
233. Ferreira, N.N.; Ferreira, L.M.; Miranda-Gonçalves, V.; Reis, R.M.; Seraphim, T.V.; Borges, J.C.; Baltazar, F.; Gremião, M.P.D. Alginate hydrogel improves anti-angiogenic bevacizumab activity in cancer therapy. *Eur. J. Pharm. Biopharm.* **2017**, *119*, 271–282. [CrossRef]
234. Liu, L.; Gao, Q.; Lu, X.; Zhou, H. In situ forming hydrogels based on chitosan for drug delivery and tissue regeneration. *Asian J. Pharm. Sci.* **2016**, *11*, 673–683. [CrossRef]
235. Li, X.; Kong, X.; Zhang, J.; Wang, Y.; Wang, Y.; Shi, S.; Guo, G.; Luo, F.; Zhao, X.; Wei, Y.; et al. Pharmaceutical nanotechnology: A novel composite hydrogel based on chitosan and inorganic phosphate for local drug delivery of camptothecin nanocolloids. *J. Pharm. Sci.* **2011**, *100*, 232–241. [CrossRef]
236. Zhu, X.; Zhang, Y.; Huang, H.; Zhang, H.; Hou, L.; Zhang, Z. Functionalized graphene oxide-based thermosensitive hydrogel for near-infrared chemo-photothermal therapy on tumor. *J. Biomater. Appl.* **2016**, *30*, 1230–1241. [CrossRef]

materials

MDPI

Review

CO-Releasing Materials: An Emphasis on Therapeutic Implications, as Release and Subsequent Cytotoxicity Are the Part of Therapy

Muhammad Faizan [1], Niaz Muhammad [2], Kifayat Ullah Khan Niazi [3], Yongxia Hu [1], Yanyan Wang [1], Ya Wu [1], Huaming Sun [1], Ruixia Liu [4], Wensheng Dong [1], Weiqiang Zhang [1,*] and Ziwei Gao [1,*]

1 Key Laboratory of Applied Surface and Colloid Chemistry MOE, School of Chemistry and Chemical Engineering, Shaanxi Normal University, Xi'an 710062, China; faizanattari@snnu.edu.cn (M.F.); huyongxia@snnu.edu.cn (Y.H.); yyw@snnu.edu.cn (Y.W.); wuya@snnu.edu.cn (Y.W.); hmsun@snnu.edu.cn (H.S.); wsdong@snnu.edu.cn (W.D.)
2 Department of Biochemistry, College of Life Sciences, Shaanxi Normal University, Xi'an 710062, China; niazpk@hotmail.com
3 School of Materials Science and Engineering, Xi'an Jiaotong University, Xi'an 710049, China; niazikifayat@stu.xjtu.edu.cn
4 Institute of Process Engineering, Chinese Academy of Science, Beijing 100190, China; rxliu@ipe.ac.cn
* Correspondence: zwq@snnu.edu.cn (W.Z.); zwgao@snnu.edu.cn (Z.G.); Tel.: +86-181-8243-8818 (W.Z.); +86-150-2929-3998 (Z.G.)

Received: 7 April 2019; Accepted: 14 May 2019; Published: 20 May 2019

Abstract: The CO-releasing materials (CORMats) are used as substances for producing CO molecules for therapeutic purposes. Carbon monoxide (CO) imparts toxic effects to biological organisms at higher concentration. If this characteristic is utilized in a controlled manner, it can act as a cell-signaling agent for important pathological and pharmacokinetic functions; hence offering many new applications and treatments. Recently, research on therapeutic applications using the CO treatment has gained much attention due to its nontoxic nature, and its injection into the human body using several conjugate systems. Mainly, there are two types of CO insertion techniques into the human body, i.e., direct and indirect CO insertion. Indirect CO insertion offers an advantage of avoiding toxicity as compared to direct CO insertion. For the indirect CO inhalation method, developers are facing certain problems, such as its inability to achieve the specific cellular targets and how to control the dosage of CO. To address these issues, researchers have adopted alternative strategies regarded as CO-releasing molecules (CORMs). CO is covalently attached with metal carbonyl complexes (MCCs), which generate various CORMs such as CORM-1, CORM-2, CORM-3, ALF492, CORM A1 and ALF186. When these molecules are inserted into the human body, CO is released from these compounds at a controlled rate under certain conditions or/and triggers. Such reactions are helpful in achieving cellular level targets with a controlled release of the CO amount. However on the other hand, CORMs also produce a metal residue (termed as i-CORMs) upon degradation that can initiate harmful toxic activity inside the body. To improve the performance of the CO precursor with the restricted development of i-CORMs, several new CORMats have been developed such as micellization, peptide, vitamins, MOFs, polymerization, nanoparticles, protein, metallodendrimer, nanosheet and nanodiamond, etc. In this review article, we shall describe modern ways of CO administration; focusing primarily on exclusive features of CORM's tissue accumulations and their toxicities. This report also elaborates on the kinetic profile of the CO gas. The comprehension of developmental phases of CORMats shall be useful for exploring the ideal CO therapeutic drugs in the future of medical sciences.

Keywords: CO administration; therapeutic agent; pharmaceutical drugs; heme oxygenase; CO-releasing materials; CO-releasing molecules; organometallic complexes; pharmacokinetic functions; pathological role; CO kinetic profile; cellular targets

1. Introduction

Carbon monoxide (CO) is considered harmful due to its toxic behavior since the last century. It has a tasteless, odorless and colorless nature. Its colorless nature allows CO to remain undetectable even at high concentrations and toxic levels, thus marked as the "silent killer" [1,2]. This poisonous CO behavior is exerted due to the formation of carboxy hemoglobin (COHb) along with oxygen present in the mainstream blood circulation. Haldane and Douglas scientifically explored it the first time through dissociation curves of CO-hemoglobin using a constant percentage of CO along with a variable percentage of oxygen at an atmospheric pressure [3,4]. The ubiquitous enzyme, heme oxygenase (HO) has been investigated in most of the biological species. In the middle of the 19th century, two scientists Tenhunen and Schmidt discovered the intracellular CO production by heme oxygenase with an enzyme being the heme catalyst [5,6]. HO is categorized under two isoforms: HO-1 (HMOX1; gene name), with its capability to remain inducible in all cell functions; and HO-2 (HMOX2; gene name), that is constitutively expressed and substantially contained in vasculature and testes [7,8]. HO-1 is identified as an element exclusively found in spleen and liver [9]; however it might be influenced by varying intensity in most biological tissues. Both HO-1 and HO-2 indicate the rate-determining step, drag-out biliverdin from the heme conversion with the CO release and Iron product associated with a tetra pyrrole ring. Biliverdin using biliverdin reductase transforms into bilirubin while generating ferritin quickly from the Iron segregate (Figure 1) [10]. The released amount of CO attaches with the Iron containing objects due to its higher diffusion rate and tendency. It tends to make itself bonded with blood in the circulatory system; and ultimately it is exhaled through lungs. CO causes a common sagginess by bringing affliction for the mammalians, completely dependent on oxygen for the blood transport system and mitochondrial respiration. Collectively, endogenously generated CO is featured in the physiological role. Generally, a low dose of CO gas endures tremendous benefits and can achieve remarkable therapeutic targets.

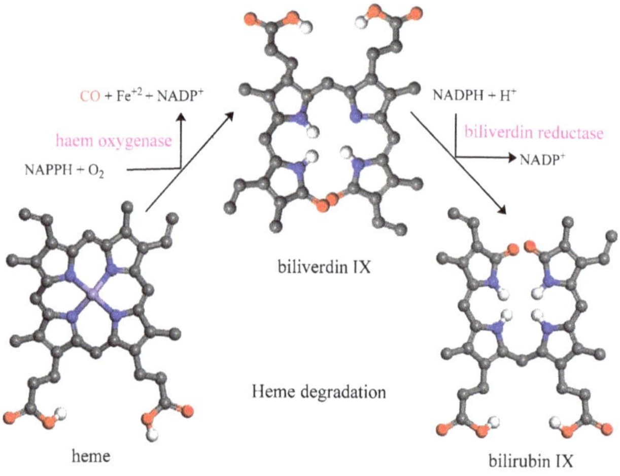

Figure 1. The intracellular carbon monoxide (CO) production by heme oxygenase (HO) in the mammalian system justifies its biological role.

CO is endogenously produced by either specific enzymes or through gas transmission into a biological system, both types exhibit the physiology and pathophysiology functions through inter- and intra- cellular interactions. The endogenously produced CO also raises their potential as a therapeutic agent. The scientists and researchers are availing this opportunity and spending their time and energies for developing modern drug techniques [11]. Their aim is to explore the modern work with the novelty of this great strategy.

1.1. CO Biological Scope

The CO gas is known for its leading role as a molecular messenger in the physiological process for the nervous system [12] and also for following some important therapeutic treatments [2,13]. It has the potential for anti-inflammatory [14], anti-proliferative [15], anti-atherogenic [16], anti-allodynia [17], anti-nociceptive [18], anti-hyperalgesia [17], and anti-apoptotic [19] effects. It is vital for vasodilatory phenomena reducing intraocular pressure [20], immunosuppressive administrated medications [21] and also has the capability to develop the pathological cellular process (Figure 2) [22]. CO also has many advantages for different biological organs: Organ transplantation [2], protection [23] and preservation; heart [24–26]; kidney [23,27–29]; liver [30,31]; lungs [32,33]; pancreatic islet [34] and the small intestine [35]. It is helpful to de-escalate the Ischemia/Reperfusion Injury (IRI) [36], mitigate the myocardial infarction and allograft rejection [37], stimulate the cytoprotective [38], and is also involved in anti-microbial [39] and anti-hypertensive activities [40]. It has a modulated utility for heme-dependent proteins like mitochondrial cytochromes and NADPH [28,41]. Moreover, the intercellular CO production by heme oxygenase has proved itself as a valuable reagent [42]. The pharmaceutical dose through endogenous CO enrichment or exogenous direct transformation is flourishing and will be attracted as a therapeutic interest lately.

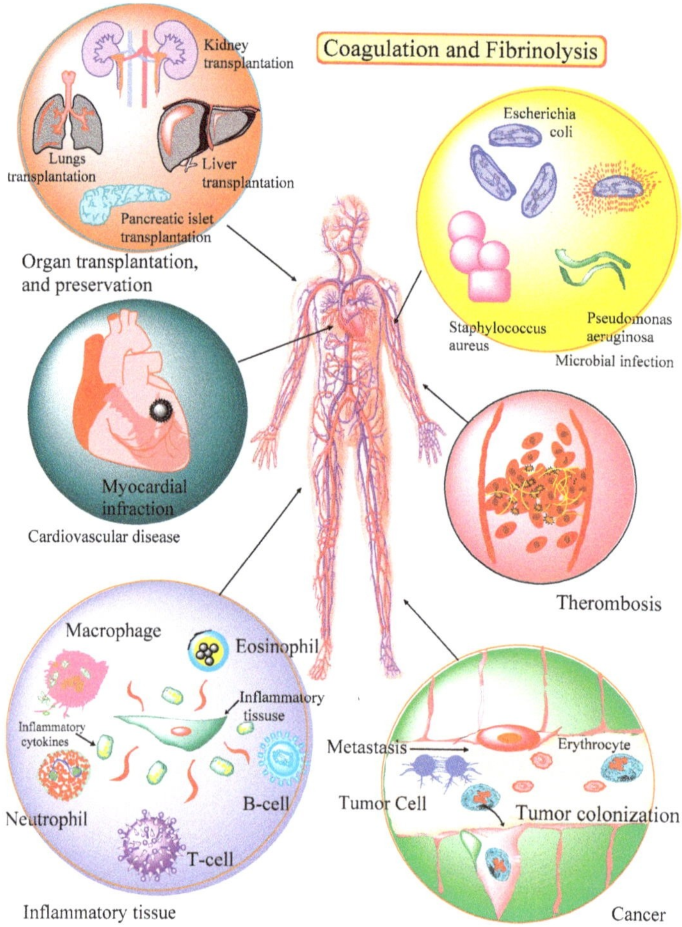

Figure 2. The coagulation and fibrinolysis scope of CO-releasing materials (CORMats).

1.2. CO Therapeutic Ways

Mainly, there are two ways to insert the CO molecules as a therapeutic agent inside the human body, i.e., direct and indirect CO insertion. The direct inhalation has not been preferred, owing to its rise in the COHb level above 10% and lack of tissue selectivity (Figure 3). Moreover, it provides a direct interaction of the CO and lungs only while detainment of CO is also observed in this method. These limitations don't allow CO to approach other biological organisms for therapy. To overcome these problems, researchers have developed an alternate strategy called "Exogenous Endeavor" for obtaining the required therapeutic actions. In the early 19th century, researchers also recognized the toxic gas NO as the nitro drug having therapeutic impacts. The nitric medications are well demonstrated as nitric oxide-releasing molecules (NORMs), and that established their well reputation afterward CORMs cogitation [43].

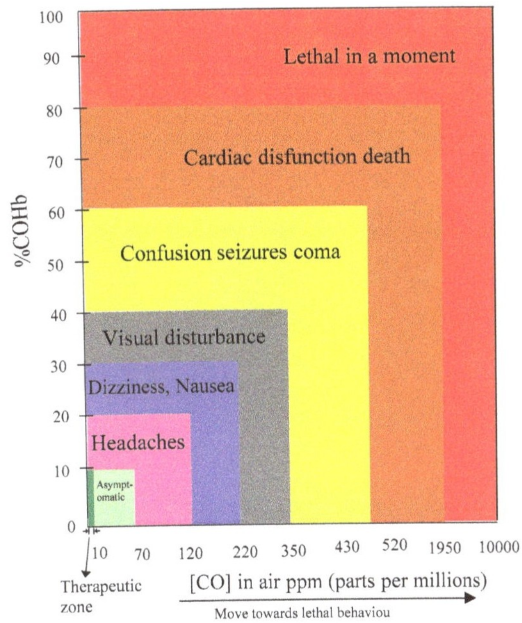

Figure 3. The carboxy hemoglobin (COHb) percentage is increasing in the direct CO inhalation beyond the therapeutic zone (~10%) during the mainstream blood circulation. (This information is based on data reported in reference [44,45]).

1.3. Why Exogenous Endeavor is Required?

The CO-releasing fragment is basically an exogenous endeavor that has opened up the paths for therapeutic treatments (Figure 4). The exogenous stakeholder CO, makes space for searching the affected sites, reaches at the diseased tissue site and makes conflict/collusion with the selected tissues for the destruction of damaged organs or/and diseased cells. If required, the CO-releasing rate can be regulated and modified according to specs. To disintegrate CORMs and CORMats into CO and metal residue, numerous activators are being administratively applied for controlling the CO liberation rate that has already been experienced in Photo-CORMs and Photo-CORMats through Ultra Violet (UV), Visible and NIR light with on/off switching facility [11,46–50]. The photon energy also has a utility to extract CO from its parent organometallic ligand. The main advantage of CO's exogenous interactions with the mammalian organism is that it reduces the CO moiety to be directly induced into blood streams for maintaining the COHb under allowable serum levels (up to 10%). Without an endogenous CO administration, it is quite challenging to get productive outcomes.

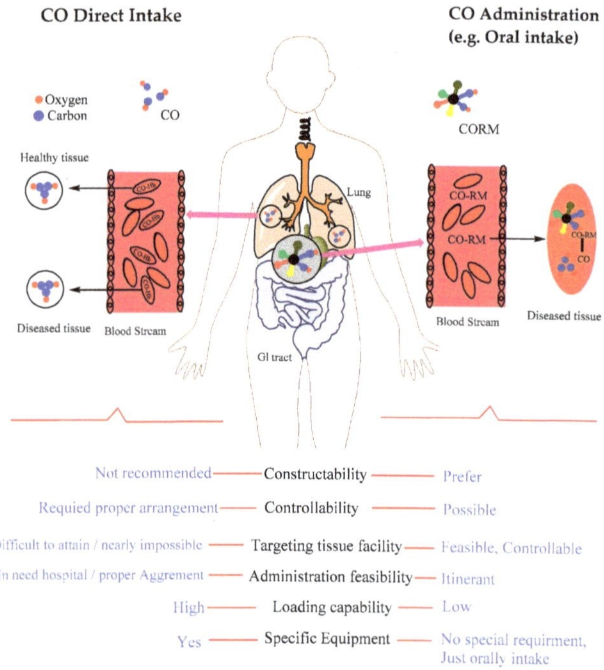

Figure 4. The feasibility analysis of the CO direct and indirect inhalation shows their different biological observance inside a human body.

1.4. Clinical Translations

In spite of its hazardous nature, a controlled CO direct inhalation has some therapeutic benefits as well. A clinical trial of controlled CO dosage was conducted on healthy volunteers for temporary paralysis of intestines known as Post-Operative Ileus (POI), and usually every patient is engaged in this POI after surgery of the abdomen. This clinical study revealed that serious POI complications could be significantly reduced if the CO dose (~250 ppm) is inhaled before and after the colon surgery (ClinicalTrials.gov identifier: NCT01050712). Another CO clinical translation test also shows a transplantation protection when the CO–saturated medium is provided for harvesting islets as it protects the cell from chronic pancreatitis (ClinicalTrials.gov identifier: NCT02567240).

This valuable intensive information about the CO therapeutic analysis plays a vital role for all the researchers' attention. Although there are few unfavorable emblems associated with releasing CO moiety, but recent outcomes of therapeutic potential helps to promote CO as preclinical stems [13]. This novel idea was initiated through clinical and pre-clinical trials for either the direct inhaled therapy [13] or oral intake of CO-releasing substances including CORMs or CORMats, which is a modern result of the professional chemistry enterprise [51].

1.5. Challenges and Demanding Features of CORMs and CORMats

Although CORMs and CORMats have a tremendous therapeutic utility but it also possesses some sort of following limitations for releasing the embedded CO.

- Availability, solubility and stability of reagents under ambient conditions.
- Feasibility to release the captured CO from in situ CORM.
- Controllability to release CO kinetics up to a desired level.
- Prone to toxicity which arises due to the transition metal foundation of metal-ligand fragments.

The abovementioned fundamentals have been discussed as a prescribed domain while exploring the CO discharge. Considering the CO gas as a therapy treatment based on CORMs and CORMats are easier to control by the transportation of gas molecules rather than the direct CO gas intake. Moreover, the rapid diffusion of these small molecules limits their ability to concentrate in specific tissues. Many challenges also arise during movability of the CO gas molecules by these strategies. Both the carbonyl transition metal and all its degradation products are biologically toxic in nature. Hence, it is difficult to manage the CO discharge with respective biological tissues. Particularly, the release of the CO molecules from CORMs also participates in depositing heavy metal ions inside the human body, which could be harmful for biological organisms.

1.6. Triggers

There are different ways to disintegrate CORMs/CORMats for the release of CO moiety by means of; peculiar physiological conditions [52], trigger by temperature [53], activation by an enzyme [54], pH alteration and increase in reactive oxygen species (ROS) concentration, accessibility to distinct wavelength of light [55,56], either using thermal degradation or ligand exchange/substitution or both [53] and prototypically activation through oxidation mechanism [11,57,58]. In order to deliver the CO molecules at a specified therapeutic rate, it is necessary to characterize the CORMs and CORMats entities, as quantification of such mechanisms might be contingent with its physical conditions like O_2, temperature, assay solution and light conditions.

1.7. CO Identifier

The authorized CO identification method consists of the following techniques: Electrochemical assays [59,60], laser infrared absorption [61] and gas chromatography [62]. Moreover, a colorimetric CO sensor facility is an alternative platform, especially for observing the CO behavior in living cellular tissues or/and organs [63,64]. The vibrational spectroscopy technique (such as IR, infrared and Raman) is one of the quickest ways to monitor the CO attached with transition metal through carbonylation at distinctive bond ranges. To date, the plethora method "Myoglobin Assay" famous as "*Gold Standard*" has been in operation for in vitro interrogation of the CO release from CORMs and CORMats. The "Myoglobin assay" is also a worthy and standard method for observing the behavior of the CO release kinetics in a biological environment and monitors the performed activity of CO (Equation (1)).

$$\text{Mb} + \text{CO} \rightarrow \text{Mb-CO} \tag{1}$$

Recently, It has been observed that binuclear Rhodium compounds, i.e., *cis*-$[Rh_2(C_6H_4PPh_2)_2(O_2CCH_3)_2](HAc)_2]$ can detect CO with a substantial selectivity and superior sensitivity [65]. The gas-phase IR spectroscopy is the most reliable and high-resolution technique for analyzing the CO release activity. For directly sensing the CO, various gas chromatography (GC) detectors have been introduced to date including the gas chromatography-mass spectrometer (GC-MS) [66], reduction gas detector (GC-RGD) [67] and thermal conductivity detector (GC-TCD). The fluorescent probe has the ability to recognize the CO-release entity even at a low concentration as compared with the myoglobin assay, but it is unable to operate in a short interval kinetic measurement (for that at least one hour interaction is required) [68,69].

1.8. The Development Phases of CORMs Motifs

1.8.1. Metal Carbonyl Complexes (MCCs)

To construct the bonding relation between the CO and low valent metal ions for producing carbonyl complexes (M-CO), the M-CO bonds must undergo an inert ambiance along with reducing conditions, which are mostly feasible in organic solvents. Irrespective of a metal physical state, the CO gas can react and develop the volatile metal carbonyl complexes (MCCs) for example, $Ni(CO)_4$ and $Fe(CO)_5$ [70]. MCC acts as a core entity for organometallic transition chemistry. The general representation for MCCs is $[M_m(CO)_xL_y]^{z\pm}[Q^{\pm}]^z$ [53], in which, M, L are the basic entity known as transition metal (B, Cr, Mn, Fe, Co, Mo, Ru, Rh, W, Re, Ir) [71], and ancillary ligand might be the C, O, P, N, S or halide ligand. Furthermore, Q and z represent the counter ion and overall complex charges. If no counter ion is available, then z will be zero. Moreover, m, y, z are calibrated as stoichiometric coefficients and x and m values should be ≥ 1 [53]. Modern and classic complexes can be distinguished by two determining factors: Low oxidation state (OS) or either very low oxidation state (LOS) and total valance electron occupied in the outer coordination sphere. For 4th, 9th, and 10th groups, the compounds were observed to have $16e^-$ configuration and the rest of the complexes were generally observed having $18e^-$ configuration. Each of these commodities must comply with the chemical, biological and physical characteristics of MCCs. Furthermore, it needs to be precisely selected in the configuration of pharmaceutical CORMs [71].

It is important to note that CORM and CORMats are stable in the aqueous medium, and it is also feasible to store it under ambient environmental conditions like the majority of other pharmaceutical drugs. Their circulation must be ensured, as it needs contact both with the diseased and damaged tissues. Moreover, the potent and non-toxic metabolites may be left behind after the CO removal. This is exactly what is required and regarded as therapeutic features. It sets the basic pattern, in the line of action for the development of such challenge-able MCCs. Mostly, during administration MCCs incorporates with the organic solvent and results in traditional oxygen-free atmosphere. These medical conditions might be different from a variety of other biological surroundings, considering that most reactants and their resulting complexes are uncertain under ambient conditions, i.e., oxygen and humidity [71]. Hypothetically, this biological activity of MCCs remains toxic in nature like $Ni(CO)_4$, and $(MeCp)Mn(CO)_3$ (MMT) as an anti-knock gasoline additive [72]. Hence, the common MCCs chemistry can act as a simple guideline for the development of pharmaceutical CORMs. Currently, this research is focused on novel strategies for establishing MCCs-CORM's activity and specifically for therapeutic purposes.

The above discussion suggested that MCCs when triggered as CORM's, become a competitor for the CO availability during its decomposition. In organometallic complexes the releasing strategy is as follows: A new incoming ligand (L′) can push itself to a metal center resulting in a new bond, which establishes and influences on the coordination number. The elevated coordination number then promotes it to elongate the M-CO bonds, and eventually it then breaks. Consequently, the CO is liberated by this method and the new L′-M bond is constructed. This information provides the foundations of the CORMs concept. The chemistry of MCCs provides per se different strategies and has an adverse impact on the CO release (Scheme 1) [53].

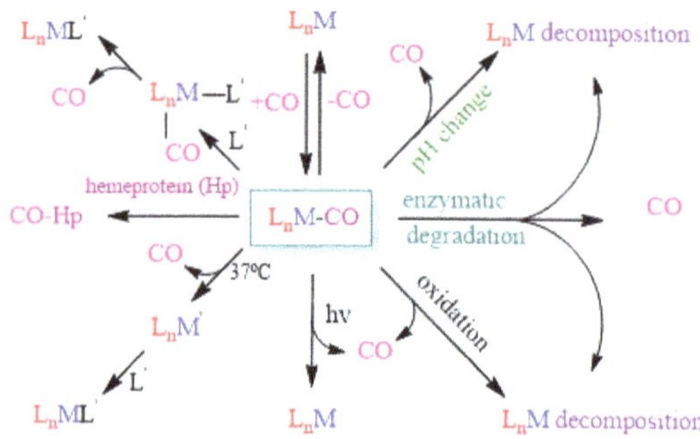

Scheme 1. CO releases from the ligand-metal CO framework (L_nM-CO).

1.8.2. Proposed Strategies for CORMs Development

CORMs based on metal-to-ligand charge transfers (MLCT) morphology. These CORMs are the elementary motifs with organometallic ingredients; corresponding to a series of MCCs occupied at the transition metal core. Other exclusive features are mentioned below:

- Structural variance of unique chemistry;
- Expected divergence with different oxidation states;
- Covalently bound with the metal center;
- Assisting alterations for the attached carbonyl ligands;
- The dynamics of co-ligands binding;
- Tendency of the outer coordination sphere.

As abovementioned, the spectroscopic nature of MCCs confirms the identification and recognition of significant trace elements like ruthenium [73,74], manganese [49,74–78], iron [37,79], cobalt [80,81], tungsten [82], osmium [83], molybdenum [82] and rhenium [84]. The developed organometallic carbonyl complexes CORMs are CORM-1, CORM-2, CORM-3, CORM-401, ALF492, CORM-A1, B_{12}-ReCORM-2, Re-CORM-1, CORMA-1-PLA and ALF186.

Along with organometallic complexes, the miscellaneous compounds can be nominated as the CORMs family. In this scenario, numerous nonmetallic compounds [85] have been accomplished by entertaining the CO release such as silica-carboxylates [86], borano-carbonates [87], borano-carbamates [88], xanthene carboxylic acid (XCA) [89], unsaturated cyclic diketones (DKs) [90], methylene chloride (MC) [91,92], meso-carboxy BODIPYs [46], hydroxy-flavones [93] (Scheme 2). Furthermore, 1,2-disubstituted ferrocenes belongs to an aldehyde family, unfavorably elicits the toxicity and its slow release mechanism restricts the researchers from developing another nonmetallic CORMs (NCORMs). The main drawbacks of NCORM are potentially a low CO content releasing, and always producing organic molecules along with the CO moiety. Anyhow, NCORM clinical traits have shown their utility to communicate with biological activity [94].

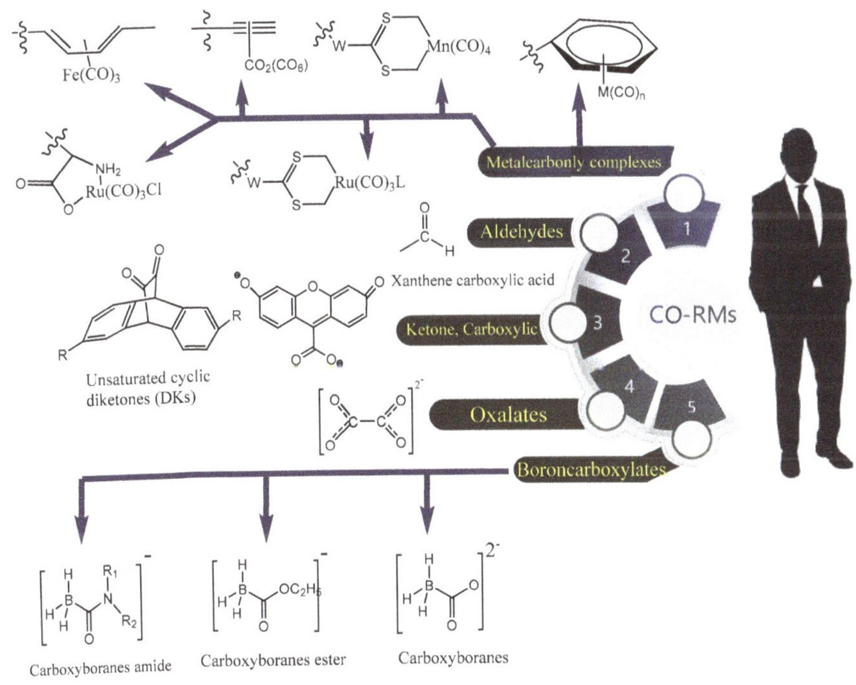

Scheme 2. Various CO-releasing molecules (CORMs) formulation associates with different functional capabilities.

1.9. CORM's Therapeutic Scope

Strategically, the synthesis route of CORMs development is not the only objective. The main theme of this CORM's innovation is to obtain the therapeutic advantage eventually. The biological significance of CORM is associated with their bacterial performance in cells lines, (i.e., standard myoglobin assay). The important biological roles of CORMs are listed below (Table 1).

Table 1. Nonmetallic and organometallics CO-releasing molecules (CORM's) fragments exhibited therapeutics activities.

List	CORMs	Therapeutic Implications	Refs
1	CORM-1	Increase coronary perfusion pressures; attenuates the L-NAME-mediated; restore unstable blood pressure and modulates vessel contractility ex-vivo in animals.	[95]
2	CORM-2	Attenuates inflammatory response in lungs and liver; induces vasorelaxation; protects against IRI; activates K^+/Ca^{+2} channels; possible for pulmonary hypertension.	[95,96]
3	CORM-3	Improves the liver & kidney functions during transplantation; Vasorelaxation induction; prevents sepsis & cardiac graft rejection; helps in bacterial infections; support rheumatoid arthritis; RBF improvement in the treatment of cynomolgus for monkeys.	[23,28,71,97–99]
4	CORM-401	Improves insulin-sensitivity and metabolic switch induces in adipocytes.	[100]
5	ALF492	In severe malaria, fully protects with artesunate combination.	[101]
6	CORM-A1	Induces the vasorelaxation; Increases RBF and reduced vascular; gives resistance in the kidney of mice; good cerebroprotective agent for epileptic seizures treatments.	[102]
7	Re-CORM-1	Anti-oxidative characteristic and protects against IRI from the affected neonatal rat of cardiomyocytes.	[103]
8	B_{12}-ReCORM-2	Protects against IRI (neonatal rat cardio-myocyte); hindrance cell mortality up to 80%; support the cardiac repairing and cardiac disease (ameliorates degenerative); anti-oxidative agent; augments and direct cardiomyogenesis.	[104,105]
9	3-hydroxyflavon CORMs	Exerting anti-oxidative activity; anti-inflammatory services and anti-cancer effects.	[46,55,89]
10	ALF 186	Protective effects for gastric ulcers and neuro protective, while IRI-induced apoptosis of retinal ganglion cells (RGC).	[106–109]
11	CORMA-1-PLA	Prevents fibroblasts and internalized into 3T3 cells during metabolic and hypoxia depletion conditions.	[110]
12	α-DK-CORMs	Absorbs in acute myeloid leukemia (AML) KG-1 cells and releases CO In-vivo upon 470nm irradiation.	[111]

1.10. Solubility

For making CORM acceptable as a pharmaceutical drug in the mammalian biological system, solubility is one of the most prominent factors where the researcher can evaluate the proficiency of the product. Solubility estimates how much CORMs and CORMats are convenient for the practical demonstration. CORM-2 is soluble in DMSO, olive oil and PEG [95,112], while CORM-3 contributes to the water compatibility with a weak acidic nature (pH = 3) [26,37]. Specifically, CORM-A1 possesses the water solubility and stability but it breaks-down immediately after liberating the CO under acidic condition (pH = 11) [87]. CORM-ALF186 can afford the disintegration in the water system and is unstable at an aerobic condition [113] and ALF062 is soluble in methanol and DMSO, while it remains unstable in the air [98,113]. Furthermore, CORM-1 has the compatibility with DMSO and ethanol [114].

Although the CORMs motif is good for releasing the CO moiety, but the tissue selectivity and targeting sites dilemma has reduced its overall biological performance and hence lost its therapeutic significance. Most importantly, the toxicity of organometallic complexes is handled very poorly, so to reduce the toxicity and increase its reactivity, it requires an exploration of all the alternative strategies. Therefore, the researchers have moved from CORMs to CORMats.

2. Research on New CO Transport Materials

As discussed, the MCCs are the admissible and professional class of (soluble) CORMs; however, it has been imperative to examine their probable shortcomings. In fact, a small number of organometallic compounds can be manipulated for pharmaceutical agents predominantly caused by the side-reaction of metals with biological chemical compounds, (e.g., nucleophilic or even electrophilic side chains of proteins) together with the toxicity of several heavy metals. Water-soluble CORMs are approaching the entire body organism and it could accelerate the toxicity against healthy tissues. The spatial and acceptable releasing rate of it into biological tissues/cells is still the utmost challenge. Furthermore, the CO-releasing activity inevitably accumulates a metal and co-ligand fragment, probably takes part in the biological activity as well. This residue (i-CORMs) can be managed through the insoluble framework. On the basis of abovementioned issues and challenges, new and compatible CO transport materials and strategies are emerging in order to get rid of the CO lethal gas dilemma and to convert it into a valuable clinical agent. CORM-1 [115], CORM-2 [116,117], CORM-3 [118] and CORM-A1 [87] have been tested in various disease models to observe their therapeutic effects and to obtain its surprisingly outcomes in typical clinical conditions [2,119]. CORM-3 has good cure-ability for inflammatory disorders like rheumatoid arthritis, osteoarthritis and collagen-induced arthritis (CIA) [97,120,121]. CORM-A1 provides ameliorated course in experimental auto-immune uveoretinitis (EAU) [122], while CORM-2 attenuates the tumor proliferation [123] and a considerable enhancement the coagulation and slow-down of the fibrinolytic bleeding [117,124] and improves survival in the liver injury affected by cecal ligation and perforation (CLP) [116,125]. CORM-3, CORM-2 and ALF-062 corroborates with antimicrobial functions [98,126–131]. The CO is encouraging the proliferation of endothelial cells, progenitor cells and regulatory T-cells [34,132,133]. There is still more interrogation required for further improvement to employ practical knowledge. So, the development of the solid CO precursor in tandem with peculiar trigger for releasing the enclosed CO gas commodity is an imperative research motive. To date, due to the unavailability of a safe delivery system for CORM; none of those prescribed formulations could have been employed in humans as a direct dose for respective damaged tissues or disease. Although a few scientific proposals have been presented in that scenario for making CORMs as a clinically viable project, but none of them exhibits the secure transportation material system for the patient's right choice. All that discussion pursued that the nanoscale and macromolecular carrier system could be exploited to obtain a selected tissue enrichment and proposed mechanism strategy for CORMs delivery (Scheme 3) [134,135].

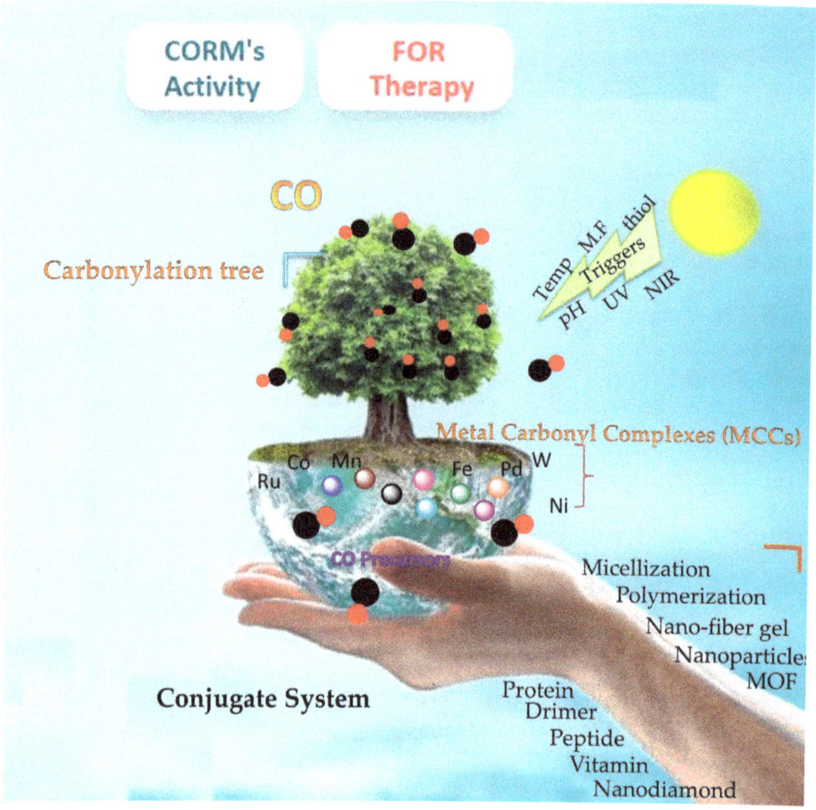

Scheme 3. Various organometallics MCCs incorporate with numerous conjugate systems to produce carbonylation complexes, i.e., CORMats for therapeutic CO release upon trigging.

The clinical trials on CORMs proved that CORMs exhibited the important biological applications, but after CORMs, the degradation metal residue (i-CORMs) also caused toxicity unfortunately [136]. The prohibited i-CORMs activity containment is a big challenge for the researchers. To reduce toxicity and capture the i-CORMs toxic moiety, scientists have explored a strategy known as CORMats. In this strategy, firstly, CO is entrapped inside the CORMats through specific administration, and then upon certain conditions the captured CO is escaped out. Several scaffolds and conjugated formulations have been introduced in this scope and it is still under investigation by compatible conjugate CORM's such as Ruthenium-MCC (Ru-MCC) and Manganese-MCCs (Mn-MCC) by different nano-transporting services such as Iron MOFs [137,138], peptide [139–144], micellization [55,59,145], protein [121,146–149], vitamins [150–153], co-polymer systems [47,154–156], nanofiber gel [142], inorganic hybrid scaffolds [157–160] and metallodendrimers [161] (Figure 5). The intrinsic toxicity control of i-CORMats is the top priority for each developed system.

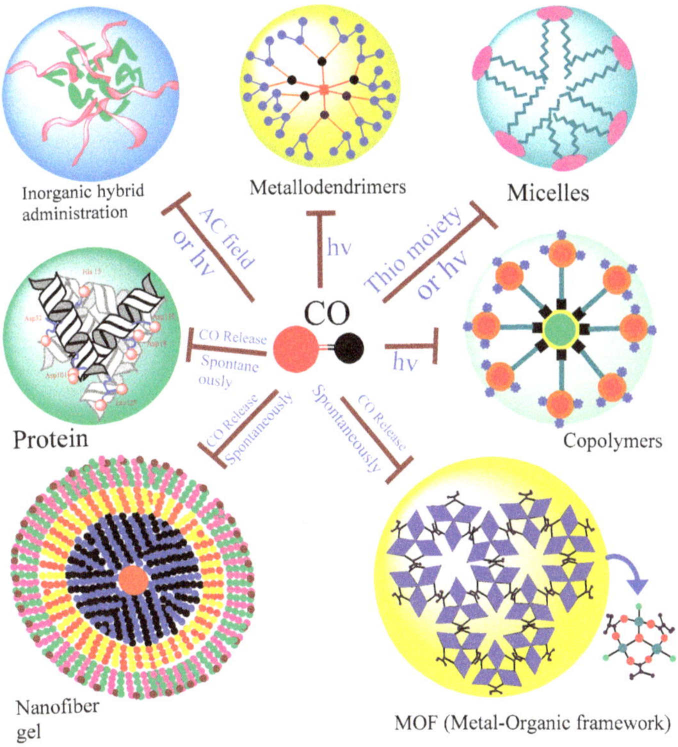

Figure 5. The CO-releasing administration with different conjugate and encapsulate strategies.

2.1. Micellization

Hubbell et al. engineered the micellization technique as the CO-producer with reduced diffusion; creditably targeted to the distal tissue draining sites [59]. Micelles were synthesized by the tri-block copolymer composed of poly(ornithine acrylamide) block and poly(ethylene glycol) block (hydrophilic nature) hosted by $[Ru(CO)_3Cl$-(ornithinate)] moieties with poly(n-butylacrylamide) block (hydrophobic nature). The CO-releasing micelles consists of a triblock copolymers (Figure 6): A hydrophilic poly(ethylene glycol) (PEG) fragment that stabilizes the micelles; a poly-OrnRu fragment that releases CO. A hydrophobic poly(n-butylacrylamide) fragment drives to construct the micelles forms.

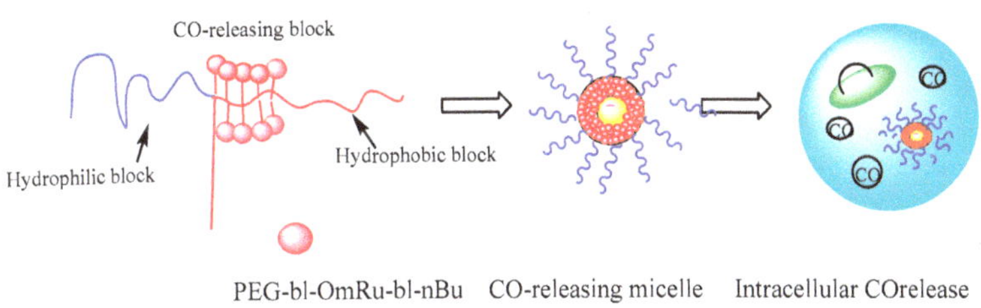

Figure 6. Triblock copolymer assembles for releasing the CO at biological sites.

The micelles polymer can be used as a pharmaceutically acceptable carrier to solubilize the poorly soluble drugs and produces the therapeutic effect against the targeting sites. Probably it promotes the reduction in toxic effects of the drugs on normal tissues and organs. Significantly, the toxicity of the $Ru(CO)_3Cl$ moiety is well-reduced in the polymer micelle due to the stealth characteristic of the PEG fragment. Moreover, the micelles moderately respond to human monocytes against the lipopolysaccharide (LPS) -induced inflammatory disease model. Importantly, poly(ethylene glycol) attenuates the toxic feature of [Ru(CO)$_3$Cl(amino acidate)] moieties. The addition of cysteine allows the release of CO from an occupied area with a slower rate as compared to [Ru(CO)$_3$Cl(glycinate)] (CORM-3). The release of CO from micelles was tested in a myoglobin assay and it has been found to be slower than CORM-3. The diffusion of the Orn-Ru substrate is facing hindrance in the cells due to the micelles stereoscopic effect. Anyhow, the mechanism approach of CO-releasing is not obvious. It has been evidently proved through experiments that thiol compounds such as cysteine, glutathione and protein are compatible to induce the CO release from micelles.

Hiroshi Maeda et al. incorporated the tricarbonyldichlororuthenium dimer (CORM-2: [RuCl(μ-Cl)(CO)$_3$]$_2$) as water-soluble styrene-maleic acid and copolymer (SMA) while gaining the optimum half-life and numerous therapeutic effects [145]. They established the micellization structure for encapsulating the CORM-2 (SMA/CORM-2) (Figure 7). The micellization has good water solubility and it is compatible with the aqueous environment. The sustain CO kinetic profile performs well in vivo bioactivity such as murine model of inflammatory colitis. The half-life of this complex was almost 35-folds compared with the free CORM-2.

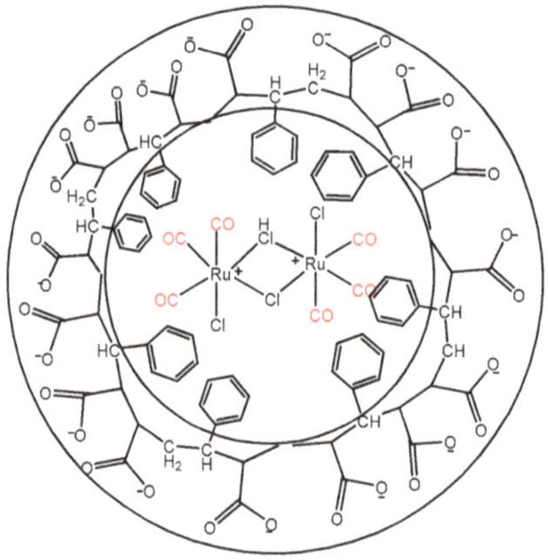

Figure 7. Tricarbonylchlororuthenium (II) dimer (CORM-2) synthesized with water-soluble styrene-maleic acid copolymer (SMA) for micellization CORMats.

In spite of the ligand exchange CO-release mechanism, photo light is able to disintegrate the CORMats moiety as the CO donor. Robert Igarashi and Yi Liao explored the micelle-based photo-CORMs synthesized by the cyclic α-diketones (α-DK) encapsulation [55]. These CORMs require visible photo light for releasing the embedded CO. This research demonstrates the therapeutic potential of CORMs. The photo-activated micelle CORMs strategy has been explained in Figure 8. During the study of these micelle CORMats on the cell proliferation, it has been found that no difference was monitored in the viability of cells in response to the micelles of DKs.

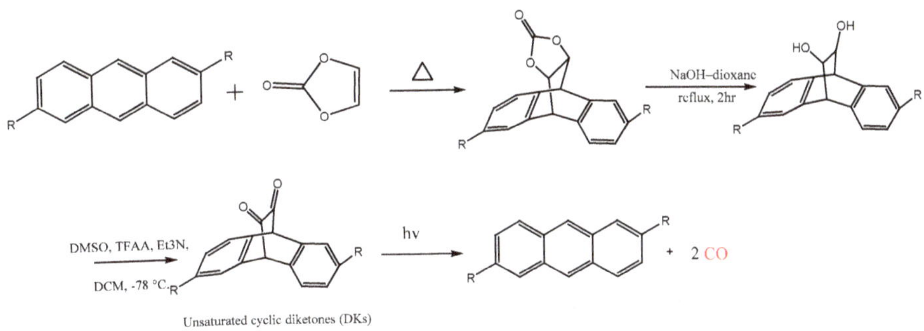

Figure 8. Synthesis route of unsaturated α-diketones (α-DKs) has been activated by photons energy.

2.2. Peptide

John reported a self-assembled amphiphilic peptide (PA) that was used to produce the CO [142]. Thereby, a covalent combination of a hydrophobic alkyl chain and a hydrophilic short sequence peptide endures the self-assembled peptide chain material. Amphiphilic peptides can spontaneously release CO and are prone to toxicity themselves. In that perspective they first designed the amphiphilic peptide PA1; which contains a β-aspartate residue to generate the NH_2-CH-RCOOH unit closely resembling the CORM-3 fragment. Next, PA1 and $[Ru(CO)_3Cl_2]_2$ were synthesized in the presence of sodium methoxide at room temperature to synthesize the CO-releasing peptide PA2 (Figure 9). The CO kinetic release curve proved that peptide P2 was synthesized in an aqueous solution like the first-order rate constant of CORM-3. The half-life of the CO released from both sources is quite identical. The half-life of CO released from CORM-3 is 2.14 ± 0.17 min, while the half-life of the CO released peptide P2 is 2.16 ± 0.05 min. In order to increase the half-life of CO released from P2, the incorporation of nanofiber gel PA2 and a strong gel PA was made. The half-life of the CO released from this nanofiber gel was significantly increased (~17.8 min) compared to PA2 and CORM-3 in an aqueous solution.

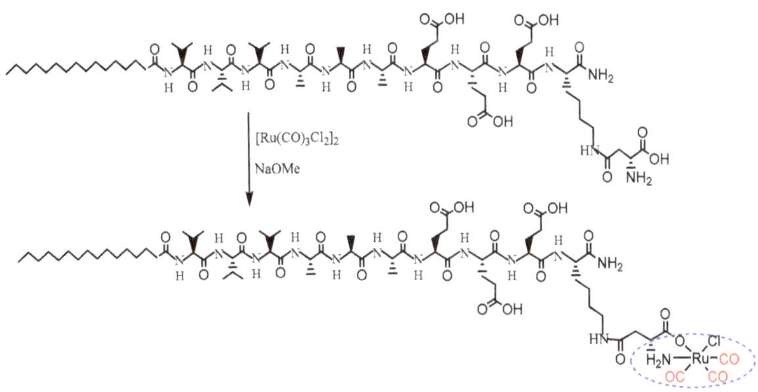

Figure 9. Synthesis of PA2 having CO-moiety for spontaneous release CO.

Rather than encapsulating a mere CO segment inside the transport materials, the CORMs entity incorporates with different functional groups of parent CORM's ligand commodity. In that scenario, peptide is linked with the Manganese-based Photo-CORM $[Mn(CO)_3]^+$ ligand tpm (tris(pyrazolyl)methane) using a Pd-catalyzed based Songashira cross-coupling mechanism and click reaction at N-terminal (azide-) and side-chain (iodoarene-) functionalization [140].

Ulrich Schatzschneider exposed the peptide linkage for the CORMats development [144]. They introduced the Mo-carbonyl $[Mo(CO)_4(bpy^{CH3,CHO})]$ associated with aldehyde functional groups at the peripheral position. The bioactive β-target peptide ligand 2,20-bipyridine (bpy) attached with molybdenum-carbonyl by *N*-terminal bonds of aminoxy acetic through catalyst-free and bio-orthogonal oxime ligation (Figure 10). The photo-activated CORMats gets activated upon 468 nm photons lights irradiations.

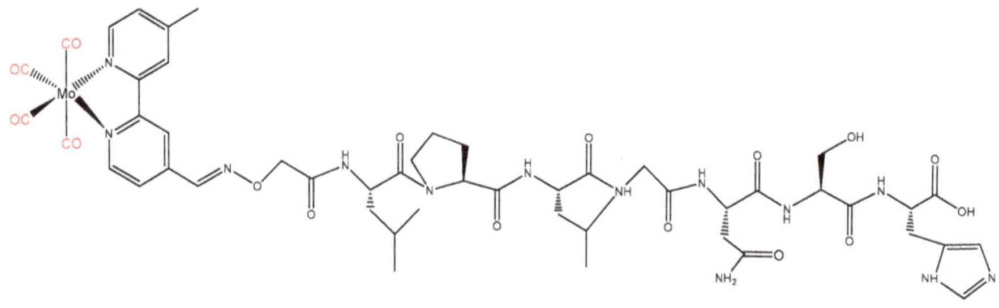

Figure 10. $[Mo(CO)_4(bpy^{CH3,CHO})]$ complex has been constructed through bio-orthogonal peptide conjugate.

Radacki and Ulrich Schatzschneider jointly synthesized the Manganese carbonyl complexes $[Mn(bpea^{NHCH2C6H4CHO})(CO)_3]PF_6$ [139]. The peptides ligand 2,2-bis(pyrazolyl)ethylamine (bpea) is bearing aminoxy, azide and N-terminal alkyne residues (Figure 11). The researchers applied the transforming growth factor β-recognizing (TGF-β) peptide sequence for developing the photo-activated delivery agent. This peptide conjugation could be utilized for further development of new CORMats.

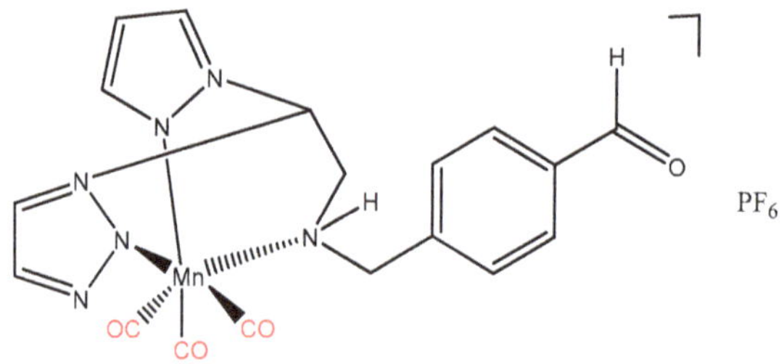

Figure 11. The synthesis route of $[Mn(bpea^{NHCH2C6H4CHO})(CO)_3]PF_6$ for Photo-CORMats.

The JJ Kodanko group successfully synthesized the ionic water-soluble compound $[Fe^{II}(CO)(N4Py)]$ by a method of continuous CO bubbling through a ligand N4Py and one equivalent of $Fe^{II}(ClO_4)_2$ under the action of the organic solvent acetone $(ClO_4)_2$ (Figure 12A) [141]. A myoglobin experiment shows that the compound is stable under dark conditions and its releasing half-life is more than one day. When it was irradiated with 365 nm ultraviolet light, the CO can be quickly released. According to MTT experimental studies, $[Fe^{II}(CO)(N4Py)](ClO_4)_2$ exhibited the effective cytotoxicity against human prostate cancer cell line (PC-3) under light-induced conditions. When the concentration reaches 10 μM the cell survival rate was monitored as 63% of the control group. In order to further investigate the CO release behavior, the carbonyl segment with acetonitrile was replaced. The UV-vis analysis found that the substitution process is very slow $[Fe^{II}\text{-}(MeCN)(N4Py)]^{2+}$ and the concentration of acetonitrile was a quick step to replace CO. Moreover, it was also found that the N4Py ligand can be modified with a peptide (Ac-Ala-Gly-OBn) to obtain a peptide-conjugated photo-induced release molecule (Figure 12B). Biological experiments have showed that the peptide chain conjugation might be evaluated for the improvement of cell-specific itself or tissue-specific CO transport properties.

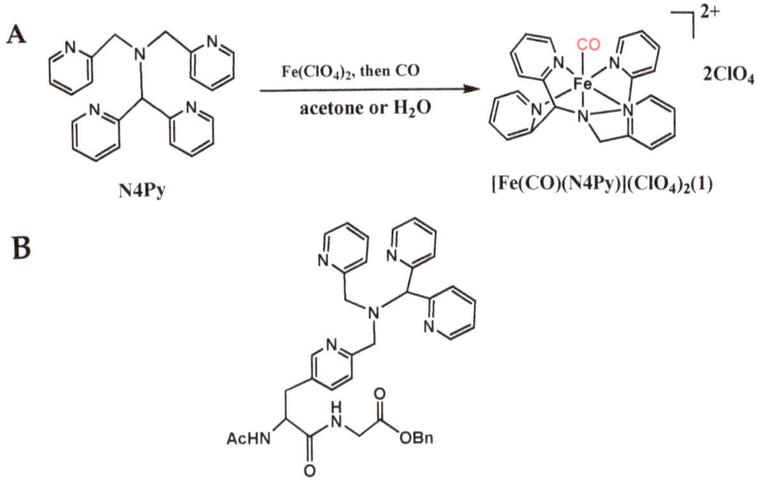

Figure 12. The ionic water-soluble Iron complexes [Fe^{II}(CO)(N4Py)] (**A**) could be modified into Photo-CORMats by the replacement of the N4Py ligand with peptide china Ac-Ala-Gly-OBn (**B**) for the improvement of the cell-specific itself or tissue-specific therapeutic properties.

In the development of Photo-CORMats, metal-coligand plays a vital role in the photoexcitation at a prescribed wavelength. This allows the photons energy to penetrate and push the CO molecules to pull out these molecules from the metal-ligand fragment. In other words, these wavelengths are providing extra energy that enables the CO molecules excitation from its parent location. The CORM-2 and CORM-3 are hydrolytically active. Synthetically, these CORMs could be transformed into the photo-activated reagent. Leone Spiccia and Ulrich Schatzschneider described the ruthenium (II) dicarbonyl complexes functionalized with 2-(2-pyridyl)pyrimidine-4-carboxylic acid (CppH) (Figure 13A) [143]. They were able to successfully construct the monomeric PNA backbone with ruthenium (II) di-carbonyl complexes to produce ruthenium (II) dicarbonyl dichloride-based PNA-like monomer [$RuCl_2$(Cpp-L-PNA)CO_2] (where PNA= peptide nucleic acid) (Figure 13B).

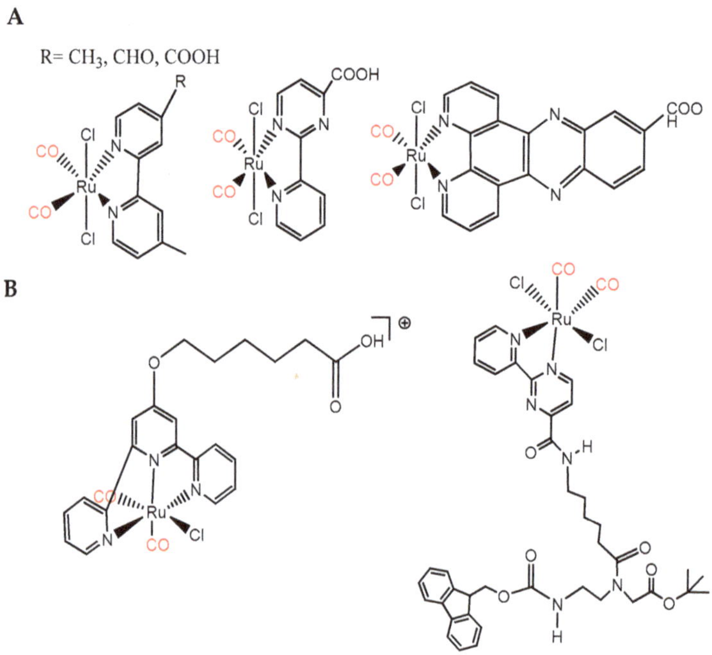

Figure 13. The hydrolytically activated Ruthenium dicarbonyl complexes CORM-2 and CORM-3 could be transformed into Photo-CORMats by peptide ligands through different functionalization: (**A**) Polypyridyl ligand of 2-(2-pyridyl)pyrimidine-4-carboxylic acide (CppH); (**B**) monomeric PNA backbone.

2.3. Proteins

G. J. Bernardes et al. disclosed that CORMs was compatible with the protein complexes transportation (Figure 14) [121]. They presented that the $Ru^{II}(CO)_2$-protein complexes using the reaction between CORM-3 and histidine fragment at the protein surface, the spontaneous CO is released to deliver in cells and mice. They also discussed that plasma protein acts as a CO carrier for in vivo by the CORM-3 formulation. Therapeutically, the controlled CO release favors in downregulation of the cytokines interleukin, i.e., IL-6 and IL-10.

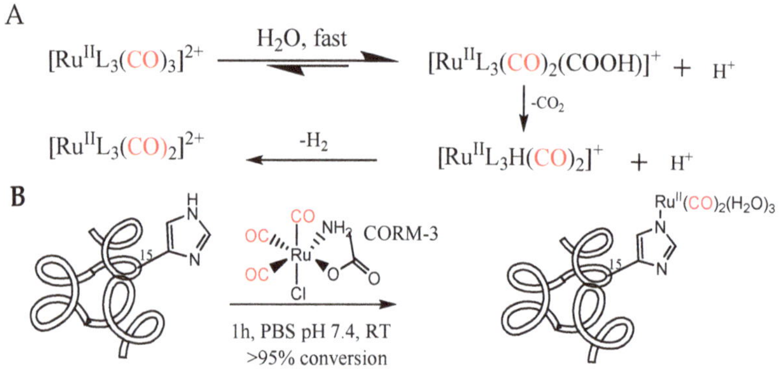

Figure 14. The reactivity of the *fac*-$[RuL_3(CO)_3]^{2+}$ complex (**A**) and CORM-3 react with single-His protein (**B**).

Another Ruthenium (II) carbonyl reagent *cis*-$[Ru(CO)_2(H_2O)_4]^{2+}$ has been reported for the spontaneous CO release in live cells using histidine (His) metalloprotein and retained at IL-8 (Figure 15a) [148]. The *cis*-$[Ru(CO)_2]^{2+}$ carbonyl segment could be produced by aqua dirutheniumcarbonyl *cis*-$[Ru(CO)_2(H_2O)_4]^{2+}$ (Figure 15b). It was also explained that metalloproteins can be modified as organometallic pro-drugs rather than catalysis. Such artificial metallohydrolase performance can be compared with the human carbonic anhydrase (CA)-II.

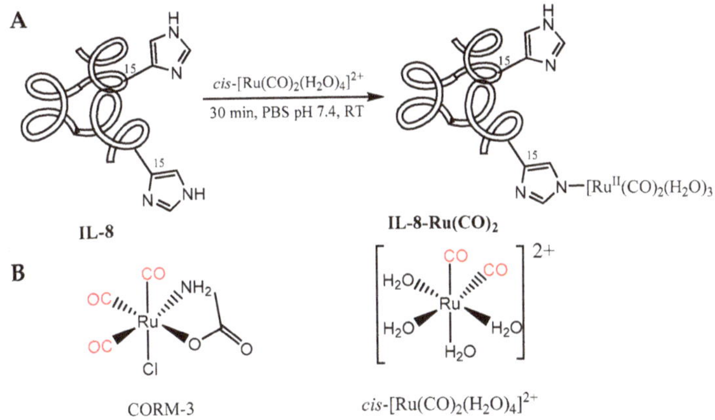

Figure 15. The spontaneous CO release by metalloprotein: (**A**) The carbonyl reagent *cis*-$[Ru(CO)_2(H_2O)_4]^{2+}$ spontaneous CO release in live cells using histidine (His) metalloprotein and retained at IL-8; (**B**) the *cis*-$[Ru(CO)_2]^{2+}$ carbonyl segment can be produced by the aqua carbonyl *cis*-$[Ru^{II}(CO)_2(H_2O)_4]^{2+}$.

In spite of the CORMs fragment incorporation with protein, Takafumi Ueno et al. explored the cages of protein for CO releasing [149]. They administrated the ferritin (Fr) cage of protein for capturing the CORM-3 moiety (Figure 16). Furthermore, it was observed that the half-life of the CO release could be enhanced; which indicates a good sign for ideal drug development. When they interrogated their performance at the biological sites, they described that the nuclear factor kappa B (NF-κB) becomes 10-times higher than the parent CORM-3. The CORM-3 protein cage is quite a unique way of CORM's engineering.

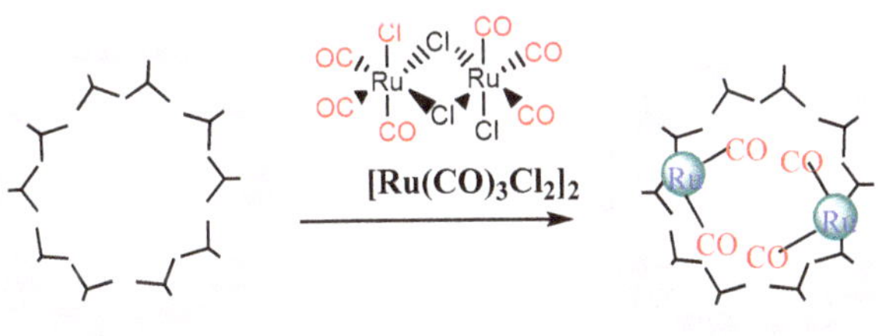

Figure 16. The recombinant L-chain apoferritin (apo-rHLFr) of Ru carbonyl complexes.

Additionally, Takafumi Ueno et al. explored the immobilization of the crosslinked hen egg white lysozyme CL-HEWL crystal deposit on MCCs for therapeutic purposes [146]. The scientist disclosed that NF-κB is remarkably high in order to respond to the pathological signals. The extra scaffold Ruthenium carbonyl moiety (Ru·CL-HEWL), was used to induce the NF-κB activation and immobilized Ruthenium carbonyl [*cis*-$Ru(CO)_2X_4]^{2-}$ moieties inside the protein cage (Ru·CL-HEWL). Optimist approaches of this transport service bear the potential of the artificial extracellular scaffold.

NF-κB can be regulated by the protein fragment. Susumu Kitagawa et al. disclosed the crystalline assembly of protein with CORM-2 in polyhedra crystals (PhC) [147]. They introduced the ruthenium carbonyls immobilized on hexahistidine. The activation of NF-κB was significantly improved up to six folds. This therapeutic research will lead to further investigation on the extracellular scaffold.

2.4. Vitamins

Anna Yu. Bogdanovab et al. presented the cyanocobalamin (B_{12}) as a biocompatible B_{12}-$Re^{II}(CO)_2$ scaffold for CORMats (Figure 17) [150]. They incorporated the *cis-trans*-$[Re^{II}(CO)_2Br_2]^0$ core having 17e⁻ dicarbonyl complexes. This research also elaborated that ReCORM-1 is compatible with B_{12} for pharmaceutical applications, and the obtained cobalamin conjugates are also feasible with aqueous aerobic media. Interestingly, after CO releasing, metal degradation is not involved in toxicity due to the exclusive configuration and metal oxidation of ReO_4^- generation.

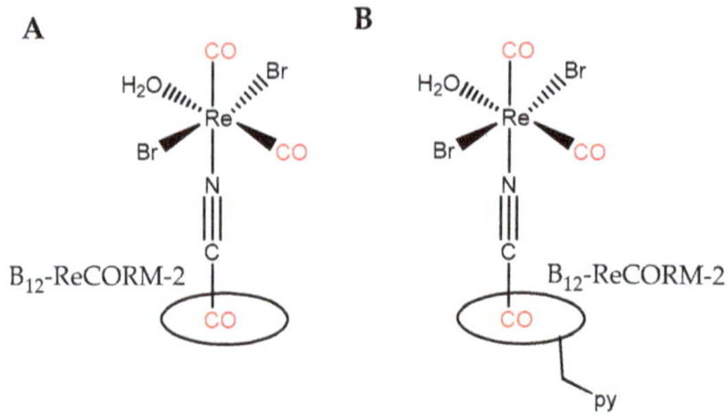

Figure 17. B_{12}-$Re^{II}(CO)_2$ CORMats conjugate: (**A**) B_{12}-ReCORM-2; (**B**) B_{12}-ReCORM-4.

The most promising anticancer drug agent is a macromolecular conjugate. The HO-1 and transcription factor Nrf2 are the prime parameters to provide resistance against inflammation and oxidative stress disease. In this analogy, biliverdin is the enzymatic activity of HO-1, while CO directs the therapeutic exploitation. Roberta Foresti et al. found that hybrid molecules were simultaneously involved in the CO liberation and Nrf2 activation [151]. The newly developed CORMats termed as hybrid-CORMats (HY-CORMats) is shown in Figure 18.

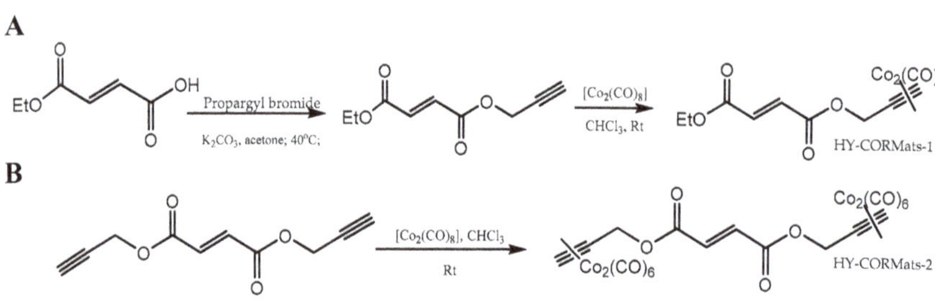

Figure 18. Synthesis route of hybrid CORMats: (**A**) HY-CORMats-1; (**B**) HY-CORMats-2.

After synthesizing the HY-CORMats, researchers further interrogated the biological activities and described the HO-1 expression along with the nuclear accumulation of Nrf2 and showed viability in different cell types (Scheme 4).

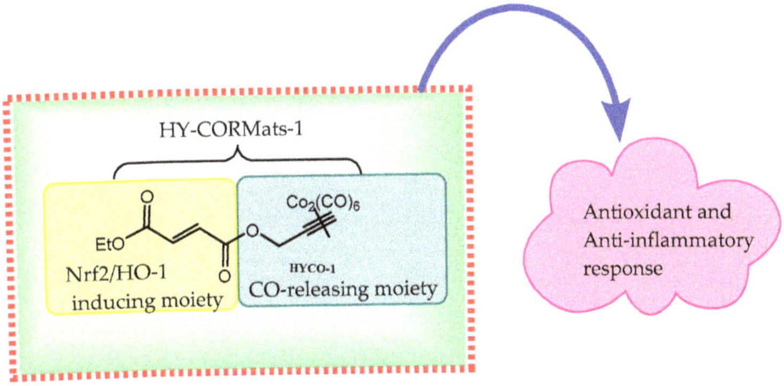

Scheme 4. The hybrid CORMats (HY-CORMats-1) has a CO moiety for various anti-inflammatory, antioxidant actions and induced nuclear accumulation of Nrf2.

A. Pamplona's group coordinated the galactose with a central metal to synthesize a polyhydric-containing water-soluble CORMats [$RuCl_2$-thiogalacto-pyranoside$(CO)_3$] (ALF492) through Sulphur bond (Figure 19) [152]. The Sulphur bond coordination of the galactose ligand with the central metal increases the water solubility and biocompatibility. This compound also exhibited the appropriate drug-like properties. The presence of the galactose ligand increases the specificity of liver glycoprotein. A myoglobin study was used to monitor the CO release kinetic profile. The target selectivity of ALF492 can be well administrated. Significantly, when ALF492 is mixed with the antimalarial drug artesunate, ALF492 responds to the effective adjuvant treatment for cerebral malaria. Collectively, this marks the outstanding potential of ALF492 in the treatment of falciparum malaria.

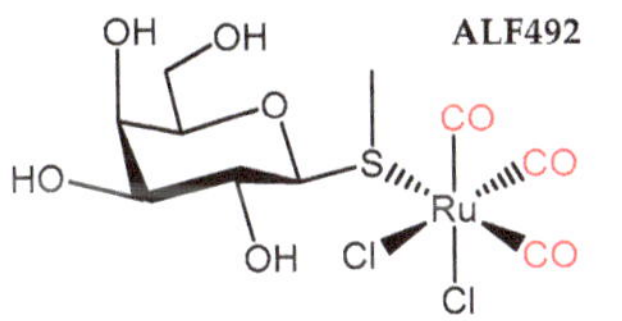

Figure 19. Galactose chelated three carbonyl ruthenium complexes.

The CC Romao group synthesized the molybdenum-based water-soluble release molecule $Mo(CO)_3(CNCR'R''CO_2H)_3$ (R′=R″=H, ALF795) (R′=R″=CH_3, ALF794) (Figure 20) [153]. Myoglobin experiments have shown that compounds were stable and did not decompose in an aerobic aqueous solution for at least 1 hr. ALF794 is less toxic and suitable for drug-like properties. It can deliver CO to acetamido phenol-induced liver in mice with acute liver failure. After 5 min, intravenous injection, the ratio of ALF794 in liver/blood and liver/kidney were reported at 5.27 and 12.58, respectively and ALF795 liver/blood and liver/kidney ratios were observed at 0.33 and 0.50, respectively.

ALF795: R'=R''=H,

ALF794: R'=R''=CH_3

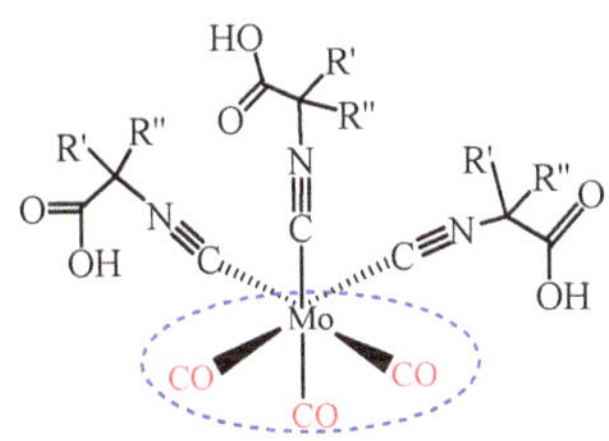

Figure 20. β-isocyanate coordinated molybdenum carbonyl complexes.

2.5. Polymers

The aforementioned most promising anticancer drug agent is a macromolecular conjugate. Ruth Duncan et al. reported for the first time a polymer-anticancer conjugate [154]. These macromolecular drugs consist of at least three (3) parts: One is a polymer carrier (HPMA), which transports metal organic drugs such as $Mn(CO)_3$ light-induced CORMs; the second is a biodegradable polymer drug connector; the third is an anti-tumor agent. The Bruckmann and Kunz groups reported that N-(2-hydroxypropyl)methacrylamide (HPMA) and pentacarbonyl bromide were heated under the re-flux of a dry acetone solution to obtain a copolymer P1 at high yield (Figure 21) [155]. Inducing CORMs can passively transport CO to release metal drugs for tumor cells or sites of inflammation. The polymer conjugate P1 releases CO at 365 nm light, which is not cytotoxic to the HCT116 human colon cancer and HepG2 liver cancer cells.

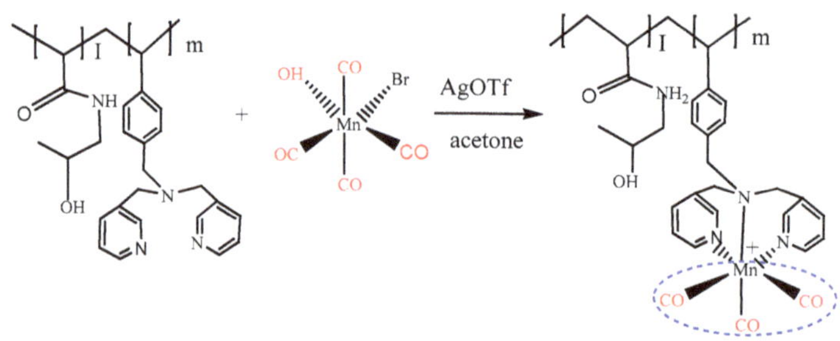

Figure 21. The copolymer P1 synthesized for releasing CO segment.

HPMA-copolymer has a great nature of water solubility. Bernhard Spingler et al. discovered the copolymer materials [156]. The bis(2-pyridylmethyl)benzylamine ligand was prepared from picolyl chloride and benzylamine, then this ligand coordinated with $Re(CO)_3$ moiety to construct the HPMA-co-bis(2-pyridylmethyl)-4-vinylbenzylamine copolymers (Figure 22). HPMA-copolymer characteristics such as the average molecular weight can be modified according to the requirement by replacing the radical starter and co-monomers. For instance, the molecular weight 52KDa is the appropriate choice for remarkably enhancing the permeability and retention (EPR) effect. The established copolymer system with the $Re(CO)_3$ fragment has an ability to diagnose the used 99mTc. Moreover, the identical behavior in IR spectra and X-ray crystallography have shown their resemblance in the binding sites of $Re(CO)_3$-labelled copolymer and bis(2-pyridylmethyl)-amine-derived complexes (solid-state structures (SSS) of $\cdot CH_2Cl_2$ and$\cdot CH_2Cl_2 \cdot H_2O$). HPMA-morphology can be efficient for targeting the tumor sites for the EPR effect.

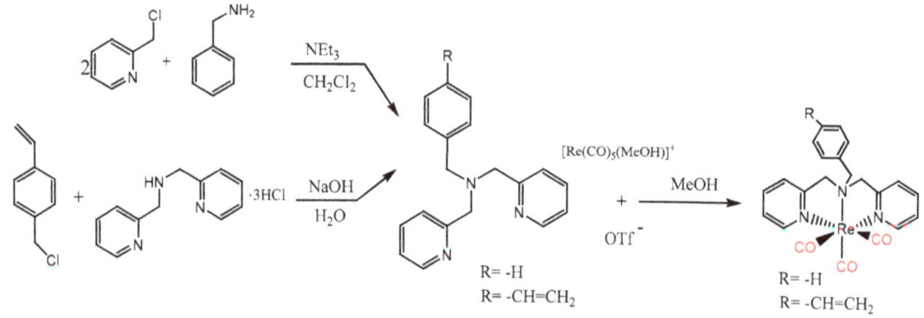

Figure 22. The copolymers HPMA-co-bis(2-pyridylmethyl)-4-vinyl-benzylamine construct through $Re(CO)_3$ moiety.

Not only solvent exchange triggers complexes have been reported but photons energy has also been explored in this research area. Pierri et al. have shown that the water-soluble photo-CORM can be controlled through NIR photons energy. They utilized the amphiphilic polymer for encapsulating the Photo-CORM *trans*-Mn(bpy)(PPh_3)(CO_2)$_2$ (Figure 23) [47]. Since CO has a strong ligand field (L.F), so its absorption bands lies in higher energy zone almost closer to the UV-region [162]. Usually Photo-CORMats belongs to the MCCs family and having CO photolysis-lability from the excited states of LF associated metal-centered [163]. Although the higher MLCT transitions obviate the ligand re-configuration, during the MLCT exhibition, the photolysis might be affected due to π-back-bonding of metal-to-CO configuration [164]. The energy gaps were found between the MLCT and LF excited states. The lower excitation states of MCLT have reduced the ligand lability as compared to LF states [163]. Thus, it is difficult to construct the Photo-induced-CO-release upon NIR or longer visible wavelengths. Up-conversion nanoparticles (UCNPs), (i.e., lanthanide ion doped) have a utility of uncaging from NIR photolysis wavelengths. UCNPs are already claimed as photodynamic therapy [165,166]. Those research analysts firstly developed the polymer matrix UNCPs@PL-PEG (an amphiphilic phospholipid-functionalized poly(ethylene glycol) then used NIR photons energy as a trigger to release the CO segment for biological purposes. PL-PEG shows a remarkable characteristic of water solubility and it has also provided space for incoming other soluble photo-organic CORMs. These organic compounds might be helpful for searching the special physiological targets.

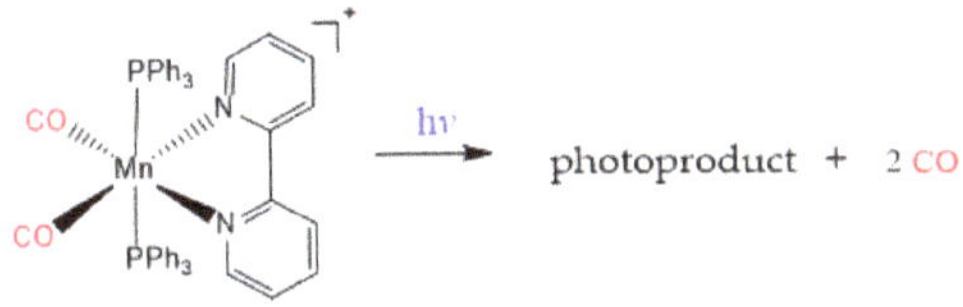

Figure 23. NIR-responsive amphiphilic polymer conjugates (PhotoCORMats).

2.6. Metal Organic Frameworks (MOFs-CORMats)

Metzler-Nolte et al. synthesized the hybrid material metal-organic framework as a class of porous coordinated polymer to encapsulate the CO segment. [138]. They established the bio-compatibility with MOFs; NH_2-MIL-88B (Fe) and MIL-88B (Fe) by capturing the CO at susceptible Fe^{II} and Fe^{III} coordinative unsaturated metal sites (CUSs). Unfavorably, it requires higher activation temperature [167] (≥ 550 K). An adventuring feature of MOF has a breathing effect while providing accommodation for CO at the adjacent site; probably having a controlled CO release, potential through the opening/closing gesture of porous MOF.

Instead, it encapsulates the mere CO segment. The research can also accommodate the CORM's fragment inside MOFs. A new class of Manganese carbonyl complex Photo-CORMs has been explored from Zr(IV) based MOF. MnBr(bpydc)$(CO)_3$ (bpydc=5,5′-dicarboxylate-2,2′-bipyridine) embedded into Zr(IV) MOFs [137]. This photo-activated commodity has been evaluated into cellular substrates along with the biocompatible polymer matrix, claimed to be controlled and efficient light-induced CO-release. After the successful CO release from CORMats, the inability in containment of metal degradation growth, this leads to scientists being reluctant for the employment of genuine medication. The CO has played its own role effectively; how other body organisms responds remains questionable.

2.7. Porous Structure Materials

Porous structure materials have already been in use for encapsulation of the CORMs commodity for the usage of the CO release. Maldonado, Elisa Barea et al. shared the amazing research work about the organometallic community by embedding the anion porous framework with the exhibition of the cation exchange strategy (Figure 24) [50]. This innovation invigorates under physiological condition while it is triggered by visible light. At first, they developed the air-stable, nontoxic, photoactive and water-soluble cation species CORM $[Mn(tacn)(CO)_3]^+$ ($ALF472^+$) then encapsulates into anionic porous matrixes belonging to the inorganic framework. Anionic silica matrix exerts the good administration by reducing the CO release kinetics and providing the control-able rate. The ALF472@hybrid silica-SO_3 material might support the on/off switch delivery management, probably experiencing the command and control on the released CO. For ensuring the safety of the CORMats, the process metal degradation fragment was examined for ALF472@MCM-41-SO_3 up to 72 h and no significant appearance was reported. This silica framework is providing an 80% CORMs metal fragment containment. In phototherapies, the CO supplied is a provocative feature in the controlled manner, such as many inflammatory skin issues and topical skin cancer treatment.

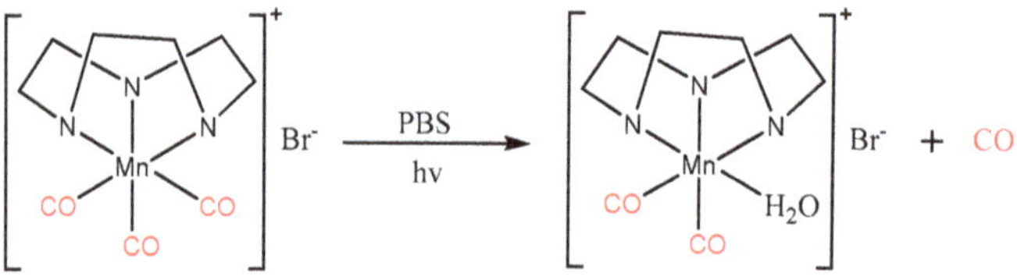

Figure 24. Photo-activated ALF472 CORM [Mn(tacn)$(CO)_3$]Br simulated under physiological parameter.

Although many of the CORMats have been explored for therapeutic purposes, but the lack of biocompatibility and inconvenient features like solubility and toxicity from organometallic compounds, it could not be used as a drug agent. To overcome this dilemma, CORM-1 has been embedded into the nonporous fibers structure poly(L-lactide-co-D/L-lactide). The bioavailability and water accessibility were confidently achieved by photo-activated electro-spinning [111]. Specifically, there was no toxicity observed during the mouse fibroblast 3T3 cells culture. This feature might be promoting the CORMats into viable drug materials in the future.

2.8. Nanaoparticles

Among the CO-carrying support system, nanoparticles are intensely focused because nanoparticles can passively target the malignant cells and actively involve with the targeted tumor cells. Ulrich Schatzschneider et al. open the gateway for the nanoparticles carriers of Silicium dioxide as photoactivatable CORMats [157]. These research analysts, first time synthesized the 3-azidopropane-functionalized SiO_2 nanoparticles, followed by [Mn$(CO)_3$(tpm-L1)] in a dimethylamide solution at room temperature. Manganese-based Photo-CORM $[Mn(CO)_3(tpm)]^+$ whose tpm ligand is linked with Silicium dioxide nanoparticles by CuAAC "click" reaction through construction of the Azido group by emulsion copolymerization of (3-azidopropyl)triethoxysilane and trimethoxymethylsilane at the surface. In this mechanism nanoparticles were functionalized with manganese tricarbonyl as light-induced CORMats (Figure 25). SiO_2 nanoparticles are stable in nature with an amazing bio-compatibly and easy to modify at the surface. It means that various target molecules could be incorporated at the surface to achieve the drug delivery objectives. Eventually, it will increase the drug delivery capacity at specific sites. So therefore, reduce the systemic risk. To reduce the side effects of chemotherapy, SiO_2 nanoparticles provide the most valuable platform for tumor drug delivery.

In the nanoparticle carrier strategy, Urara Hasegawa et al. introduced the CO-releasing polymeric nanoparticles (CONPs) through phenylboronic acid-catechol complexation of catechol-bearing CO-donor $Ru(CO)_3Cl$(L-DOPA) and phenylboronic acid-containing framboidal nanoparticles. It has been testified in a biological organism and gives feedback to cysteine, and subdues the pro-inflammatory mediator's IL-6 [168].

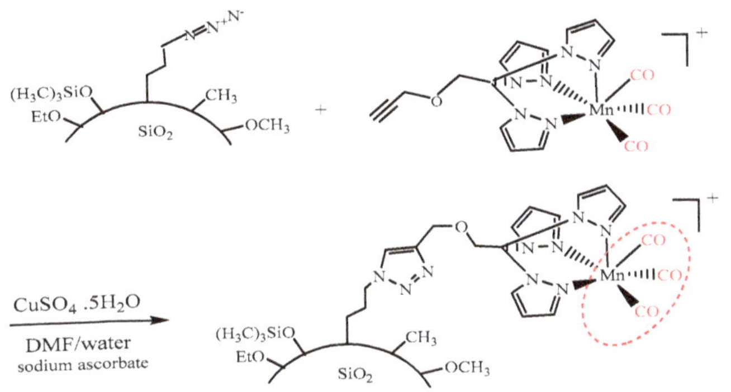

Figure 25. The mechanism of Manganese tricarbonyl functionalized with silica nanoparticles.

2.9. Nanosheets

The encapsulation of CORM's commodity shows promising features. X. Chen et al. worked out to cage the Manganese-carbonyl CORM inside the small MnCO-graphene oxide (PEG-BPY[$MnBr(CO)_3$]-GO) nanosheet, recruited as a drug carrier trigger by NIR light energy with on-demand CO release for photochemical CORMats [48]. They successfully constructed the novel combinations of the CO-release mechanism but triggered facility provided inconvenience for therapeutic purposes. However, we also need to pay attention to its complete management system, as in some cases big trouble may be faced for its state of being clinically applied or not. Merely an advantage for the CO releasing behavior is not enough, as it will always be difficult to handle it and nearly impossible for remote area patients, thus it raises a concern for being a practically viable medicine or not. Moreover, there might be no concern to what happens to metal degradation and leftover residue, which actually needs to be addressed properly too. Apart from this, how can we provide this medication to the patient using NIR trigging, so its mobility will remain the utmost challenge. In reality, the novel production in terms of its laboratory scale is quite different from its practical application as a cure agent.

2.10. Metallodendrimers

Metallodendrimers has a monodisperse nature with facile preparation. This exclusive feature is demonstrated by Smith et al. to entrepreneur the Photo-CORMats (Figure 26) [161]. For this purpose, Photo-CORM $Mn(CO)_3$ moiety scaffolds with polypyridyl dendritic. The general representation of polypyridyl dendritic is [DAB-PPI-{MnBr($bpy^{CH3,CH=N}$)$(CO)_3$}n] (whereas DAB–1,4-diaminobutane, PPI=poly(propyleneimine), bpy=bipyridyl). Photo-activated metallodendritic CORMats has been observed to liberate CO molecules upon 410 nm visible light photons penetration.

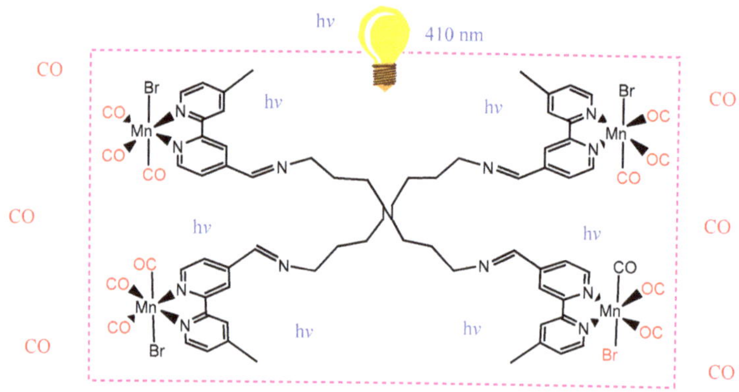

Figure 26. Metallodendrimers photoactivated CORMats.

2.11. Nanodiamond (ND)

The tpm ligand of Photo-CORM $[Mn(CO)_3(tpm)]^+$ peptide material and nanoparticle $[Mn(CO)_3(tpm\text{-}L1)]$ could further be constructed to explore the azide-modified nanodiamond (ND) by CuAAC (copper-catalyzed 1,3-dipolar azide–alkyne cycloaddition) "click" formation as Manganese-MCCs Photo-CORMats (Figure 27) [159]. Dordelmann, G. et al. introduced the first-time CuAAC coupling to attach the CO-liberating agent with ND as a biocompatible supporter. Photoactivatable CORM $[Mn(CO)_3(tpm)]^+$ retained at the ND's surface for CO biological services and therapeutic purposes and were evaluated through standard myoglobin assay.

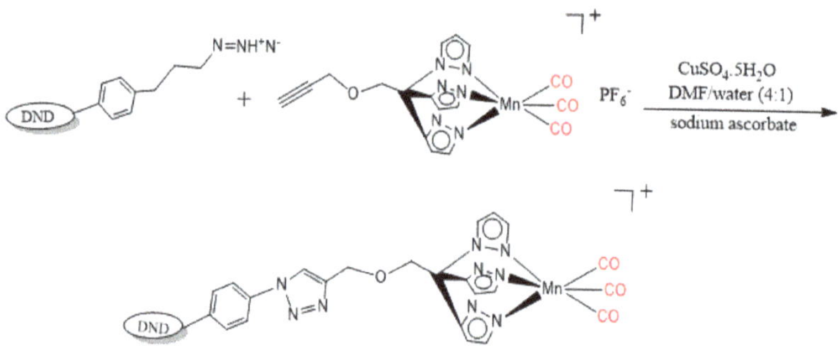

Figure 27. The $[Mn(CO)_3(tpm)]^+$-functionalized nanodiamond (ND) immobilized on azidemodified ND'surface through CuAAC "click" reaction.

Different CORMats are compatible with special cellular environments and are free to perform their therapeutic activities. Certain conditions restrict CORMats activities; definitely it would directly affect the therapeutic performance. CORMats therapeutic potential relies on the material's nature such as solubility, compatibility and activation mechanism. Another advantage of CORMats is to modify the CORMats assembly according to respective disease cells, which could be more helpful in searching the selective targets. For cancer treatment, the redirected T-cells, i.e., chimeric antigen receptor (CAR) T-cell might be providing a governing principle for cancer therapy [169]. CAR-Tcell is easier to find its own therapeutic targets from the peculiar receptor configuration. This exclusive feature facilitates gene-therapy. Similarly, this morphology can be applied to the CORMats development for special tissue selectivity. Numerous CORMats with their biological significance are described in Table 2.

Table 2. Conjugate strategies for therapeutic CO release.

Sr. #	CORMats	Therapeutic Implications	Refs
1	Micellization	Bioactive in a murine model of inflammatory colitis; potential for curing the ROS affected inflammatory disease; In response to human monocytes and attenuates the LPS-induced inflammatory.	[55,59,145]
2	Proteins	Regulation the cytokines IL-6 and IL-10; artificial metallohydrolase performance and elevated NF-κB factor (10 folds).	[121,146–149]
3	Vitamins	Shows HO-1 expression; inducing nuclear accumulation of Nrf2; antimalarial drug artesunate and acute liver failure.	[150–153]
4	Polymers	HCT116 human colon cancer and HepG2 liver cancer cells; enhance the EPR effect and targeting the tumor sites; achieve special and selective physiological targets.	[47,154–156]
5	Porous structure materials	Inflammatory skin issues and topical skin cancer treatment. Surprisingly, no toxicity was found in mouse fibroblast 3T3 cells.	[50,111].
6	Nanoparticles	Cysteine and subdues feedback to pro-inflammatory mediator's IL-6; cardiovascular therapy and relax the rat aorta muscle rings.	[157,168]
7	Peptide	Human prostate cancer cell line (PC-3) and supported cardiomyocyte viability.	[139–144]
8	Nano-sheets	Controllable CO release (e.g. GO-MnCORMats) suitable for inflammatory diseases after LPS stimulation and responsive intracellular CO release.	[48]
9	Nano-diamond	Nano-diamond precursor compatible with photons, hopefully, could be modified for special cell targeting.	[159]
10	MOFs	Inflammatory bowel disease and expected pharmacological applications by downsizing the MOF crystals to the nanoscale.	[137,138]
11	Metallodendrimers	Potential for inflammatory disease and cancer cells.	[161]

Briefly if the above intensive research discussion is summarized; it is evident that Ru-MCCs and Mn-MCCs are the right choices for nano-medicine due to bio-compatibility and tremendous prescribed feasibility analysis especially Ru-MCCs due to its lessened toxicity. Scrolling down from micellization to nanodiamond, all CO-prescriptions have CO liberation capability, but none of them could be claimed as safe therapeutic management and can't be directly applied for exogenous CO-prodrug. During the CORMats administration, rather than focus on the exploration of new advanced materials, the researchers might be considering already existing pharmaceutical materials. A pharmaceutical drug like substance such as crystalline smectite clay; is one of the promising biocompatible and pharmaceutical composites, which could be transformed into CORMats after proper formulation and careful administration. There are two types of strategies that have been introduced for the CORMats production. One is exploiting from the already developed CORMs with biocompatible materials as CO carriers. Another analogy is captured from the CO moiety in the vicinity of material specific akin to MOFs [138]. Already developed CO incorporating strategies are generalized in Table 3.

Table 3. Summarize the CO-releasing substrate along with their association.

Strategies	CO-Release Mechanism	Molecules/Materials
CO-releasing molecules (CORMs)	organometallics	CORM-1, CORM-2, CORM-3, ALF492, CORM-A1, B_{12}-ReCORM-2, Re-CORM-1, CORMA-1-PLA and ALF186.
	nonmetallic	Silica-carboxylates, boranocarbonates, boranocarbamates, xanthene carboxylic acid (XCA), hydroxy-flavones, 1,2-disubstituted ferrocenes, methylene chloride, meso-carboxy BODIPYs, unsaturated cyclic and diketones (DKs).
CO-releasing materials (CORMats)	conjugated systems	Micellization, peptide, vitamins, proteins, polymers, metal organic framework, nanoparticles, nano-sheets, porous structure materials, metallodendrimer and nano-diamond.

3. CO-Releasing Kinetic Profile

CORMs and CORMats must have a CO utility to deliver in response to the biological system soon after trigger. The specified trigger plays a decisive role for therapeutic applications. Their kinetics is highly dependent on the trigger facility at which they are applied for. The CO discharging rate was exclusively committed for searching the affected sites of selected targets. The half-life ($t_{1/2}$) of CORMs/CORMats ($t_{1/2}$ is defined in time duration as half of the introduced CORMs/CORMats amount will be dis-integrated) is the key parameter for examining the CORMs/CORMats stability and sustainability. The fast CO-releasing rate is difficult to attain predetermined clinical objectives.

Just an illustration [170], the half-life ($t_{1/2}$) of CORM-3 is 3.6 min only when anticipated with the human plasma. At that moment, CORM-3 dissolves in plasma configuration and suddenly reacting with albumin, supplies CO_2 and $Ru(CO)_2$ segment; and also makes an alliance with protein in vivo circulation, where CO serves slowly and nonspecifically [171]. Non-technical CO release is unable to deliver the necessary pharmaceutical features. Likewise slow and fast CO release molecules would be engineered to accommodate the distinct clinical trials (Figure 28).

The half-life of CORM-1 and CORM-2 is about 1 min in PBS (*phosphate buffered saline*) at 37 temperature with pH~7.4 [37,172]. Such types of half-lives are considered very short intervals. The CORMs and CORMats deliberation must be regulated along with integral body fluids in order to communicate with victim organs and/or tissues before CORMs/CORMats (as CO-producer) consume entire CO quantity [173]. To improve the sustainability of CO carriers, the half-life ($t_{1/2}$) should be extended for few minutes, but somehow few seconds and milliseconds extension will be more beneficial. A different designated strategy has promoted the transient CO releases. These mechanisms will be observed through ion-channel kinetic studies [174]. The extended pharmacokinetic qualities containing nanomaterials (NPs) and macromolecular models could be exploited for the management of CO transporters or CO carriers.

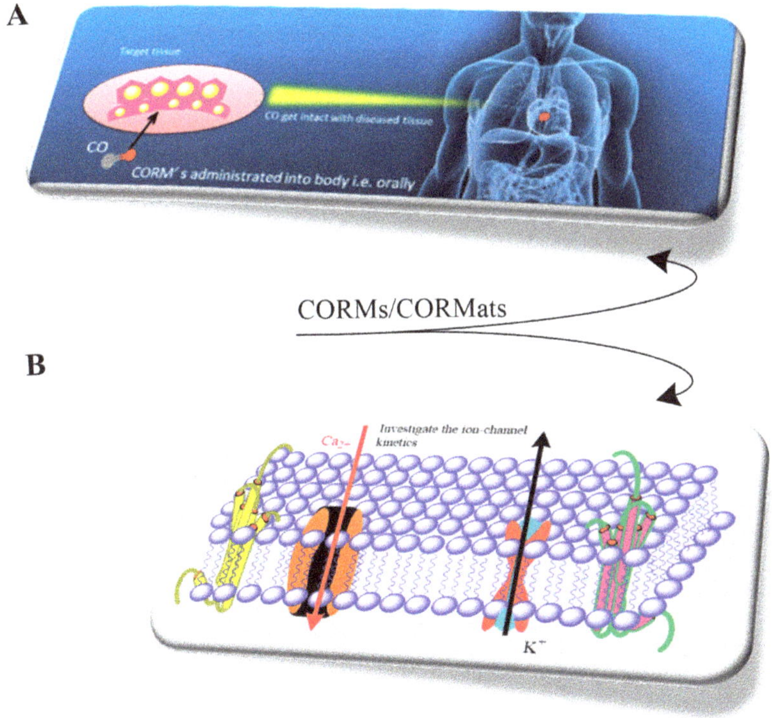

Figure 28. The CO-releasing rate profile reflects the different characteristics: (**a**) Slow CO release has therapeutic significance; (**b**) fast CO release demonstrates the path of ion-channel kinetics.

4. CORMs/CORMats Cytotoxicity and Tissue Accumulation

In addition to CORMs and CORMats pharmaceutical advantages, it delivers some adverse effects too because of their toxicological profile or even proliferation of toxic metal residues (i-CORMs/i-CORMats) resulting soon after CORMs/CORMats launch the CO into the biological environment [95,175–177]. Subsequently, a particular deficiency of CORMs/CORMats is usually observed after the CO excretion; their CO-missing analogues tend to prevail in situ administration. Therefore the transition-heavy metal core usually harbors cofactors and is involved in some uncontrolled reactions/activities with neighboring tissues/cells, thereby contributing serious cellular impairment (Figure 29).

Wang et al. studied the in vivo toxicity, cytotoxicity, metabolism and bio-distribution of two carbonyl metal CORMs series including $Ru(CO)_3ClnL$ and $M(CO)_5L$ (M= Cr, Mo, W) [178]. The cytotoxic effect was monitored on murine macrophages through MTT colorimetric assay with respect to IC_{50} and LD_{50} values; the severely damaged kidney and liver were observed to picture both morphological and functional aspects. The cell culture RAW264.7 was incubated with CORMs/CORMats while examining cytotoxicity and demonstrates their bactericidal activity against a variety of microbes, including *Staphylococcus aureus, Escherichia coli, and Pseudomonas aeruginosa* [126,127]. In this study, they found the uneven distribution of metal complexes in organs and tissues that subsequent damage through metal ions oxidation such as Ruthenium complexes. It was oxidized from Ru^{II} to Ru^{III} by P450 enzymes. This toxicity issue was elucidated by Winburn et al., they also performed the CORM's toxicological profile [179]. CORM-2 and its depleted form i-CORM-2 were studied and compared in two kidney cell lines (MDCK and HeK lines) and primary rat cardiomyocytes. This study explained that the CORM-2 cytoprotective concentration (<20mM) is approaching to cytotoxic value (>100 mM). Moreover, both CORM-2 and i-CORM-2 exerted the cellular toxicity by means of the abnormal cell cytology, cell cycle arrest, reduced cell viability, increased apoptosis and inhibited mitochondrial enzyme activity [136]. These particular consequences were observed through the metal-core mediated toxicity. Different studies have also been explored that intensifying the polarity of CORMs/CORMats would be possibly limiting their penetration over the cellular membrane, and thus attenuating their toxicity [136].

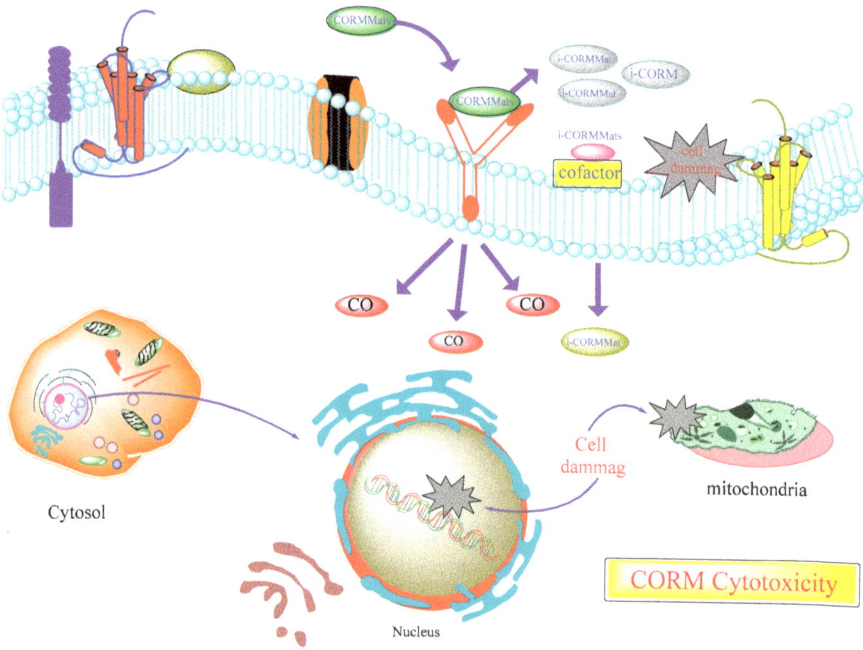

Figure 29. The CORMs/CORMats integration exhibited the cytotoxicity of metal residue (i-CORMs/CORMats).

5. Concluding Remarks

CO is generally infamous for its toxicity, but a controlled dose of CO shows useful biological impacts. The CO's detail analysis exhibits endogenous production by heme oxygenase and explores the therapeutic scope. This scientific study not only confirms the endogenous generation of CO, which has important potential in pathological tissues but also guarantees exogenously released CO's therapeutic impacts. Therefore, the challenges for the pharmaceutical drug chemists have always been and are continuing still, for the development of a risk-free and more convenient strategy to deliver therapeutic CO dosage. The CO administration with biological system suggests their therapeutic potential. This CO administration relies on MCCs for CO liberation. Thus, CORMats were developed by MCCs with different conjugate/scaffold systems. The CORMats have been covalently assembled with different nanomaterial including polymers, silica nanoparticles, proteins cages, vitamins, metallodendrimer, micelles, nanodiamond and nanofiber gel (peptide amphiphilic) or even incorporated with magnetic nanoparticles (maghemite), tablets, non-woven, or either MOFs for following features: To enhance sustainability and stability; to approach the special cellular tissues/organs; to reduce the toxicity; to attain the EPR- effect; or to permit special triggers facilities. CORM has the capability to deliver the CO to tissues and cells in vivo, in-fact constitute the most appropriate scheme to accomplish the therapeutic outcomes. This proof-of-concept refers to the medicinal chemists to endeavor modern

CORMats furnished with ADME (CORMats characteristics: Administration, Distribution, Metabolism, Excretion), prerequisite for the clinical utility. As a prodrug, these developed mechanisms are highly dependent on in vivo performance. It has been worth mentioning that many pharmaceutical materials were also claimed to be non-toxic such as smectite clay that might be transformed into CORMats for promising therapeutic benefits. Probably crystalline smectite clays are the best choice for the CORMats development due to its con-comitant administration. Additionally, their layered structure exfoliations and cation exchange capacity (CEC) have been encouraging for developing the new class of CORMats. Furthermore, it is mandatory to investigate the metal residues (remaining fragments) after CO liberation, if any side effect of newly developed materials is reported should try to minimize it by modern carrier designs. The aim of the controlled CO delivery management was sponsored by tissue selection and distribution. The CORMats activation with different triggers did not permit to develop "universal" CORMats for every disease model. The method of CORMs/CORMats trigger or even CO activation is used to disintegrate the MCCs through photo, thermal, enzyme, pH, oxidation and solvent trigger CORMs/CORMats bearing ligand exchange strategies. These CORMs/CORMats strategies are promising candidates of the therapeutic potential and deserve exclusive attention for thorough therapeutic investigations.

The toxicity of the CO precursor is still a big challenge for the researchers. The CORMs/CORMats toxicity was in-action during and after the CO release with depleted metal residues abbreviated as i-CORMs/CORMats. The fast CO release helps to study the ion-channel path, while the slow release favors tissue targets. It is mandatory that CORMs/i-CORMs and CORMats/i-CORMats (before/after CO release) should not be participating in any toxic activity. Otherwise it will not be possible to prescribe for patients; as the safety of human organs is the utmost priority.

The above discussion confirms that CORMs and CORMats are accountable for the CO-produce being the active ingredient. It should be noted that CORMats did not technically modify specific receptors but only provided a transport and discharge services for the CO gas. Therefore, the therapeutic impacts of CORMats under physiological conditions to employ CO preferentially and professionally against damaged biological tissues/organism must prevail and ensure the quick release of loaded CO upon trigger.

Author Contributions: Conceptualization, W.Z. and Y.W. (Yanyan Wang); Methodology, Y.W. (Yanyan Wang), M.F., Y.H. and W.Z.; Writing —Original Draft Preparation, M.F., H.S., Y.H., N.M., K.N. and Y.W. (Yanyan Wang); Writing —Review and Editing, Y.H., N.M., Y.W. (Ya Wu), W.D., R.L., K.N., W.Z. and M.F.; Supervision, Z.G. and W.Z.; Project Administration, W.Z. and Z.G.

Funding: This research was supported by the grant from National Natural Science Foundation of China (21771122, 21571121), the 111 Project (B14041), the Program for Chang Jiang Scholars and Innovative Research Team in University (IRT_14R33), Key Research and Development Project of Shaanxi Science and Technology Department (2017SF-064).

Acknowledgments: This work was performed within the "211 Project" of Ministry of Education (MOE) China at the Key Laboratory of Applied Surface and Colloid Chemistry with the support of China Scholarship Council program at Shaanxi Normal University, Xi'an, China.

Conflicts of Interest: The authors declare that there is no conflict of interests regarding the publication of the paper.

References

1. Pitto-Barry, A.; Lupan, A.; Ellingford, C.; Attia, A.A.A.; Barry, N.P.E. New Class of Hybrid Materials for Detection, Capture, and "On-Demand" Release of Carbon Monoxide. *ACS Appl. Mater. Interfaces* **2018**, *10*, 13693–13701. [CrossRef] [PubMed]
2. Motterlini, R.; Otterbein, L.E. The therapeutic potential of carbon monoxide. *Nat. Rev. Drug Discov.* **2010**, *9*, 728–743. [CrossRef] [PubMed]
3. Haldane, J.B.S. Carbon monoxide as a tissue poison. *Biochem. J.* **1927**, *21*, 1068–1075. [CrossRef] [PubMed]
4. Douglas, C.G.; Haldane, J.S.; Haldane, J.B.S. The laws of combination of haemoglobin with carbon monoxide and oxygen. *J. Physiol.* **1912**, *44*, 275–304. [CrossRef]

5. Tenhunen, R.; Marver, H.S.; Schmid, R. Microsomal heme oxygenase. Characterization of the enzyme. *J. Biol. Chem.* **1969**, *244*, 6388–6394.
6. Tenhunen, R.; Marver, H.S.; Schmid, R. The enzymatic conversion of heme to bilirubin by microsomal heme oxygenase. *Proc. Natl. Acad. Sci. USA* **1968**, *61*, 748–755. [CrossRef]
7. Kocer, G.; Nasircilar Ulker, S.; Senturk, U.K. The contribution of carbon monoxide to vascular tonus. *Microcirculation* **2018**, *25*, e12495. [CrossRef]
8. Peng, J.; Hu, T.; Li, J.; Du, J.; Zhu, K.; Cheng, B.; Li, K. Shepherd's Purse Polyphenols Exert Its Anti-Inflammatory and Antioxidative Effects Associated with Suppressing MAPK and NF-kappaB Pathways and Heme Oxygenase-1 Activation. *Oxidative Med. Cell. Longev.* **2019**, *2019*, 7202695. [CrossRef]
9. Dercho, R.A.; Nakatsu, K.; Wong, R.J.; Stevenson, D.K.; Vreman, H.J. Determination of in vivo carbon monoxide production in laboratory animals via exhaled air. *J. Pharmacol. Toxicol. Methods* **2006**, *54*, 288–295. [CrossRef]
10. Heinemann, S.H.; Hoshi, T.; Westerhausen, M.; Schiller, A. Carbon monoxide—Physiology, detection and controlled release. *Chem. Commun.* **2014**, *50*, 3644–3660. [CrossRef]
11. Yang, S.; Chen, M.; Zhou, L.; Zhang, G.; Gao, Z.; Zhang, W. Photo-activated CO-releasing molecules (PhotoCORMs) of robust sawhorse scaffolds [μ^2-OOCR^1, η^1-NH_2CHR^2(C=O] OCH_3, $Ru(I)_2CO_4$]. *Dalton Trans.* **2016**, *45*, 3727–3733. [CrossRef]
12. Stupfel, M.; Bouley, G. Physiological and biochemical effects on rats and mice exposed to small concentrations of carbon monoxide for long periods. *Ann. N. Y. Acad. Sci.* **1970**, *174*, 342–368. [CrossRef]
13. Ling, K.; Men, F.; Wang, W.C.; Zhou, Y.Q.; Zhang, H.W.; Ye, D.W. Carbon Monoxide and Its Controlled Release: Therapeutic Application, Detection, and Development of Carbon Monoxide Releasing Molecules (CORMs). *J. Med. Chem.* **2018**, *61*, 2611–2635. [CrossRef] [PubMed]
14. Otterbein, L.E. Carbon monoxide: Innovative anti-inflammatory properties of an age-old gas molecule. *Antioxid. Redox Signal.* **2002**, *4*, 309–319. [CrossRef]
15. Otterbein, L.E.; Soares, M.P.; Yamashita, K.; Bach, F.H. Heme oxygenase-1: Unleashing the protective properties of heme. *Trends Immunol.* **2003**, *24*, 449–455. [CrossRef]
16. Zamani, M.; Aleyasin, A.; Fakhrzadeh, H.; Kiavar, M.; Raoufzadeh, S.; Larijani, B.; Mahmoodi, E. Heme Oxigenase 2 Gene Polymorphisms as Genetic Risk Factor in Atherosclerosis in Iranian Patients. *Iran. Red Crescent Med. J.* **2010**, *12*, 559–563.
17. Joshi, H.P.; Kim, S.B.; Kim, S.; Kumar, H.; Jo, M.J.; Choi, H.; Kim, J.; Kyung, J.W.; Sohn, S.; Kim, K.T.; et al. Nanocarrier-mediated Delivery of CORM-2 Enhances Anti-allodynic and Anti-hyperalgesic Effects of CORM-2. *Mol. Neurobiol.* **2019**, *56*. [CrossRef]
18. Motterlini, R.; Haas, B.; Foresti, R. Emerging concepts on the anti-inflammatory actions of carbon monoxide-releasing molecules (CO-RMs). *Med. Gas Res.* **2012**, *2*, 28. [CrossRef]
19. Al-Huseini, L.M.; Aw Yeang, H.X.; Hamdam, J.M.; Sethu, S.; Alhumeed, N.; Wong, W.; Sathish, J.G. Heme oxygenase-1 regulates dendritic cell function through modulation of p38 MAPK-CREB/ATF1 signaling. *J. Biol. Chem.* **2014**, *289*, 16442–16451. [CrossRef]
20. Stagni, E.; Privitera, M.G.; Bucolo, C.; Leggio, G.M.; Motterlini, R.; Drago, F. A water-soluble carbon monoxide-releasing molecule (CORM-3) lowers intraocular pressure in rabbits. *Br. J. Ophthalmol.* **2009**, *93*, 254–257. [CrossRef]
21. Allanson, M.; Reeve, V.E. Ultraviolet A (320-400 nm) modulation of ultraviolet B (290-320 nm)-induced immune suppression is mediated by carbon monoxide. *J. Investig. Dermatol.* **2005**, *124*, 644–650. [CrossRef] [PubMed]
22. Mazzola, S.; Forni, M.; Albertini, M.; Bacci, M.L.; Zannoni, A.; Gentilini, F.; Lavitrano, M.; Bach, F.H.; Otterbein, L.E.; Clement, M.G. Carbon monoxide pretreatment prevents respiratory derangement and ameliorates hyperacute endotoxic shock in pigs. *FASEB J.* **2005**, *19*, 2045–2047. [CrossRef]
23. Bagul, A.; Hosgood, S.A.; Kaushik, M.; Nicholson, M.L. Carbon monoxide protects against ischemia-reperfusion injury in an experimental model of controlled nonheartbeating donor kidney. *Transplantation* **2008**, *85*, 576–581. [CrossRef] [PubMed]
24. Sato, K.; Balla, J.; Otterbein, L.; Smith, R.N.; Brouard, S.; Lin, Y.; Csizmadia, E.; Sevigny, J.; Robson, S.C.; Vercellotti, G.; et al. Carbon monoxide generated by heme oxygenase-1 suppresses the rejection of mouse-to-rat cardiac transplants. *J. Immunol.* **2001**, *166*, 4185–4194. [CrossRef]

25. Chen, B.; Guo, L.; Fan, C.; Bolisetty, S.; Joseph, R.; Wright, M.M.; Agarwal, A.; George, J.F. Carbon monoxide rescues heme oxygenase-1-deficient mice from arterial thrombosis in allogeneic aortic transplantation. *Am. J. Pathol.* **2009**, *175*, 422–429. [CrossRef] [PubMed]
26. Clark, J.E.; Naughton, P.; Shurey, S.; Green, C.J.; Johnson, T.R.; Mann, B.E.; Foresti, R.; Motterlini, R. Cardioprotective actions by a water-soluble carbon monoxide-releasing molecule. *Circ. Res.* **2003**, *93*, e2–e8. [CrossRef]
27. Yoshida, J.; Ozaki, K.S.; Nalesnik, M.A.; Ueki, S.; Castillo-Rama, M.; Faleo, G.; Ezzelarab, M.; Nakao, A.; Ekser, B.; Echeverri, G.J.; et al. Ex vivo application of carbon monoxide in UW solution prevents transplant-induced renal ischemia/reperfusion injury in pigs. *Am. J. Transplant.* **2010**, *10*, 763–772. [CrossRef]
28. Sandouka, A.; Fuller, B.J.; Mann, B.E.; Green, C.J.; Foresti, R.; Motterlini, R. Treatment with CO-RMs during cold storage improves renal function at reperfusion. *Kidney Int.* **2006**, *69*, 239–247. [CrossRef]
29. Nakao, A.; Faleo, G.; Shimizu, H.; Nakahira, K.; Kohmoto, J.; Sugimoto, R.; Choi, A.M.; McCurry, K.R.; Takahashi, T.; Murase, N. Ex vivo carbon monoxide prevents cytochrome P450 degradation and ischemia/reperfusion injury of kidney grafts. *Kidney Int.* **2008**, *74*, 1009–1016. [CrossRef] [PubMed]
30. Pizarro, M.D.; Rodriguez, J.V.; Mamprin, M.E.; Fuller, B.J.; Mann, B.E.; Motterlini, R.; Guibert, E.E. Protective effects of a carbon monoxide-releasing molecule (CORM-3) during hepatic cold preservation. *Cryobiology* **2009**, *58*, 248–255. [CrossRef] [PubMed]
31. Kaizu, T.; Ikeda, A.; Nakao, A.; Tsung, A.; Toyokawa, H.; Ueki, S.; Geller, D.A.; Murase, N. Protection of transplant-induced hepatic ischemia/reperfusion injury with carbon monoxide via MEK/ERK1/2 pathway downregulation. *Am. J. Physiol. Gastrointest. Liver Physiol.* **2008**, *294*, G236–G244. [CrossRef]
32. Kohmoto, J.; Nakao, A.; Sugimoto, R.; Wang, Y.; Zhan, J.; Ueda, H.; McCurry, K.R. Carbon monoxide-saturated preservation solution protects lung grafts from ischemia-reperfusion injury. *J. Thorac. Cardiovasc. Surg.* **2008**, *136*, 1067–1075. [CrossRef]
33. Minamoto, K.; Harada, H.; Lama, V.N.; Fedarau, M.A.; Pinsky, D.J. Reciprocal regulation of airway rejection by the inducible gas-forming enzymes heme oxygenase and nitric oxide synthase. *J. Exp. Med.* **2005**, *202*, 283–294. [CrossRef] [PubMed]
34. Lee, S.S.; Gao, W.; Mazzola, S.; Thomas, M.N.; Csizmadia, E.; Otterbein, L.E.; Bach, F.H.; Wang, H. Heme oxygenase-1, carbon monoxide, and bilirubin induce tolerance in recipients toward islet allografts by modulating T regulatory cells. *FASEB J.* **2007**, *21*, 3450–3457. [CrossRef] [PubMed]
35. Nakao, A.; Toyokawa, H.; Tsung, A.; Nalesnik, M.A.; Stolz, D.B.; Kohmoto, J.; Ikeda, A.; Tomiyama, K.; Harada, T.; Takahashi, T.; et al. Ex vivo application of carbon monoxide in University of Wisconsin solution to prevent intestinal cold ischemia/reperfusion injury. *Am. J. Transplant.* **2006**, *6*, 2243–2255. [CrossRef] [PubMed]
36. Bojakowski, K.; Gaciong, Z.; Grochowiecki, T.; Szmidt, J. Carbon monoxide may reduce ischemia reperfusion injury: A case report of complicated kidney transplantation from a carbon monoxide poisoned donor. *Transplant. Proc.* **2007**, *39*, 2928–2929. [CrossRef]
37. Motterlini, R.; Mann, B.E.; Foresti, R. Therapeutic applications of carbon monoxide releasing molecules. *Expert Opin. Investig. Drugs* **2005**, *14*, 1305–1318. [CrossRef]
38. Ryter, S.W.; Choi, A.M. Cytoprotective and anti-inflammatory actions of carbon monoxide in organ injury and sepsis models. *Novartis Found. Symp.* **2007**, *280*, 165–175.
39. Nobre, L.S.; Jeremias, H.; Romao, C.C.; Saraiva, L.M. Examining the antimicrobial activity and toxicity to animal cells of different types of CO-releasing molecules. *Dalton Trans.* **2016**, *45*, 1455–1466. [CrossRef] [PubMed]
40. Motterlini, R.; Gonzales, A.; Foresti, R.; Clark, J.E.; Green, C.J.; Winslow, R.M. Heme oxygenase-1-derived carbon monoxide contributes to the suppression of acute hypertensive responses in vivo. *Circ. Res.* **1998**, *83*, 568–577. [CrossRef] [PubMed]
41. Taille, C.; El-Benna, J.; Lanone, S.; Boczkowski, J.; Motterlini, R. Mitochondrial respiratory chain and NAD(P)H oxidase are targets for the antiproliferative effect of carbon monoxide in human airway smooth muscle. *J. Biol. Chem.* **2005**, *280*, 25350–25360. [CrossRef]
42. Siow, R.C.; Sato, H.; Mann, G.E. Heme oxygenase-carbon monoxide signalling pathway in atherosclerosis: Anti-atherogenic actions of bilirubin and carbon monoxide? *Cardiovasc. Res.* **1999**, *41*, 385–394. [CrossRef]
43. Crespy, D.; Landfester, K.; Schubert, U.S.; Schiller, A. Potential photoactivated metallopharmaceuticals: From active molecules to supported drugs. *Chem. Commun.* **2010**, *46*, 6651–6662. [CrossRef]

44. Otterbein, L.E. The evolution of carbon monoxide into medicine. *Respir. Care* **2009**, *54*, 925–932. [CrossRef]
45. Omaye, S.T. Metabolic modulation of carbon monoxide toxicity. *Toxicology* **2002**, *180*, 139–150. [CrossRef]
46. Palao, E.; Slanina, T.; Muchová, L.; Šolomek, T.; Vítek, L.; Klán, P. Transition-Metal-Free CO-Releasing BODIPY Derivatives Activatable by Visible to NIR Light as Promising Bioactive Molecules. *J. Am. Chem. Soc.* **2016**, *138*, 126–133. [CrossRef] [PubMed]
47. Pierri, A.E.; Huang, P.J.; Garcia, J.V.; Stanfill, J.G.; Chui, M.; Wu, G.; Zheng, N.; Ford, P.C. A photoCORM nanocarrier for CO release using NIR light. *Chem. Commun.* **2015**, *51*, 2072–2075. [CrossRef] [PubMed]
48. He, Q.; Kiesewetter, D.O.; Qu, Y.; Fu, X.; Fan, J.; Huang, P.; Liu, Y.; Zhu, G.; Liu, Y.; Qian, Z.; et al. NIR-Responsive On-Demand Release of CO from Metal Carbonyl-Caged Graphene Oxide Nanomedicine. *Adv. Mater.* **2015**, *27*, 6741–6746. [CrossRef]
49. Mede, R.; Hoffmann, P.; Klein, M.; Goerls, H.; Schmitt, M.; Neugebauer, U.; Gessner, G.; Heinemann, S.H.; Popp, J.; Westerhausen, M. A Water-Soluble Mn(CO)3-Based and Non-Toxic PhotoCORM for Administration of Carbon Monoxide Inside of Cells. *Z. Anorg. Allg. Chem.* **2017**, *643*, 2057–2062. [CrossRef]
50. Carmona, F.J.; Rojas, S.; Sanchez, P.; Jeremias, H.; Marques, A.R.; Romao, C.C.; Choquesillo-Lazarte, D.; Navarro, J.A.; Maldonado, C.R.; Barea, E. Cation Exchange Strategy for the Encapsulation of a Photoactive CO-Releasing Organometallic Molecule into Anionic Porous Frameworks. *Inorg. Chem.* **2016**, *55*, 6525–6531. [CrossRef]
51. Foresti, R.; Bani-Hani, M.G.; Motterlini, R. Use of carbon monoxide as a therapeutic agent: Promises and challenges. *Intensive Care Med.* **2008**, *34*, 649–658. [CrossRef]
52. Kueh, J.T.B.; Stanley, N.J.; Hewitt, R.J.; Woods, L.M.; Larsen, L.; Harrison, J.C.; Rennison, D.; Brimble, M.A.; Sammut, I.A.; Larsen, D.S. Norborn-2-en-7-ones as physiologically-triggered carbon monoxide-releasing prodrugs. *Chem. Sci.* **2017**, *8*, 5454–5459. [CrossRef]
53. Romão, C.C.; Blättler, W.A.; Seixas, J.D.; Bernardes, G.J.L. Developing drug molecules for therapy with carbon monoxide. *Chem. Soc. Rev.* **2012**, *41*, 3571–3583. [CrossRef] [PubMed]
54. Stamellou, E.; Storz, D.; Botov, S.; Ntasis, E.; Wedel, J.; Sollazzo, S.; Krämer, B.K.; van Son, W.; Seelen, M.; Schmalz, H.G.; et al. Different design of enzyme-triggered CO-releasing molecules (ET-CORMs) reveals quantitative differences in biological activities in terms of toxicity and inflammation. *Redox Biol.* **2014**, *2*, 739–748. [CrossRef] [PubMed]
55. Peng, P.; Wang, C.; Shi, Z.; Johns, V.K.; Ma, L.; Oyer, J.; Copik, A.; Igarashi, R.; Liao, Y. Visible-light activatable organic CO-releasing molecules (PhotoCORMs) that simultaneously generate fluorophores. *Org. Biomol. Chem.* **2013**, *11*, 6671–6674. [CrossRef] [PubMed]
56. Schatzschneider, U. Photoactivated Biological Activity of Transition-Metal Complexes. *Eur. J. Inorg. Chem.* **2010**, *2010*, 1451–1467. [CrossRef]
57. Seixas, J.D.; Mukhopadhyay, A.; Santos-Silva, T.; Otterbein, L.E.; Gallo, D.J.; Rodrigues, S.S.; Guerreiro, B.H.; Gonçalves, A.M.L.; Penacho, N.; Marques, A.R.; et al. Characterization of a versatile organometallic pro-drug (CORM) for experimental CO based therapeutics. *Dalton Trans.* **2013**, *42*, 5985–5998. [CrossRef]
58. Queiroga, C.S.F.; Vercelli, A.; Vieira, H.L.A. Carbon monoxide and the CNS: Challenges and achievements. *Br. J. Pharmacol.* **2015**, *172*, 1533–1545. [CrossRef] [PubMed]
59. Hasegawa, U.; van der Vlies, A.J.; Simeoni, E.; Wandrey, C.; Hubbell, J.A. Carbon monoxide-releasing micelles for immunotherapy. *J. Am. Chem. Soc.* **2010**, *132*, 18273–18280. [CrossRef]
60. Park, S.S.; Kim, J.; Lee, Y. Improved electrochemical microsensor for the real-time simultaneous analysis of endogenous nitric oxide and carbon monoxide generation. *Anal. Chem.* **2012**, *84*, 1792–1796. [CrossRef]
61. Morimoto, Y.; Durante, W.; Lancaster, D.G.; Klattenhoff, J.; Tittel, F.K. Real-time measurements of endogenous CO production from vascular cells using an ultrasensitive laser sensor. *Am. J. Physiol. Heart Circ. Physiol.* **2001**, *280*, H483–H488. [CrossRef]
62. Marks, G.S.; Vreman, H.J.; McLaughlin, B.E.; Brien, J.F.; Nakatsu, K. Measurement of endogenous carbon monoxide formation in biological systems. *Antioxid. Redox Signal.* **2002**, *4*, 271–277. [CrossRef]
63. Barbe, J.M.; Canard, G.; Brandes, S.; Guilard, R. Selective chemisorption of carbon monoxide by organic-inorganic hybrid materials incorporating cobalt(III) corroles as sensing components. *Chemistry* **2007**, *13*, 2118–2129. [CrossRef]
64. McLean, S.; Mann, B.E.; Poole, R.K. Sulfite species enhance carbon monoxide release from CO-releasing molecules: Implications for the deoxymyoglobin assay of activity. *Anal. Biochem.* **2012**, *427*, 36–40. [CrossRef]

65. Esteban, J.; Ros-Lis, J.V.; Martinez-Manez, R.; Marcos, M.D.; Moragues, M.; Soto, J.; Sancenon, F. Sensitive and selective chromogenic sensing of carbon monoxide by using binuclear rhodium complexes. *Angew. Chem.* **2010**, *49*, 4934–4937. [CrossRef]
66. Nandi, C.; Debnath, R.; Debroy, P. Intelligent Control Systems for Carbon Monoxide Detection in IoT Environments. In *Guide to Ambient Intelligence in the IoT Environment: Principles, Technologies and Applications*; Mahmood, Z., Ed.; Springer International Publishing: Cham, Switerzland, 2019; pp. 153–176.
67. Vreman, H.J.; Stevenson, D.K. Heme oxygenase activity as measured by carbon monoxide production. *Anal. Biochem.* **1988**, *168*, 31–38. [CrossRef]
68. Yuan, L.; Lin, W.; Tan, L.; Zheng, K.; Huang, W. Lighting up carbon monoxide: Fluorescent probes for monitoring CO in living cells. *Angew. Chem.* **2013**, *52*, 1628–1630. [CrossRef]
69. Michel, B.W.; Lippert, A.R.; Chang, C.J. A Reaction-Based Fluorescent Probe for Selective Imaging of Carbon Monoxide in Living Cells Using a Palladium-Mediated Carbonylation. *J. Am. Chem. Soc.* **2012**, *134*, 15668–15671. [CrossRef]
70. Bauschlicher, C.W., Jr; Bagus, P.S. The metal–carbonyl bond in $Ni(CO)_4$ and $Fe(CO)_5$: A clear–cut analysis. *J. Chem. Phys.* **1984**, *81*, 5889–5898. [CrossRef]
71. Schatzschneider, U. Novel lead structures and activation mechanisms for CO-releasing molecules (CORMs). *Br. J. Pharmacol.* **2015**, *172*, 1638–1650. [CrossRef]
72. Kitazawa, M.; Wagner, J.R.; Kirby, M.L.; Anantharam, V.; Kanthasamy, A.G. Oxidative stress and mitochondrial-mediated apoptosis in dopaminergic cells exposed to methylcyclopentadienyl manganese tricarbonyl. *J. Pharmacol. Exp. Ther.* **2002**, *302*, 26–35. [CrossRef]
73. Gessner, G.; Sahoo, N.; Swain, S.M.; Hirth, G.; Schonherr, R.; Mede, R.; Westerhausen, M.; Brewitz, H.H.; Heimer, P.; Imhof, D.; et al. CO-independent modification of K(+) channels by tricarbonyldichlororuthenium(II) dimer (CORM-2). *Eur. J. Pharmacol.* **2017**, *815*, 33–41. [CrossRef]
74. Mansour, A.M. Rapid green and blue light-induced CO release from bromazepam Mn(I) and Ru(II) carbonyls: Synthesis, density functional theory and biological activity evaluation. *Appl. Organomet. Chem.* **2017**, *31*, e3564. [CrossRef]
75. Carmona, F.J.; Jimenez-Amezcua, I.; Rojas, S.; Romao, C.C.; Navarro, J.A.R.; Maldonado, C.R.; Barea, E. Aluminum Doped MCM-41 Nanoparticles as Platforms for the Dual Encapsulation of a CO-Releasing Molecule and Cisplatin. *Inorg. Chem.* **2017**, *56*, 10474–10480. [CrossRef]
76. Wareham, L.K.; McLean, S.; Begg, R.; Rana, N.; Ali, S.; Kendall, J.J.; Sanguinetti, G.; Mann, B.E.; Poole, R.K. The Broad-Spectrum Antimicrobial Potential of [Mn(CO)4(S2CNMe(CH2CO2H))], a Water-Soluble CO-Releasing Molecule (CORM-401): Intracellular Accumulation, Transcriptomic and Statistical Analyses, and Membrane Polarization. *Antioxid. Redox Signal.* **2018**, *28*, 1286–1308. [CrossRef] [PubMed]
77. Mansour, A.M.; Shehab, O.R. Reactivity of visible-light induced CO releasing thiourea-based Mn(I) tricarbonyl bromide (CORM-NS1) towards lysozyme. *Inorg. Chim. Acta* **2018**, *480*, 159–165. [CrossRef]
78. Aucott, B.J.; Ward, J.S.; Andrew, S.G.; Milani, J.; Whitwood, A.C.; Lynam, J.M.; Parkin, A.; Fairlamb, I.J.S. Redox-tagged carbon monoxide-releasing molecules (CORMs): Ferrocene-containing [Mn(C-N)(CO)4] complexes as a promising new CORM class. *Inorg. Chem.* **2017**, *56*, 5431–5440. [CrossRef]
79. Kretschmer, R.; Gessner, G.; Goerls, H.; Heinemann, S.H.; Westerhausen, M. Dicarbonyl-bis(cysteamine)iron(II): A light induced carbon monoxide releasing molecule based on iron (CORM-S1). *J. Inorg. Biochem.* **2011**, *105*, 6–9. [CrossRef]
80. Li, J.; Zhang, J.; Zhang, Q.; Bai, Z.; Zhao, Q.; Wang, Z.; Chen, Y.; Liu, B. Synthesis and biological activities of carbonyl cobalt CORMs with selectively inhibiting cyclooxygenase-2. *J. Organomet. Chem.* **2018**, *874*, 49–62. [CrossRef]
81. Gong, Y.; Zhang, T.; Li, M.; Xi, N.; Zheng, Y.; Zhao, Q.; Chen, Y.; Liu, B. Toxicity, bio-distribution and metabolism of CO-releasing molecules based on cobalt. *Free Radic. Biol. Med.* **2016**, *97*, 362–374. [CrossRef]
82. Wong, J.; MacDonald, N.; Mottillo, C.; Hiskic, I.; Butler, I.S.; Friscic, T. *Synthesis of Organometallic C0-Releasing Molecules (CORMs) in the Absence of a Bulk Organic Solvent*; American Chemical Society: Denver, CO, USA, 2015; INOR-785.
83. Finze, M.; Bernhardt, E.; Willner, H.; Lehmann, C.W.; Aubke, F. Homoleptic, sigma-bonded octahedral superelectrophilic metal carbonyl cations of iron(II), ruthenium(II), and osmium(II). Part 2: Syntheses and characterizations of [M(CO)(6)][BF(4)](2) (M = Fe, Ru, Os). *Inorg. Chem.* **2005**, *44*, 4206–4214. [CrossRef]

84. Kottelat, E.; Chabert, V.; Crochet, A.; Fromm, K.M.; Zobi, F. Towards Cardiolite-Inspired Carbon Monoxide Releasing Molecules—Reactivity of d4, d5 Rhenium and d6 Manganese Carbonyl Complexes with Isocyanide Ligands. *Eur. J. Inorg. Chem.* **2015**, *2015*, 5628–5638. [CrossRef]
85. Abeyrathna, N.; Washington, K.; Bashur, C.; Liao, Y. Nonmetallic carbon monoxide releasing molecules (CORMs). *Org. Biomol. Chem.* **2017**, *15*, 8692–8699. [CrossRef]
86. Friis, S.D.; Taaning, R.H.; Lindhardt, A.T.; Skrydstrup, T. Silacarboxylic acids as efficient carbon monoxide releasing molecules: Synthesis and application in palladium-catalyzed carbonylation reactions. *J. Am. Chem. Soc.* **2011**, *133*, 18114–18117. [CrossRef] [PubMed]
87. Motterlini, R.; Sawle, P.; Hammad, J.; Bains, S.; Alberto, R.; Foresti, R.; Green, C.J. CORM-A1: A new pharmacologically active carbon monoxide-releasing molecule. *FASEB J.* **2005**, *19*, 284–286. [CrossRef] [PubMed]
88. Klein, M.; Neugebauer, U.; Schmitt, M.; Popp, J. Elucidation of the CO-Release Kinetics of CORM-A1 by Means of Vibrational Spectroscopy. *Chemphyschem A Eur. J. Chem. Phys. Phys. Chem.* **2016**, *17*, 985–993. [CrossRef]
89. Antony, L.A.; Slanina, T.; Sebej, P.; Solomek, T.; Klan, P. Fluorescein analogue xanthene-9-carboxylic acid: A transition-metal-free CO releasing molecule activated by green light. *Org. Lett.* **2013**, *15*, 4552–4555. [CrossRef]
90. Mondal, R.; Okhrimenko, A.N.; Shah, B.K.; Neckers, D.C. Photodecarbonylation of alpha-diketones: A mechanistic study of reactions leading to acenes. *J. Phys.Chem. B* **2008**, *112*, 11–15. [CrossRef] [PubMed]
91. Martins, P.N.A.; Reuzel-Selke, A.; Jurisch, A.; Atrott, K.; Pascher, A.; Pratschke, J.; Buelow, R.; Neuhaus, P.; Volk, H.D.; Tullius, S.G. Induction of carbon monoxide in the donor reduces graft immunogenicity and chronic graft deterioration. *Transplant. Proc.* **2005**, *37*, 379–381. [CrossRef]
92. Chauveau, C.; Bouchet, D.; Roussel, J.C.; Mathieu, P.; Braudeau, C.; Renaudin, K.; Tesson, L.; Soulillou, J.P.; Iyer, S.; Buelow, R.; et al. Gene Transfer of Heme Oxygenase-1 and Carbon Monoxide Delivery Inhibit Chronic Rejection. *Am. J. Transplant.* **2002**, *2*, 581–592. [CrossRef]
93. Anderson, S.N.; Richards, J.M.; Esquer, H.J.; Benninghoff, A.D.; Arif, A.M.; Berreau, L.M. A Structurally-Tunable 3-Hydroxyflavone Motif for Visible Light-Induced Carbon Monoxide-Releasing Molecules (CORMs). *ChemistryOpen* **2015**, *4*, 590–594. [CrossRef]
94. Petrovski, Ž.; Norton de Matos, M.R.P.; Braga, S.S.; Pereira, C.C.L.; Matos, M.L.; Gonçalves, I.S.; Pillinger, M.; Alves, P.M.; Romão, C.C. Synthesis, characterization and antitumor activity of 1,2-disubstituted ferrocenes and cyclodextrin inclusion complexes. *J. Organomet. Chem.* **2008**, *693*, 675–684. [CrossRef]
95. Motterlini, R.; Clark, J.E.; Foresti, R.; Sarathchandra, P.; Mann, B.E.; Green, C.J. Carbon monoxide-releasing molecules: Characterization of biochemical and vascular activities. *Circ. Res.* **2002**, *90*, E17–E24. [CrossRef]
96. Steiger, C.; Luhmann, T.; Meinel, L. Oral drug delivery of therapeutic gases—Carbon monoxide release for gastrointestinal diseases. *J. Control. Release* **2014**, *189*, 46–53. [CrossRef] [PubMed]
97. Ferrandiz, M.L.; Maicas, N.; Garcia-Arnandis, I.; Terencio, M.C.; Motterlini, R.; Devesa, I.; Joosten, L.A.; van den Berg, W.B.; Alcaraz, M.J. Treatment with a CO-releasing molecule (CORM-3) reduces joint inflammation and erosion in murine collagen-induced arthritis. *Ann. Rheum. Dis.* **2008**, *67*, 1211–1217. [CrossRef]
98. Nobre, L.S.; Seixas, J.D.; Romao, C.C.; Saraiva, L.M. Antimicrobial action of carbon monoxide-releasing compounds. *Antimicrob. Agents Chemother.* **2007**, *51*, 4303–4307. [CrossRef] [PubMed]
99. Bikiel, D.E.; Gonzalez Solveyra, E.; Di Salvo, F.; Milagre, H.M.; Eberlin, M.N.; Correa, R.S.; Ellena, J.; Estrin, D.A.; Doctorovich, F. Tetrachlorocarbonyliridates: Water-soluble carbon monoxide releasing molecules rate-modulated by the sixth ligand. *Inorg. Chem.* **2011**, *50*, 2334–2345. [CrossRef] [PubMed]
100. Braud, L.; Pini, M.; Wilson, J.L.; Czibik, G.; Sawaki, D.; Derumeaux, G.; Foresti, R.; Motterlini, R. A carbon monoxide-releasing molecule (CORM-401) induces a metabolic switch in adipocytes and improves insulin-sensitivity on high fat diet-induced obesity in mice. *Arch. Cardiovasc. Dis. Suppl.* **2018**, *10*, 188. [CrossRef]
101. Wareham, L.K.; Poole, R.K.; Tinajero-Trejo, M. CO-releasing Metal Carbonyl Compounds as Antimicrobial Agents in the Post-antibiotic Era. *J. Biol. Chem.* **2015**, *290*, 18999–19007. [CrossRef]
102. Zhang, W.Q.; Atkin, A.J.; Thatcher, R.J.; Whitwood, A.C.; Fairlamb, I.J.; Lynam, J.M. Diversity and design of metal-based carbon monoxide-releasing molecules (CO-RMs) in aqueous systems: Revealing the essential trends. *Dalton Trans.* **2009**, 4351–4358. [CrossRef] [PubMed]

103. Seixas, J.D.; Santos, M.F.; Mukhopadhyay, A.; Coelho, A.C.; Reis, P.M.; Veiros, L.F.; Marques, A.R.; Penacho, N.; Goncalves, A.M.; Romao, M.J.; et al. A contribution to the rational design of Ru(CO)3Cl2L complexes for in vivo delivery of CO. *Dalton Trans.* **2015**, *44*, 5058–5075. [CrossRef]
104. Suliman, H.B.; Zobi, F.; Piantadosi, C.A. Heme Oxygenase-1/Carbon Monoxide System and Embryonic Stem Cell Differentiation and Maturation into Cardiomyocytes. *Antioxid. Redox Signal.* **2016**, *24*, 345–360. [CrossRef]
105. Musameh, M.D.; Green, C.J.; Mann, B.E.; Fuller, B.J.; Motterlini, R. Improved myocardial function after cold storage with preservation solution supplemented with a carbon monoxide-releasing molecule (CORM-3). *J. Heart Lung Transplant.* **2007**, *26*, 1192–1198. [CrossRef]
106. Ulbrich, F.; Hagmann, C.; Buerkle, H.; Romao, C.C.; Schallner, N.; Goebel, U.; Biermann, J. The Carbon monoxide releasing molecule ALF-186 mediates anti-inflammatory and neuroprotective effects via the soluble guanylate cyclase ß1 in rats' retinal ganglion cells after ischemia and reperfusion injury. *J. Neuroinflamm.* **2017**, *14*, 130. [CrossRef]
107. Zhang, W.Q.; Whitwood, A.C.; Fairlamb, I.J.; Lynam, J.M. Group 6 carbon monoxide-releasing metal complexes with biologically-compatible leaving groups. *Inorg. Chem.* **2010**, *49*, 8941–8952. [CrossRef]
108. Long, L.; Jiang, X.; Wang, X.; Xiao, Z.; Liu, X. Water-soluble diiron hexacarbonyl complex as a CO-RM: Controllable CO-releasing, releasing mechanism and biocompatibility. *Dalton Trans.* **2013**, *42*, 15663–15669. [CrossRef]
109. Zobi, F.; Degonda, A.; Schaub, M.C.; Bogdanova, A.Y. CO releasing properties and cytoprotective effect of cis-trans-[Re(II)(CO)2Br2L2]n complexes. *Inorg. Chem.* **2010**, *49*, 7313–7322. [CrossRef]
110. Schatzschneider, U. PhotoCORMs: Light-triggered release of carbon monoxide from the coordination sphere of transition metal complexes for biological applications. *Inorg. Chim. Acta* **2011**, *374*, 19–23. [CrossRef]
111. Bohlender, C.; Gläser, S.; Klein, M.; Weisser, J.; Thein, S.; Neugebauer, U.; Popp, J.; Wyrwa, R.; Schiller, A. Light-triggered CO release from nanoporous non-wovens. *J. Mater. Chem. B* **2014**, *2*, 1454–1463. [CrossRef]
112. Lomont, J.P.; Nguyen, S.C.; Harris, C.B. Exploring the utility of tandem thermal-photochemical CO delivery with CORM-2. *Organometallics* **2014**, *33*, 6179–6185. [CrossRef]
113. Bannenberg, G.L.; Vieira, H.L. Therapeutic applications of the gaseous mediators carbon monoxide and hydrogen sulfide. *Expert Opin. Ther. Pat.* **2009**, *19*, 663–682. [CrossRef]
114. Motterlini, R.; Mann, B.E.; Johnson, T.R.; Clark, J.E.; Foresti, R.; Green, C.J. Bioactivity and pharmacological actions of carbon monoxide-releasing molecules. *Curr. Pharm. Des.* **2003**, *9*, 2525–2539. [CrossRef]
115. Fiumana, E.; Parfenova, H.; Jaggar, J.H.; Leffler, C.W. Carbon monoxide mediates vasodilator effects of glutamate in isolated pressurized cerebral arterioles of newborn pigs. *Am. J. Physiol. Heart Circ. Physiol.* **2003**, *284*, H1073–H1079. [CrossRef]
116. Cepinskas, G.; Katada, K.; Bihari, A.; Potter, R.F. Carbon monoxide liberated from carbon monoxide-releasing molecule CORM-2 attenuates inflammation in the liver of septic mice. *Am. J. Physiol. Gastrointest. Liver Physiol.* **2008**, *294*, G184–G191. [CrossRef]
117. Nielsen, V.G.; Chawla, N.; Mangla, D.; Gomes, S.B.; Arkebauer, M.R.; Wasko, K.A.; Sadacharam, K.; Vosseller, K. Carbon monoxide-releasing molecule-2 enhances coagulation in rabbit plasma and decreases bleeding time in clopidogrel/aspirin-treated rabbits. *Blood Coagul. Fibrinolysis Int. J. Haemost. Thromb.* **2011**, *22*, 756–759. [CrossRef]
118. Santos-Silva, T.; Mukhopadhyay, A.; Seixas, J.D.; Bernardes, G.J.; Romao, C.C.; Romao, M.J. Towards improved therapeutic CORMs: Understanding the reactivity of CORM-3 with proteins. *Curr. Med. Chem.* **2011**, *18*, 3361–3366. [CrossRef]
119. Maines, M.D. Heme oxygenase: Function, multiplicity, regulatory mechanisms, and clinical applications. *FASEB J.* **1988**, *2*, 2557–2568. [CrossRef]
120. Freitas, A.; Alves-Filho, J.C.; Secco, D.D.; Neto, A.F.; Ferreira, S.H.; Barja-Fidalgo, C.; Cunha, F.Q. Heme oxygenase/carbon monoxide-biliverdin pathway down regulates neutrophil rolling, adhesion and migration in acute inflammation. *Br. J. Pharmacol.* **2006**, *149*, 345–354. [CrossRef]
121. Chaves-Ferreira, M.; Albuquerque, I.S.; Matak-Vinkovic, D.; Coelho, A.C.; Carvalho, S.M.; Saraiva, L.M.; Romao, C.C.; Bernardes, G.J. Spontaneous CO release from Ru(II)(CO)2-protein complexes in aqueous solution, cells, and mice. *Angew. Chem.* **2015**, *54*, 1172–1175. [CrossRef]

122. Chora, A.A.; Fontoura, P.; Cunha, A.; Pais, T.F.; Cardoso, S.; Ho, P.P.; Lee, L.Y.; Sobel, R.A.; Steinman, L.; Soares, M.P. Heme oxygenase-1 and carbon monoxide suppress autoimmune neuroinflammation. *J. Clin. Investig.* **2007**, *117*, 438–447. [CrossRef]
123. Vitek, L.; Gbelcova, H.; Muchova, L.; Vanova, K.; Zelenka, J.; Konickova, R.; Suk, J.; Zadinova, M.; Knejzlik, Z.; Ahmad, S.; et al. Antiproliferative effects of carbon monoxide on pancreatic cancer. *Dig. Liver Dis.* **2014**, *46*, 369–375. [CrossRef]
124. Chlopicki, S.; Lomnicka, M.; Fedorowicz, A.; Grochal, E.; Kramkowski, K.; Mogielnicki, A.; Buczko, W.; Motterlini, R. Inhibition of platelet aggregation by carbon monoxide-releasing molecules (CO-RMs): Comparison with NO donors. *Naunyn Schmiedeberg's Arch. Pharmacol.* **2012**, *385*, 641–650. [CrossRef]
125. Chung, S.W.; Liu, X.; Macias, A.A.; Baron, R.M.; Perrella, M.A. Heme oxygenase-1-derived carbon monoxide enhances the host defense response to microbial sepsis in mice. *J. Clin. Investig.* **2008**, *118*, 239–247. [CrossRef]
126. Desmard, M.; Davidge, K.S.; Bouvet, O.; Morin, D.; Roux, D.; Foresti, R.; Ricard, J.D.; Denamur, E.; Poole, R.K.; Montravers, P.; et al. A carbon monoxide-releasing molecule (CORM-3) exerts bactericidal activity against Pseudomonas aeruginosa and improves survival in an animal model of bacteraemia. *FASEB J.* **2009**, *23*, 1023–1031. [CrossRef]
127. Desmard, M.; Foresti, R.; Morin, D.; Dagouassat, M.; Berdeaux, A.; Denamur, E.; Crook, S.H.; Mann, B.E.; Scapens, D.; Montravers, P.; et al. Differential antibacterial activity against Pseudomonas aeruginosa by carbon monoxide-releasing molecules. *Antioxid. Redox Signal.* **2012**, *16*, 153–163. [CrossRef]
128. Murray, T.S.; Okegbe, C.; Gao, Y.; Kazmierczak, B.I.; Motterlini, R.; Dietrich, L.E.; Bruscia, E.M. The carbon monoxide releasing molecule CORM-2 attenuates Pseudomonas aeruginosa biofilm formation. *PLoS ONE* **2012**, *7*, e35499. [CrossRef]
129. Wegiel, B.; Larsen, R.; Gallo, D.; Chin, B.Y.; Harris, C.; Mannam, P.; Kaczmarek, E.; Lee, P.J.; Zuckerbraun, B.S.; Flavell, R.; et al. Macrophages sense and kill bacteria through carbon monoxide-dependent inflammasome activation. *J. Clin. Investig.* **2014**, *124*, 4926–4940. [CrossRef]
130. Otterbein, L.E.; May, A.; Chin, B.Y. Carbon monoxide increases macrophage bacterial clearance through Toll-like receptor (TLR)4 expression. *Cell. Mol. Biol.* **2005**, *51*, 433–440.
131. Tavares, A.F.; Teixeira, M.; Romao, C.C.; Seixas, J.D.; Nobre, L.S.; Saraiva, L.M. Reactive oxygen species mediate bactericidal killing elicited by carbon monoxide-releasing molecules. *J. Biol. Chem.* **2011**, *286*, 26708–26717. [CrossRef]
132. Hu, C.M.; Lin, H.H.; Chiang, M.T.; Chang, P.F.; Chau, L.Y. Systemic expression of heme oxygenase-1 ameliorates type 1 diabetes in NOD mice. *Diabetes* **2007**, *56*, 1240–1247. [CrossRef]
133. Wegiel, B.; Gallo, D.J.; Raman, K.G.; Karlsson, J.M.; Ozanich, B.; Chin, B.Y.; Tzeng, E.; Ahmad, S.; Ahmed, A.; Baty, C.J.; et al. Nitric oxide-dependent bone marrow progenitor mobilization by carbon monoxide enhances endothelial repair after vascular injury. *Circulation* **2010**, *121*, 537–548. [CrossRef] [PubMed]
134. Nguyen, D.; Boyer, C. Macromolecular and Inorganic Nanomaterials Scaffolds for Carbon Monoxide Delivery: Recent Developments and Future Trends. *ACS Biomater. Sci. Eng.* **2015**, *1*, 895–913. [CrossRef]
135. Kautz, A.C.; Kunz, P.C.; Janiak, C. CO-releasing molecule (CORM) conjugate systems. *Dalton Trans.* **2016**, *45*, 18045–18063. [CrossRef]
136. Schallner, N.; Otterbein, L.E. Friend or foe? Carbon monoxide and the mitochondria. *Front. Physiol.* **2015**, *6*, 17. [CrossRef]
137. Diring, S.; Carné-Sánchez, A.; Zhang, J.; Ikemura, S.; Kim, C.; Inaba, H.; Kitagawa, S.; Furukawa, S. Light responsive metal–organic frameworks as controllable CO-releasing cell culture substrates. *Chem. Sci.* **2017**, *8*, 2381–2386. [CrossRef] [PubMed]
138. Ma, M.; Noei, H.; Mienert, B.; Niesel, J.; Bill, E.; Muhler, M.; Fischer, R.A.; Wang, Y.; Schatzschneider, U.; Metzler-Nolte, N. Iron metal-organic frameworks MIL-88B and NH2-MIL-88B for the loading and delivery of the gasotransmitter carbon monoxide. *Chemistry* **2013**, *19*, 6785–6790. [CrossRef]
139. Pai, S.; Radacki, K.; Schatzschneider, U. Sonogashira, CuAAC, and Oxime Ligations for the Synthesis of MnI Tricarbonyl PhotoCORM Peptide Conjugates. *Eur. J. Inorg. Chem.* **2014**, *2014*, 2886–2895. [CrossRef]
140. Pfeiffer, H.; Rojas, A.; Niesel, J.; Schatzschneider, U. Sonogashira and Click reactions for the N-terminal and side-chain functionalization of peptides with [Mn(CO)3(tpm)]+-based CO releasing molecules (tpm = tris(pyrazolyl)methane). *Dalton Trans.* **2009**, 4292–4298. [CrossRef]

141. Jackson, C.S.; Schmitt, S.; Dou, Q.P.; Kodanko, J.J. Synthesis, characterization, and reactivity of the stable iron carbonyl complex [Fe(CO)(N4Py)](ClO4)2: Photoactivated carbon monoxide release, growth inhibitory activity, and peptide ligation. *Inorg. Chem.* **2011**, *50*, 5336–5338. [CrossRef]
142. Matson, J.B.; Webber, M.J.; Tamboli, V.K.; Weber, B.; Stupp, S.I. A peptide-based material for therapeutic carbon monoxide delivery. *Soft Matter* **2012**, *8*, 6689–6692. [CrossRef]
143. Bischof, C.; Joshi, T.; Dimri, A.; Spiccia, L.; Schatzschneider, U. Synthesis, spectroscopic properties, and photoinduced CO-release studies of functionalized ruthenium(II) polypyridyl complexes: Versatile building blocks for development of CORM-peptide nucleic acid bioconjugates. *Inorg. Chem.* **2013**, *52*, 9297–9308. [CrossRef] [PubMed]
144. Pfeiffer, H.; Sowik, T.; Schatzschneider, U. Bioorthogonal oxime ligation of a Mo(CO)4(N–N) CO-releasing molecule (CORM) to a TGF β-binding peptide. *J. Organomet. Chem.* **2013**, *734*, 17–24. [CrossRef]
145. Yin, H.; Fang, J.; Liao, L.; Nakamura, H.; Maeda, H. Styrene-maleic acid copolymer-encapsulated CORM2, a water-soluble carbon monoxide (CO) donor with a constant CO-releasing property, exhibits therapeutic potential for inflammatory bowel disease. *J. Control. Release* **2014**, *187*, 14–21. [CrossRef] [PubMed]
146. Tabe, H.; Fujita, K.; Abe, S.; Tsujimoto, M.; Kuchimaru, T.; Kizaka-Kondoh, S.; Takano, M.; Kitagawa, S.; Ueno, T. Preparation of a Cross-Linked Porous Protein Crystal Containing Ru Carbonyl Complexes as a CO-Releasing Extracellular Scaffold. *Inorg. Chem.* **2015**, *54*, 215–220. [CrossRef]
147. Tabe, H.; Shimoi, T.; Fujita, K.; Abe, S.; Ijiri, H.; Tsujimoto, M.; Kuchimaru, T.; Kizaka-Kondo, S.; Mori, H.; Kitagawa, S.; et al. Design of a CO-releasing Extracellular Scaffold Using in Vivo Protein Crystals. *Chem. Lett.* **2015**, *44*, 342–344. [CrossRef]
148. Albuquerque, I.S.; Jeremias, H.F.; Chaves-Ferreira, M.; Matak-Vinkovic, D.; Boutureira, O.; Romão, C.C.; Bernardes, G.J.L. An artificial CO-releasing metalloprotein built by histidine-selective metallation. *Chem. Commun.* **2015**, *51*, 3993–3996. [CrossRef]
149. Fujita, K.; Tanaka, Y.; Sho, T.; Ozeki, S.; Abe, S.; Hikage, T.; Kuchimaru, T.; Kizaka-Kondoh, S.; Ueno, T. Intracellular CO release from composite of ferritin and ruthenium carbonyl complexes. *J. Am. Chem. Soc.* **2014**, *136*, 16902–16908. [CrossRef]
150. Zobi, F.; Blacque, O.; Jacobs, R.A.; Schaub, M.C.; Bogdanova, A.Y. 17 e−rhenium dicarbonyl CO-releasing molecules on a cobalamin scaffold for biological application. *Dalton Trans.* **2012**, *41*, 370–378. [CrossRef]
151. Wilson, J.L.; Fayad Kobeissi, S.; Oudir, S.; Haas, B.; Michel, B.; Dubois Rande, J.L.; Ollivier, A.; Martens, T.; Rivard, M.; Motterlini, R.; et al. Design and synthesis of new hybrid molecules that activate the transcription factor Nrf2 and simultaneously release carbon monoxide. *Chemistry* **2014**, *20*, 14698–14704. [CrossRef]
152. Pena, A.C.; Penacho, N.; Mancio-Silva, L.; Neres, R.; Seixas, J.D.; Fernandes, A.C.; Romao, C.C.; Mota, M.M.; Bernardes, G.J.; Pamplona, A. A novel carbon monoxide-releasing molecule fully protects mice from severe malaria. *Antimicrob. Agents Chemother.* **2012**, *56*, 1281–1290. [CrossRef]
153. Marques, A.R.; Kromer, L.; Gallo, D.J.; Penacho, N.; Rodrigues, S.S.; Seixas, J.D.; Bernardes, G.J.L.; Reis, P.M.; Otterbein, S.L.; Ruggieri, R.A.; et al. Generation of Carbon Monoxide Releasing Molecules (CO-RMs) as Drug Candidates for the Treatment of Acute Liver Injury: Targeting of CO-RMs to the Liver. *Organometallics* **2012**, *31*, 5810–5822. [CrossRef]
154. Duncan, R. Polymer conjugates as anticancer nanomedicines. *Nat. Rev. Cancer* **2006**, *6*, 688–701. [CrossRef]
155. Brückmann, N.E.; Wahl, M.; Reiß, G.J.; Kohns, M.; Wätjen, W.; Kunz, P.C. Polymer Conjugates of Photoinducible CO-Releasing Molecules. *Eur. J. Inorg. Chem.* **2011**, *2011*, 4571–4577. [CrossRef]
156. Kunz, P.C.; Brückmann, N.E.; Spingler, B. Towards Polymer Diagnostic Agents—Copolymers of N-(2-Hydroxypropyl)methacrylamide and Bis(2-pyridylmethyl)-4-vinylbenzylamine: Synthesis, Characterisation and Re(CO)3-Labelling. *Eur. J. Inorg. Chem.* **2007**, *2007*, 394–399. [CrossRef]
157. Dördelmann, G.; Pfeiffer, H.; Birkner, A.; Schatzschneider, U. Silicium Dioxide Nanoparticles As Carriers for Photoactivatable CO-Releasing Molecules (PhotoCORMs). *Inorg. Chem.* **2011**, *50*, 4362–4367. [CrossRef]
158. Gonzales, M.A.; Han, H.; Moyes, A.; Radinos, A.; Hobbs, A.J.; Coombs, N.; Oliver, S.R.J.; Mascharak, P.K. Light-triggered carbon monoxide delivery with Al-MCM-41-based nanoparticles bearing a designed manganese carbonyl complex. *J. Mater. Chem. B* **2014**, *2*, 2107–2113. [CrossRef]

159. Dordelmann, G.; Meinhardt, T.; Sowik, T.; Krueger, A.; Schatzschneider, U. CuAAC click functionalization of azide-modified nanodiamond with a photoactivatable CO-releasing molecule (PhotoCORM) based on [Mn(CO)3(tpm)]+. *Chem. Commun.* **2012**, *48*, 11528–11530. [CrossRef]
160. Kunz, P.C.; Meyer, H.; Barthel, J.; Sollazzo, S.; Schmidt, A.M.; Janiak, C. Metal carbonyls supported on iron oxide nanoparticles to trigger the CO-gasotransmitter release by magnetic heating. *Chem. Commun.* **2013**, *49*, 4896–4898. [CrossRef]
161. Govender, P.; Pai, S.; Schatzschneider, U.; Smith, G.S. Next generation PhotoCORMs: Polynuclear tricarbonylmanganese(I)-functionalized polypyridyl metallodendrimers. *Inorg. Chem.* **2013**, *52*, 5470–5478. [CrossRef]
162. Cadranel, A.; Pieslinger, G.E.; Tongying, P.; Kuno, M.K.; Baraldo, L.M.; Hodak, J.H. Spectroscopic signatures of ligand field states in {Ru(II)(imine)} complexes. *Dalton Trans.* **2016**, *45*, 5464–5475. [CrossRef]
163. Gonzales, M.A. Iron, Manganese and Ruthenium Metal Carbonyls as Photoactive Carbon Monoxide Releasing Molecules (photoCORMS): Ligand Design Strategies, Syntheses and Structure Characterizations. Ph.D. Thesis, University of California, Santa Cruz, CA, USA, 2013.
164. Khramov, D.M.; Lynch, V.M.; Bielawski, C.W. N-Heterocyclic Carbene−Transition Metal Complexes: Spectroscopic and Crystallographic Analyses of π-Back-bonding Interactions. *Organometallics* **2007**, *26*, 6042–6049. [CrossRef]
165. Wang, C.; Cheng, L.; Liu, Z. Drug delivery with upconversion nanoparticles for multi-functional targeted cancer cell imaging and therapy. *Biomaterials* **2011**, *32*, 1110–1120. [CrossRef]
166. Zhao, Z.; Han, Y.; Lin, C.; Hu, D.; Wang, F.; Chen, X.; Chen, Z.; Zheng, N. Multifunctional core-shell upconverting nanoparticles for imaging and photodynamic therapy of liver cancer cells. *Chem. Asian J.* **2012**, *7*, 830–837. [CrossRef] [PubMed]
167. Yoon, J.W.; Seo, Y.-K.; Hwang, Y.K.; Chang, J.-S.; Leclerc, H.; Wuttke, S.; Bazin, P.; Vimont, A.; Daturi, M.; Bloch, E.; et al. Controlled Reducibility of a Metal–Organic Framework with Coordinatively Unsaturated Sites for Preferential Gas Sorption. *Angew. Chem. Intl. Ed.* **2010**, *49*, 5949–5952. [CrossRef]
168. van der Vlies, A.J.; Inubushi, R.; Uyama, H.; Hasegawa, U. Polymeric Framboidal Nanoparticles Loaded with a Carbon Monoxide Donor via Phenylboronic Acid-Catechol Complexation. *Bioconjugate Chem.* **2016**, *27*, 1500–1508. [CrossRef]
169. Muhammad, N.; Mao, Q.; Xia, H. CAR T-cells for cancer therapy. *Biotechnol. Genet. Eng. Rev.* **2017**, *33*, 190–226. [CrossRef]
170. Johnson, T.R.; Mann, B.E.; Teasdale, I.P.; Adams, H.; Foresti, R.; Green, C.J.; Motterlini, R. Metal carbonyls as pharmaceuticals? [Ru(CO)3Cl(glycinate)], a CO-releasing molecule with an extensive aqueous solution chemistry. *Dalton Trans.* **2007**, 1500–1508. [CrossRef] [PubMed]
171. Narayan, S.P.; Choi, C.H.; Hao, L.; Calabrese, C.M.; Auyeung, E.; Zhang, C.; Goor, O.J.; Mirkin, C.A. The Sequence-Specific Cellular Uptake of Spherical Nucleic Acid Nanoparticle Conjugates. *Small (Weinh. Bergstr. Ger.)* **2015**, *11*, 4173–4182. [CrossRef]
172. Nguyen, D.; Adnan, N.N.; Oliver, S.; Boyer, C. The Interaction of CORM-2 with Block Copolymers Containing Poly(4-vinylpyridine): Macromolecular Scaffolds for Carbon Monoxide Delivery in Biological Systems. *Macromol. Rapid Comm.* **2016**, *37*, 739–744. [CrossRef] [PubMed]
173. Garcia-Gallego, S.; Bernardes, G.J. Carbon-monoxide-releasing molecules for the delivery of therapeutic CO in vivo. *Angew. Chem.* **2014**, *53*, 9712–9721. [CrossRef] [PubMed]
174. Dallas, M.L.; Scragg, J.L.; Peers, C. Modulation of hTREK-1 by carbon monoxide. *Neuroreport* **2008**, *19*, 345–348. [CrossRef]
175. Lundvig, D.M.; Immenschuh, S.; Wagener, F.A. Heme oxygenase, inflammation, and fibrosis: The good, the bad, and the ugly? *Front. Pharmacol.* **2012**, *3*, 81. [CrossRef]
176. Szeremeta, M.; Petelska, A.D.; Kotynska, J.; Niemcunowicz-Janica, A.; Figaszewski, Z.A. The effect of fatal carbon monoxide poisoning on the surface charge of blood cells. *J. Membr. Biol.* **2013**, *246*, 717–722. [CrossRef]
177. Petelska, A.D.; Kotynska, J.; Figaszewski, Z.A. The effect of fatal carbon monoxide poisoning on the equilibria between cell membranes and the electrolyte solution. *J. Membr. Biol.* **2015**, *248*, 157–161. [CrossRef]

178. Wang, P.; Liu, H.; Zhao, Q.; Chen, Y.; Liu, B.; Zhang, B.; Zheng, Q. Syntheses and evaluation of drug-like properties of CO-releasing molecules containing ruthenium and group 6 metal. *Eur. J. Med. Chem.* **2014**, *74*, 199–215. [CrossRef]
179. Winburn, I.C.; Gunatunga, K.; McKernan, R.D.; Walker, R.J.; Sammut, I.A.; Harrison, J.C. Cell damage following carbon monoxide releasing molecule exposure: Implications for therapeutic applications. *Basic Clin. Pharmacol. Toxicol.* **2012**, *111*, 31–41. [CrossRef]

MDPI
St. Alban-Anlage 66
4052 Basel
Switzerland
Tel. +41 61 683 77 34
Fax +41 61 302 89 18
www.mdpi.com

Materials Editorial Office
E-mail: materials@mdpi.com
www.mdpi.com/journal/materials

www.ingramcontent.com/pod-product-compliance
Lightning Source LLC
Chambersburg PA
CBHW040838250926
51850CB00007BA/1351
* 9 7 8 3 0 3 6 5 1 0 5 8 3 *